ESSENTIALS
OF MULTIPHASE FLOW
AND TRANSPORT
IN POROUS MEDIA

ESSENTIALS OF MULTIPHASE FLOW AND TRANSPORT IN POROUS MEDIA

George F. Pinder

University of Vermont

William G. Gray

University of North Carolina at Chapel Hill

A JOHN WILEY & SONS, INC., PUBLICATION

Library of Congress Cataloging-in-Publication Data:

ISBN 978-0-470-31762-4

Printed in the United States of America.

10 9 8 7 6 5 4 3 2 1

To the Future: Jasmine, Ryan, Charlotte, Benjamin, and Lucy

CONTENTS

Preface   **xi**

Acknowledgments   **xiii**

1 Setting the Stage   **1**

  1.1   Introduction / 1
  1.2   Phases and Porous Media / 2
  1.3   Grain and Pore Size Distributions / 6
  1.4   The Concept of Saturation / 12
  1.5   The Concept of Pressure / 13
  1.6   Surface Tension Considerations / 16
  1.7   Concept of Concentration / 30
  1.8   Summary / 32
  1.9   Exercises / 32
        Bibliography / 33

2 Mass Conservation Equations   **35**

  2.1   Introduction / 35
  2.2   Microscale Mass Conservation / 38
  2.3   Integral Forms of Mass Conservation / 39
  2.4   Integral Theorems / 44
        2.4.1   Divergence Theorem / 45
        2.4.2   Transport Theorem / 45
  2.5   Point Forms of Mass Conservation / 46
  2.6   The Macroscale Perspective / 48
        2.6.1   The Representative Elementary Volume / 49
        2.6.2   Global and Local Coordinate Systems / 50
        2.6.3   Macroscopic Variables / 53
        2.6.4   Definitions of Macroscale Quantities / 56
        2.6.5   Summary of Macroscale Quantities / 62
  2.7   The Averaging Theorems / 63
        2.7.1   Spatial Averaging Theorem / 64
        2.7.2   Temporal Averaging Theorem / 66
  2.8   Macroscale Mass Conservation / 67
        2.8.1   Macroscale Point Forms / 67
        2.8.2   Integral Forms / 71

2.9   Applications / 73
    2.9.1   Integral Analysis / 74
    2.9.2   Point Analysis / 76
2.10 Summary / 79
2.11 Exercises / 79
    Bibliography / 81

3   Flow Equations   **83**

3.1   Introduction / 83
3.2   Darcy's Experiments / 85
3.3   Fluid Properties / 88
3.4   Equations of State for Fluids / 89
    3.4.1   Mass Fraction / 89
    3.4.2   Mass Density and Pressure / 90
    3.4.3   Fluid Viscosity / 92
3.5   Hydraulic Potential / 93
    3.5.1   Hydrostatic Force and Hydraulic Head / 93
    3.5.2   Derivatives of Hydraulic Head / 97
3.6   Single-Phase Fluid Flow / 98
    3.6.1   Darcy's Law / 99
    3.6.2   Hydraulic Conductivity and Permeability / 102
    3.6.3   Derivation of Groundwater Flow Equation / 106
    3.6.4   Recapitulation of the Derivation / 111
    3.6.5   Initial and Boundary Conditions / 113
    3.6.6   Two-Dimensional Flow / 116
3.7   Two-Phase Immiscible Flow / 121
    3.7.1   Derivation of Flow Equations / 121
    3.7.2   Observations on the p^c-s^w Relationship / 127
    3.7.3   Formulas for the p^c-s^w Relationship / 135
    3.7.4   Observations of the k_{rel}^α-s^w Relationship / 143
    3.7.5   Formulas for the k_{rel}^α-s^w Relation / 146
    3.7.6   Special Cases of Multiphase Flow / 149
3.8   The Buckley-Leverett Analysis / 155
    3.8.1   Fractional Flow / 155
    3.8.2   Derivation of the Buckley-Leverett Equation / 157
    3.8.3   Solution of the Buckley-Leverett Equation / 158
3.9   Summary / 160
3.10 Exercises / 161
    Bibliography / 162

4   Mass Transport Equations   **165**

4.1   Introduction / 165
4.2   Velocity in the Species Transport Equations / 167
    4.2.1   Direct Approach / 168
    4.2.2   Rigorous Approach / 169

4.2.3 Distribution Approach / 172
4.2.4 Summary / 175

4.3 Closure Relations for the Dispersion Vector / 176
4.4 Chemical Reaction Rates / 180
4.5 Interphase Transfer Terms / 182

4.5.1 Kinetic Formulation / 183
4.5.2 Equilibrium Formulation / 187
4.5.3 Summary: Kinetic vs. Equilibrium Formulations / 194

4.6 Initial and Boundary Conditions / 195
4.7 Conclusion / 196
4.8 Exercises / 197
Bibliography / 198

5 Simulation   **199**

5.1 1-D Simulation of Air-Water Flow / 199

5.1.1 Drainage in a Homogeneous Soil / 201
5.1.2 Drainage in a Heterogeneous Soil / 205
5.1.3 Imbibition in Homogeneous Soil / 206

5.2 1-D Simulation of DNAPL-Water Flow / 207

5.2.1 Primary DNAPL Imbibition in Homogeneous Soil / 208
5.2.2 Density Effect / 208
5.2.3 DNAPL Drainage in Homogeneous Soil / 209
5.2.4 Secondary Imbibition of DNAPL in Homogeneous Soil / 210
5.2.5 Secondary Drainage in Homogeneous Soil / 211
5.2.6 Primary Imbibition in Heterogeneous Soil / 212

5.3 2-D Simulation of DNAPL-Water Flow / 213

5.3.1 DNAPL Descent into a Water-Saturated Reservoir / 213

5.4 Simulation of Multiphase Flow and Transport / 216

5.4.1 1-D Two-Phase Flow and Transport / 217
5.4.2 2-D Two-Phase Flow and Transport / 218

5.5 2-D Single-Phase Flow and Transport / 224

5.5.1 Base Case / 228
5.5.2 Effect of Inflow / 228
5.5.3 Impact of Well Discharge / 230
5.5.4 Effect of Adsorption / 231
5.5.5 Effect of a Low Transmissivity Region / 232
5.5.6 Effect of a High Transmissivity Region / 234
5.5.7 Effect of Rate of Reaction / 235

5.6 3-D Single-Phase Flow and Transport / 236
5.7 2-D Three-Phase Flow / 239
5.8 Summary / 244
Bibliography / 245

Select Symbols   **247**

Index   **253**

PREFACE

This book was prepared in response to a realization by the authors that many students and practicing professionals accept as valid, and use with confidence, equations describing multiphase flow and transport in porous media, although they have little knowledge of the underlying physical-chemical processes and simplifying assumptions implicit in these equations. The purpose of this text is to build the mathematical equations describing porous media processes from first principles, in a stepwise, coherent, rigorous, and comprehensive manner.

Experience gained in teaching the physics of flow through porous media over a 35 year period provides the pedagogical approach reflected in the structure of the book. Chapter 1 introduces intuitive and fundamental concepts that must be considered in porous media systems. These serve to provide a framework within which the careful study of porous media must be structured. This framework is filled in later chapters using concepts borrowed from the study of continuous media. Chapter 2 provides information about conservation of mass. The equations are developed for species within a phase and for the entire phase itself. The initial presentation is for point and system equations for a single phase where the other phases present will define the boundary of the volume being studied. At this scale, the phases are juxtaposed. The mathematical tools for changing the equations in these forms to appropriate forms at a larger scale are presented. At this larger scale, the phases are modeled as overlapping continua each occupying a fraction of the space. The mass conservation equations are then developed at this larger scale. One might expect a conservation of momentum equation to be posed in conjunction with the mass balance equations. However, porous media study typically replaces a theoretically derived momentum conservation equation with the experimentally based correlation known as Darcy's law. This equation is the topic of Chapter 3, wherein the nuances of using Darcy's law for single-phase flow as well as for multiphase flow are examined. Chapter 4 expands on the material presented in Chapter 2 by considering the transport equations for chemical species in detail. Supplementary conditions needed to account for species transport, such as expressions for dispersion, chemical reactions, and interphase transport, are developed for incorporation into the equations of chemical species movement. These equations can be solved in conjunction with the conditions for total mass conservation and Darcy's law to obtain velocity and concentration fields. As we have stated, the main objective of this text is to provide information on the underpinnings of simulation models that account properly for system physics. Chapter 5 demonstrates the implementation of some of the developed equations for simulation of a variety of porous media systems. The goal here is not to develop numerical codes but to demonstrate how the various terms that appear in the governing equations contribute to modeled system behavior. Indeed, this text does not consider problems involving heat transfer and does not provide insights into the development of numerical solution algorithms. The objective of the book is to provide insights into the essential elements that must be accounted for in quantifying the behavior of flow and chemical transport in porous media.

The resultant text is a presentation that is designed to meet the needs of a student at the senior undergraduate or graduate level who has a fundamental knowledge of fluid mechanics. Those who model subsurface systems will also benefit from the careful examination of the features of the flow and transport equations that are foundational to their simulations. A background in groundwater hydrology, soil mechanics, soil physics, or oil and gas reservoir engineering will provide context for the reader motivated by a desire to learn more about the elements of the theoretical description of porous media systems.

George F. Pinder
William G. Gray

ACKNOWLEDGMENTS

Over the years the first author conducted research while with the U.S. Geological Survey and taught at Princeton University and the University of Vermont. Robert Farvolden, John Bredehoeft, and Hilton Cooper have been especially helpful in providing advice and support during his career. He has learned from as well as lectured to students interested in the physics of flow and transport through porous media. Both authors appreciate the collaboration with and counsel they have received from Ahmet S. Cakmak throughout their careers. The second author has benefited greatly from his experiences on the faculty at Princeton University and the University of North Carolina at Chapel Hill. Indeed the insights of students and colleagues both at these institutions and elsewhere have provided him with extraordinary opportunities to learn about porous media. He is particularly grateful to Robert L. Irvine, George F. Galland, and Cass T. Miller for their contributions in fostering environments in which academic pursuits are possible and valued.

Both authors have been particularly fortunate to have received the unwavering support of their wives, Phyllis Pinder and Genetha Gray, in all their academic, professional, and personal endeavors.

In general, it is difficult to identify all the individuals who make contributions to a work. However, we must especially thank all the students who have attended our lectures, sat through our courses, and nudged us away from theoretical and conceptual errors. Unfortunately, we must still accept responsibility for those errors we have been unable to purge from our understanding and which may show their warts in these pages.

George F. Pinder
William G. Gray

1

SETTING THE STAGE

1.1 INTRODUCTION

The purpose of this text is to introduce the fundamental concepts that underlie the physics of multiphase flow and transport through porous media. This first chapter introduces some of the qualitative physical characteristics of porous media. Parameters are introduced that provide quantitative measures of the characteristics that arise in modeling fluid flow and chemical transport in the system of interest. Some simple elementary equations are employed that are helpful in initiating the translation of a qualitative understanding to a quantitative description. In the second chapter, the equations of conservation of mass are developed. In Chapter 3 appropriate constitutive relationships[1] are introduced that provide information needed to complete the mathematical definition of the physical systems involving fluid flow. Chapter 4 is dedicated to developing the equations that describe mass transport. Finally, in the fifth chapter, example physical problems involving multiphase flow and transport through porous media are detailed.

The approach of this presentation is to progress from observations of system behavior and characteristics to a mathematical description of those observations. This approach involves three steps: (i) description of experiments that reveal various phenomena; (ii) development and presentation of the governing equations; and (iii) application of the resulting equations to physical systems of interest.

---

[1] Constitutive, or closure, relationships are typically correlations between fluxes and physical variables. The correlations are motivated by experimental observations or from simplified theoretical considerations. They are not universal principles but are appropriate for some systems under certain operating conditions. Constitutive relations provide specific information that makes it possible to apply conservation equations to problems.

Essentials of Multiphase Flow and Transport in Porous Media, by George F. Pinder and William G. Gray
Copyright © 2008 by John Wiley & Sons, Inc.

1.2   PHASES AND POROUS MEDIA

A *phase* is a liquid, solid, or gas that is separated from another solid, liquid, or gas by an identifiable boundary. An example is an oil bubble or oil globule submerged in water, where the oil and the water are each phases and the physical demarcation between the two liquid phases is an interface. Some transfer of material, momentum, and energy may occur between phases; a phase need not have a homogeneous composition or temperature. Thus, although gradients of properties may exist within a phase, sharp discontinuities in composition at an identifiable boundary are considered to be interfaces between phases. A second example of a two-phase system is raindrops falling through air. A raindrop is a liquid phase while the air is a gas phase, and transfer of water to the air may occur by evaporation across the boundary of the raindrop. Because of evaporation, gradients in humidity may exist in the gas. An important attribute of this system is that the gas phase is continuous in that every point in the gas phase may be reached by a physical path without entering into the liquid phase. On the other hand, the liquid phase, comprised of raindrops, is an assemblage in which the properties of each drop may be distinctly different from those of a nearby drop. Modeling of a discontinuous phase as a unit requires some approximations or simplifications that are not needed when describing a continuous phase. As a third example, dry sand is actually a mixture of solid sand grains and air. The behavior of this two-phase mixture will be very different when air is pumped through a packed column of essentially immobile sand from when the air entrains the sand grains, imparting momentum and energy to them and causing them to move at significant velocity in a cloud. Thus identification of the components of a system is not sufficient for determining how to model it. Multiphase models must be formulated to account for the modes of transfer of chemical constituents, momentum, and energy within each phase and across the phase interfaces.

Porous media are considered herein to exhibit a specific set of physical attributes that distinguish them from general multiphase systems. The most notable of these are the requirements that more than one phase exist within a specified control volume, that one of these be a relatively immobile solid, and that at least one of these phases be fluid (either a liquid or a gas). Furthermore, the definition of a mixture of phases as a porous medium requires that the solid phase contain multiply-connected spaces that are accessible to the fluid.

Although the definition of a porous medium requires that the solid be "relatively" immobile, a precise specification of the degree of solid mobility or deformation that is allowable by this definition is not possible. At one extreme, an immobile solid, such as well-consolidated sand or a block of granite, may form the solid phase of a porous medium. At the other extreme, a solid such as sand scoured from the bottom of the ocean and carried in the waves or grain flowing out of a grain elevator is a solid phase mixed with fluid in a system that is not a porous medium. For a porous medium, the velocity of the solid phase with respect to the boundary of the system is much less than the velocity of the fluid that can flow within the porous system.

In natural porous media systems, some consolidation of the solid phase may occur as flow moves through the pore space. This can be accounted for under the theoretical framework of porous media flow. Infiltration of rainwater into a soil and movement of subsurface water through a geologic formation are examples where porous medium considerations apply. Situations where the withdrawal of water from the subsurface causes the ground to subside over a period of years may also be analyzed within a porous medium framework because the movement of the solid is very slow in comparison to the water movement. A system composed of snow, air, and melt-water may be studied as a porous medium consisting of a solid and two fluids if the rate of melting is small enough that the snow particles respond as a unit, are relatively immobile, and are not carried off as solid particles within the flowing water. The precise specification of the conditions under which a fluid-solid system cannot be studied in a meaningful way as a porous medium is elusive. The study of flow of water in a sponge is another system that may or may not fall under the umbrella of traditional porous media studies depending on the degree of deformation of the solid structure for the conditions of interest.

Despite the fact that it is not possible to define precisely a porous medium, we will persevere and identify additional attributes of porous systems under study here. To be amenable to porous media flow modeling, the pore space within the solid must, in general, be continuous. For example, Styrofoam is composed of a solid phase in which air bubbles are encapsulated and separated. These air pockets are disconnected, and thus the behavior of the air in the Styrofoam cannot be studied under the guise of porous media analysis. At a larger scale, Swiss cheese is composed of both gas and solid phases. However, the gas phase is contained in isolated, generally disconnected void spaces within the cheese. Gas in one void space cannot readily travel to another void space. Although the absence of a connected void space precludes the scientific study of Swiss cheese as a porous medium, experimental sampling of this system remains a highly rewarding and widely practiced endeavor.

In porous media to be studied here, the individual fluid phases must, under suitable conditions, have the potential to be continuous. That is, the possibility must exist for fluid to flow from one location to another within the porous medium. Thus the structure of the pore space within the solid must be such that pathways exist that connect the regions of the system. For a single fluid phase in such a system, the fluid region will certainly be connected. However, when more than one fluid is present in the connected pore space, one of the fluids may divide into a number of separate disconnected elements. Systems in which a phase becomes disconnected are very difficult to model as the physics of each disconnected region of fluid must be accounted for.

The study of porous media typically assumes that the solid phase is connected. However, the definition of "connection" for the solid is imprecise and the discussion can disintegrate to what it means for grains of sand to be "touching." Nevertheless, we can make the somewhat satisfying observation that for the solid in a porous medium to possess the necessary degree of immobility, any individual grains must be in contact with other grains with points of mechanical interaction between grains changing very slowly relative to the rate of change of fluid molecules that interact with a particular point on a grain. This observation does not preclude the possibility that grains will reorganize or deform in response to various stresses placed on the solid system, although it does suggest that treatment of consolidated media, in which the grains are essentially cemented together, may be somewhat more straightforward. Cases where the solid deforms more quickly and chaotically, as in grain flow or an avalanche, cannot be modeled as porous media.

As an illustration that reveals some of the fundamental concepts that arise in porous media, consider the following simple experiment. Into a graduated cylinder of 2.0 cm diameter and a height of 30 cm pour sand until the top of the sand is located at the 10.0 cm mark (see Figure 1.1). The sand-air mixture in the graduated cylinder constitutes a porous medium because the air phase that is intermixed with the sand is continuous, identifiable interfaces exist between the air and sand phases, and the sand is essentially immobile relative to the bounding walls of the cylinder. Of course, shaking the cylinder can mobilize the sand and invalidate the porous medium assumption, but this will not be the situation in this example.

The portion of the sample that is not sand is called the *pore space*. In the present case, the pore space is occupied completely by air, and the air-sand mixture constitutes a porous medium.

From a device capable of measuring the amount of fluid dispensed, add water to the graduated cylinder until the water level is located at the 15.0 cm mark, 5.0 cm above the top of the sand surface. Assume that the experimental technique employed is such that the sand in the cylinder is not disturbed by the addition of water and that all of the pore space in the sand is filled with water with no air being trapped. Under these conditions, the porous medium is said to be saturated with water. We note that achievement of saturation is difficult as air tends to become trapped in the system. Nevertheless, taking advantage of the fact that this is an illustrative example, we happily discount this experimental complication. Assume the water dispenser indicates that the total volume of water added to the graduated cylinder, V_T^w, is 25.0 cubic centimeters (cm³). The cylinder is now occupied by a sand-water mixture that is a porous medium, and this porous medium is overlain by a water phase that is connected to the water in the medium.

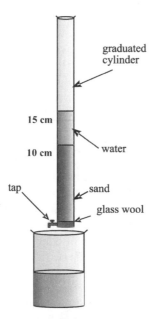

Figure 1.1: Diagrammatic presentation of experiment to show concepts of phase and porous media.

The fraction of the porous medium that is pore space can be determined by analyzing this experiment. First, determine the total volume of the porous medium consisting of the fluid and solid mixture. Based on the equation for the volume of a cylinder of radius r and height h, the volume of the porous medium in the cylinder with radius of 1.0 cm and a height of 10.0 cm is:

$$V = V^s + V^w = \pi r^2 h = \pi \times (1.0\,\text{cm})^2 \times 10.0\,\text{cm} = 31.4\,\text{cm}^3 \tag{1.1}$$

where V is the total volume of the porous medium composed of sand and water, V^s is the volume of sand in the porous medium, and V^w is the volume of water in the porous medium. In this case, where water completely fills the pore space, V^w is also the volume of pore space. Although we have calculated the total volume of porous medium, the distribution of this volume between sand and pore space is not yet known. Consideration of the total amount of water dispensed will lead to this information.

The 25.0 cm³ dispensed into the cylinder fills the pore space and the volume that extends 5.0 cm above the sand. The water volume in this 5-cm region within the cylinder, V_C^w, is easily calculated from the equation for a cylinder as:

$$V_C^w = \pi r^2 h = \pi \times (1.0\,\text{cm})^2 \times 5.0\,\text{cm} = 15.7\,\text{cm}^3 \tag{1.2}$$

Thus the volume of water in the pore space of the sample is:

$$V^w = V_T^w - V_C^w = 25.0\,\text{cm}^3 - 15.7\,\text{cm}^3 = 9.3\,\text{cm}^3 \tag{1.3}$$

Combination of equations (1.1) and (1.3) also provides the volume of sand in the system:

$$V^s = V - V^w = 31.4\,\text{cm}^3 - 9.3\,\text{cm}^3 = 22.1\,\text{cm}^3 \tag{1.4}$$

The fraction of the porous medium that is pore space, ε, is thus obtained as:

$$\varepsilon = \frac{V^w}{V^s + V^w} = \frac{9.3\,\text{cm}^3}{31.4\,\text{cm}^3} = 0.30 \tag{1.5}$$

This fraction ε is called the *porosity* or *void fraction* of the sample. In general, for a sample of porous medium of size V, the porosity is defined in terms of the size of the volume sample and the volume of solid in the sample as:

$$\varepsilon = 1 - \left(\frac{1}{V} \int_{V^s} dV \right) \tag{1.6}$$

Realize that the porosity calculated in this experiment provides a value that is characteristic of the entire sample. It provides no information as to how the pore volume is distributed within the sample. If half the sand were removed from cylinder, the value of porosity obtained from the remainder of the sample could be different from that for the entire sample. Certainly if one removes sand such that there are only a few grains left in the cylinder, the porosity for that sample could be quite different from that for the full sample, or even meaningless as a quantity intended to characterize the system. These observations introduce the notion that when one is studying porous media, the length scale at which observations are made can be an important factor that influences the values of variables measured.

Furthermore, although the total volume of pore space within a sample can be measured, in general the geometry and volume of individual pores cannot be measured. Some specific information at this small scale can be obtained for samples of size on the order of $1 \, cm^3$ using advanced imaging techniques, but, at best, only statistical distributions of the pore sizes of larger samples can be determined. Despite the fact that no single accepted physical measure of the size of a single pore exists, the concept of pore size is widely used. Each naturally occurring pore will have a variable cross section, and *grain size* is commonly used as a surrogate for the size of the pore. Methods for describing the distribution of grain sizes and pore sizes are discussed in the next section.

1.3 GRAIN AND PORE SIZE DISTRIBUTIONS

While the volume of a pore does depend upon the size of the grains in some sense, the relationship can be complex because of the influence of *grain packing*. For example, the packing of spherical grains of uniform size in Figure 1.2 is referred to as *cubic* and the porosity is 0.48. On the other hand the packing of the same grains in Figure 1.3 is *rhombohedral* and has a porosity of 0.26. The pore space is fully connected in both cases, and the pore space can be identified as the region between adjacent spheres. Although the precise specification of what constitutes a pore is not obvious, the pore space does illustrate the channels of fluid flow. If any consistent measure of a pore is selected in both figures, the volume and pore diameter of an individual pore in Figure 1.2 is larger than that in Figure 1.3. A random *packing* of uniform spheres will result in different values of porosity depending on the looseness and organization of the spheres. A loose random packing of spheres will generally generate porosities from 0.32 to 0.35 [8]. Addition to this mix of solid spherical particles with a range of sizes and of nonspherical, arbitrarily shaped grains adds complexity to the identification of "pore size" and to the range of porosities that

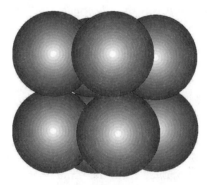

Figure 1.2: Cubic packing of spheres generates a porosity of 0.48 [6].

Figure 1.3: Rhombohedral packing of spheres generates a porosity of 0.26 [6].

Figure 1.4: Small grains tend to occupy spaces left between larger grains to yield smaller porosities.

can be achieved. In nature one is not likely to find spheres of the same size or, for that matter, spherical particles at all. Normally one will find a variety of grain sizes with the smaller grains occupying the spaces between the larger grains as conceptualized in Figure 1.4.

Because of the difficulty in characterizing pore space, the grain size distribution is used as a surrogate. For unconsolidated media, grain size is easily measured through the use of *sieves*. A classical sieve is composed of a metal cylinder approximately 5 cm in length and approximately 20 cm in diameter. It is open at one end and contains a metal screen at the other. Sieves are normally stacked with the sieve with the smallest screen size opening, or *mesh size*, at the base of the stack. Below the last sieve is a pan to collect those grains smaller than the smallest screen size (see Figure 1.5).

Sieve sizes are designated in two principal ways. Some sieves provide the sieve diameter in inches or millimeters. Others designate the sieve by a standard number that is not directly related to the mesh size but indicates the number of openings per inch. For example, a number 20 sieve has 20 openings per inch, or 400 openings per square inch. Typical sieve sizes are shown in Table 1.1.

To sieve a sample of soil, a known weight of the soil is placed in the uppermost sieve. This sieve is covered and a shaking apparatus is used to vibrate the column

Figure 1.5: Screens are stacked sequentially from the finest mesh at the bottom to the coarsest at the top.

Table 1.1: Typical sieve sizes

U.S. Standard Test Sieves (ASTM) [1]			
Sieve Designation		Nominal Sieve Opening	
Standard	Alternative	Inches	Millimeters
25.0 mm	1 in	1	25.7
11.2 mm	7/16 in	0.438	11.2
4.75 mm	No. 4	0.187	4.76
1.70 mm	No. 12	0.0661	1.68
0.075 mm	No. 200	0.0029	0.063

of sieves while it remains approximately vertical. The grains smaller than the opening in the top sieve eventually pass to the next lower sieve. This sieve, in turn, retains those grains with a diameter larger than its mesh size and smaller than the mesh size of the upper sieve. This process continues until the grains retained in the container at the bottom of the column are smaller than the diameter of the sieve with the smallest mesh. Each soil fraction is then weighed and the results plotted as weight vs. sieve size.

Soils are classified as coarse-grained when they are composed of sand and gravel. Typically, less than 50% of coarse-grained material will pass through the No. 200 mesh. Fine-grained soils are composed of silt and clay. Fifty percent or more of a fine-grained soil will pass the No. 200 mesh.

Normally, solid material with components smaller than those captured by the No. 200 mesh screen is very difficult to screen further and is therefore analyzed via a "wet" method that exploits the dependence of the settling rate of a small particle in a quiescent fluid on its size. Wet methods make use of Stokes' law, which states that the diameter of a spherical particle falling through a fluid is related to the velocity according to[2]:

$$D = \sqrt{\frac{18\mu v}{g(\rho_s - \rho_f)}} \tag{1.7}$$

where D is the particle diameter, μ is the dynamic viscosity, v is the settling velocity, g is gravity, ρ_s is the solid particle density, and ρ_f is the density of the fluid. In the analysis, ρ_s is assumed to be a constant independent of particle size, and the velocity, v, is considered to be reached as soon as settling begins. The general idea in employing a wet method is to begin with a mixture of particles uniformly distributed in a fluid consisting of water and a dispersing agent, such as hexametaphosphate, added to the mixture to ensure that the particles do not aggregate. The maximum size particle will be the largest size that passes through a number 200 sieve, D_{200}, approximately $70\,\mu m$ based on Table 1.1. Then measurement of the evolution of the density profile of the mixture due to the different settling rates of the particles provides information concerning the distribution of particle sizes. Two principal *wet methods* are employed. One is the *pipette method* while the second is the *hydrometer method*.

The pipette method involves collection of samples of the solid-fluid mixture in a cylinder at various times and depths. Initially, the distribution of particles in the fluid is uniform with a mass per volume of m_0. If the solution is dilute enough so that collisions between particles are not significant, at a depth L at time t, all particles with diameter greater than that given by equation (1.7) with v replaced by L/t will have settled to a depth below L. Suppose a small sample of the mixture collected at this time and location using a pipette has a mass of sediment per volume of $m(L, t)$. Then $1 - m/m_0$ is the mass fraction of particles with diameter, D, in the range:

$$\sqrt{\frac{18\mu L}{g(\rho_s - \rho_f)t}} < D < D_{200} \tag{1.8}$$

Collection of samples at various times and depths in the mixture allows the distribution of particle diameters to be constructed.

The hydrometer method also exploits the differential settling characteristics of a dilute mixture of particles. By this approach, a hydrometer is inserted into the settling solution at various times and the depth of flotation as well as the density of the mixture associated with that flotation are recorded. If the density of the solid fluid mixture as would be measured by the hydrometer is initially ρ_{h0} and the density reading obtained at some later time is $\rho_h(L, t)$, then:

$$\frac{m(L,t)}{m_0} = \left(\frac{\rho_h - \rho_f}{\rho_{h0} - \rho_f}\right) \tag{1.9}$$

---

[2]An important assumption that is made in Stokes' law is that the grains are spherical. While this may be appropriate for sand particles, clay particles tend to be platelike and some calibration of the procedure may be necessary.

Table 1.2: Experimental results from a wet method experiment for determining fine grain size distribution

Grain Size D (mm)	Weight with Diameter $< D$ (g)	Mass Ratio m/m_0
0.070	150.0	1.00
0.040	147.0	0.98
0.010	127.5	0.85
0.005	91.5	0.61
0.002	42.2	0.28
0.001	22.5	0.15

where m/m_0 is the mass fraction of particles with diameter less than D calculated using equation (1.7). The distribution of particle sizes may be constructed using this data collected at a sequence of times.

Although the methods outlined above are conceptually very simple, they are complicated by the need to compensate for temperature effects, for the time intervals for insertion of the pipette into the solution, the initial concentration of particles in the solution, the method of obtaining the initial uniform particle distribution in the fluid, and other protocols. Details of implementation of these methods have been standardized, for example in [1] and [5]. For purposes of subsequent discussion here, an example of a set of data that could be obtained from the pipette or hydrometer method is provided in Table 1.2.

The information gained from sieve and wet method analyses reveals more than just the range of grain sizes. It can help to classify the soil as to its type, e.g., sand, silt, silty sand, etc. Particle sizes smaller than 0.002 mm are considered to be clay or clay-sized fractions. In addition the data reveal the degree of sorting of the soil. A course-grained soil for which all the grains are approximately the same size is called well sorted (or poorly graded). A soil that exhibits a wide range of grain sizes is designated as poorly sorted (or well graded). The shape of the resulting *grain size distribution curve* can also reveal information regarding the history of the soil.[3]

The grain size distribution curves for two soil samples are plotted in Figure 1.6. Along the horizontal axis is plotted the grain size. On the vertical axis is plotted the percent weight finer than the indicated grain size. For example, the percent by weight of grains with diameter smaller than 0.01 mm in the clayey sandy-silt sample is approximately 40%. Similarly, in the case of the silty fine-sand sample, approximately 25% of the grains have diameters smaller than 0.1 mm. It should be kept in mind that the process of sieving measures the smallest cross-sectional diameter of the grain. A needle-shaped grain will be categorized as having a size equal to its width rather than its length, assuming of course it does not get lodged crosswise in the sieve. Thus only spherical particles that have the same measure of size regardless of orientation are uniquely identified by sieving. Nevertheless sieving is applied widely to soils containing grains of all shapes.

Figure 1.6 demonstrates that the clayey sandy-silt sample is finer grained than the silty fine-sand sample. In fact, by referring to the soil classification found beneath the distribution curve, it is evident how these samples received their classification.

---

[3] We will consider this in more detail in the next chapter.

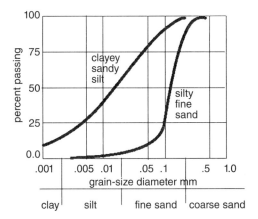

Figure 1.6: The grain size distribution indicates the soil classification of a sample and its degree of gradation.

Additional information may be obtained from the shape of the distribution curves. Note that the largest slope of the silty fine-sand curve is much steeper than that of the clayey sandy-silt curve. This indicates that the silty fine sand has a more uniform size distribution. In other words, the silty fine sand is considered to be better *sorted* or more poorly *graded* than the clayey sandy silt.

Two measures have been developed to describe the range in grain sizes of a soil sample. One is called the *coefficient of uniformity* and is defined as:

$$C_u = \frac{D_{60}}{D_{10}} \tag{1.10}$$

where D_n refers to the grain size greater than or equal to n% of the grains by weight. For example, 60% of the grains by weight are smaller than D_{60}. The denominator designated as D_{10} is also known as the *effective grain size*. The second measure is the *coefficient of curvature* calculated as:

$$C_c = \frac{D_{30}^2}{D_{10}D_{60}} \tag{1.11}$$

A well-graded soil has a coefficient of curvature between 1.0 and 3.0. Additionally, the coefficient of uniformity is greater than 4.0 for a well-graded gravel and greater than 6.0 for sands. A soil whose coefficient of uniformity is less than 2.0 is a *uniform soil*. A poorly graded soil violates at least one of these criteria, and a soil is said to be uniform if its coefficient of uniformity is less than or equal to 2.0 [7]. For the data of Figure 1.6, the coefficients of uniformity and curvature of the clayey sandy-silt sample are, respectively:

$$C_u = \frac{D_{60}}{D_{10}} = \frac{0.02\,\text{mm}}{0.001\,\text{mm}} = 20.0 \tag{1.12}$$

$$C_c = \frac{D_{30}^2}{D_{10}D_{60}} = \frac{(0.008\,\text{mm})^2}{(0.001\,\text{mm})(0.02\,\text{mm})} = 3.2 \tag{1.13}$$

For the silty fine sand, these coefficients are calculated as:

$$C_u = \frac{D_{60}}{D_{10}} = \frac{0.15\,\text{mm}}{0.05\,\text{mm}} = 3.0 \tag{1.14}$$

$$C_c = \frac{D_{30}^2}{D_{10}D_{60}} = \frac{(0.1\,\text{mm})^2}{(0.05\,\text{mm})(0.15\,\text{mm})} = 1.3 \tag{1.15}$$

The fact that C_u for clayey sandy silt is greater than C_u for silty fine sand confirms the previous observation that the clayey sandy silt is a better graded, or less uniform, soil[4]

1.4   THE CONCEPT OF SATURATION

Consider again the experimental apparatus consisting of a graduated cylinder containing 10 cm of a porous medium composed of sand and water covered by 5 cm of water. Suppose that a tap at the base of the graduated cylinder covered by a piece of glass wool is opened to allow the water to drain out while preventing the sand from escaping. The water that drains out is collected in a glass beaker. When the water has drained sufficiently such that no water remains above the porous medium, what remains in the bottom 10 cm of the graduated cylinder is a mixture of air, water, and sand. This is a three-phase porous media system. The fraction of pore space occupied by the water phase is called the *saturation*. Mathematically, the water saturation is defined as:

$$s^w = \frac{1}{V^v}\int_{V^w} dV = \frac{V^w}{V^v} \tag{1.16}$$

where V^v is the pore volume within the sample volume of porous medium, V, and is defined as:

$$V^v = V - V^s \tag{1.17}$$

Notice that in equation (1.16) the size of the sample is important. For example, if the entire porous medium is used, a single average value of saturation is obtained for the medium at any time. If smaller subvolumes within the porous system are considered, one may produce a spatially variable saturation field that characterizes the system at any instant. If the sample size is too small, the value of s^w calculated will change with small changes in the sample size. The saturation can take on values ranging from 0, when no water is present in the sample such that the void space is

---

[4] See the Earth Manual [3] for more information regarding grain size distributions and soil classification.

occupied completely by air, to 1, when the water occupies all of the void space and no air phase is present in the porous medium sample. From equation (1.15), the porosity is the pore volume divided by the total volume occupied by the porous medium such that $\varepsilon = V^v/V$. Substitution of this expression into equation (1.16) to eliminate V^v and multiplication by the porosity yields:

$$\varepsilon s^w = \frac{1}{V} \int_{V^w} dV = \frac{V^w}{V} \qquad (1.18)$$

This relation motivates definition of the *water content* as the ratio of the volume of water to the total volume denoted by θ, such that:

$$\theta = \varepsilon s^w = \frac{V^w}{V} \qquad (1.19)$$

After a sufficiently long period of time (where the definition of "sufficiently" is itself an interesting question usually answered as being somewhere from minutes to several hours), the drainage from the sand will stop, although some water will remain held in the sand. This water is called the *residual saturation*, and it exists as isolated droplets, is immobile, and will not normally drain.

While the concept of residual saturation is useful in theory, it is actually an imprecise quantity that is difficult to define unambiguously either physically or mathematically. In fact, if the graduated cylinder with the sample is kept in a low humidity environment for a number of days after residual saturation has been reached, the sand will continue to dry out gradually, and the saturation will decrease below residual. The reason for this is that the water will evaporate into the air in the pore space and then move out of the sand, primarily by vapor diffusion. The transfer of vapor from the liquid water phase to the vapor phase is a *phase transformation*. Water changes from liquid to gas as it moves across the interface from the liquid to the gas phase. Residual saturation will be encountered again when the topic of constitutive relationships is explored.

Consider, again, the drainage of the sand saturated with water. Suppose that while the drainage is occurring, some olive oil is poured onto the sand such that it seeps into some of the space between the grains. For this scenario, four phases comprise the porous medium: solid grains, water, the vapor, and oil. Although identifiable interfaces exist between each pair of phases, some material will transfer across the interfaces. Over time, some of the oil will dissolve into the water, and some of the water will dissolve into the oil. This transfer of molecules between phases is an example of *interphase transfer* or *interfacial transport* between two liquid phases across their interface. Despite this interphase transfer, the distinct interface remains as a location where material properties undergo a sharp transformation.

1.5   THE CONCEPT OF PRESSURE

Pressure, by its definition, is the magnitude of a force acting normal to a surface per unit area. It is also a measure of energy per unit volume. The concept is most easily understood for a static system. Consider the simple case of a graduated cylinder filled to the point where the water depth is 15 cm as shown in Figure 1.7. As a reference define the pressure at the top of the water column to be zero, and define a coordinate axis z that is positive downward and also equal to zero at the top of the water column. For this small system, the water density can be considered constant throughout the column. The force due to water acting downward at any cross section of the cylinder is equal to the weight of water above the cross section. This weight, W, is equal to the density of water times the gravitational acceleration times the volume. Thus, at a distance z from the top of the water column, the downward force exerted by the water is:

$$W = \rho_w g V = \rho_w g \pi r^2 z \tag{1.20}$$

where ρ_w is the density of water and r is the radius of the cylinder. Pressure is the force per unit area; thus division of equation (1.20) by the cross-sectional area, πr^2, gives the water pressure, p_w, as:

$$p_w = \frac{W}{A} = \rho_w g z \tag{1.21}$$

The pressure is independent of the size or shape of the cross section of the cylinder; it depends only on the distance from the water surface. At the bottom of the cylinder, under 15 cm of water, the pressure is:

$$p_w = \rho_w g z_{bot} = 1.0 \frac{g}{cm^3} \times 980 \frac{cm}{sec^2} \times 15\,cm = 1.47 \times 10^4 \frac{dynes}{cm^2} \tag{1.22}$$

Pressure is also energy per volume. Thus the energy in the cylinder due to water, E_p, can be calculated by integrating the pressure over the cylinder of fluid. Since the cross-sectional area is constant, this integral is:

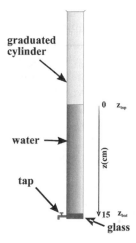

Figure 1.7: Definition sketch for discussion of pressure. Note that the axis is positive downwards with a value of zero at the top of the water column.

$$E_p = \int\limits_{z_{top}}^{z_{bot}} \pi r^2 p_w \, dz = \int\limits_{z_{top}}^{z_{bot}} \pi r^2 \rho_w g z \, dz = \pi r^2 \rho_w g \frac{z_{bot}^2 - z_{top}^2}{2} \tag{1.23}$$

where z_{bot} is the coordinate of the bottom of the cylinder (15 cm in the present example) and z_{top} is the coordinate of the top of the cylinder (0 cm in the present example). Since the volume of the fluid in the cylinder is $\pi r^2(z_{bot} - z_{top})$, the energy per volume is a pressure denoted as p_w such that:

$$\frac{E_p}{V} = p_w = \rho_w g \frac{z_{bot} + z_{top}}{2} \tag{1.24}$$

Thus p_w, the pressure obtained as energy per volume is equal to the pressure calculated as force per area using equation (1.21) if the area is located at the mid-height of the water column, the centroid.

The quantity p_w is also seen to be a volume-averaged pressure for the region of interest. For the cylindrical geometry considered here, the pressure obtained by averaging over the cylindrical volume of water can easily be shown to be equal to the pressure obtained by averaging the pressure over the surface of the cylinder. For an arbitrarily shaped region, and even for a cylinder whose axis is not aligned with the direction of gravity, this is not the case. This observation introduces the important concept that even if the point values of a quantity are well-described, average values of the quantity can be different depending on the averaging procedure used.

The preceding discussion of pressure is relatively transparent because the water is static and the geometry is simple. Most important, one can measure pressure in a fluid, p_w, based on the above concepts whenever a static column of water (or any other liquid of known density) can be created and placed in contact with a location where it is desired to measure the pressure. To demonstrate this fact and to provide a foundation for a subsequent presentation of constitutive theory, another experiment is discussed next based on the design illustrated in Figure 1.8.

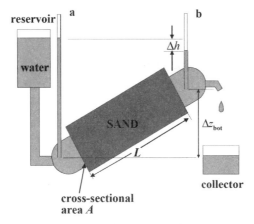

Figure 1.8: Apparatus used to demonstrate the concept of pressure measurements and later to describe the measurement of hydraulic conductivity.

In this experiment, the water moves parallel to the cylinder axis through the sand; as it does so, it loses energy. One measure of this loss of energy is the pressure decrease. While it is quite challenging to measure the pressure of the moving water in the sand directly, the pressure can be easily measured indirectly by creating a static column of water that accesses the moving fluid. The two vertical standpipes, or *manometers*, that are attached to the experimental apparatus in Figure 1.8 facilitate the measurement of the pressure in the moving water. Water rises in the manometers to the level consistent with the pressure at the access point at the base of the manometer tube. Given the height of the water in the tube, one can determine the pressure at the base of the tube using equation (1.21) and therefore in the moving fluid at that point of entry into the manometer.

If there were no flow occurring in the apparatus depicted in Figure 1.8, the water levels in the manometers would be equal such that $\Delta h = 0$. One must be careful to note that this observation does not mean that the pressure at the bases of the two manometers are equal. In fact, when there is no flow the pressures at the bases of the two manometers differ by $\rho_g \Delta z_{bot}$. A pressure difference between two points will cause flow to occur only if it is sufficient to overcome the force of gravity. The reservoir serves to provide this additional force. For the case indicated when flow is occurring, the water height in the right manometer is less than that in the left by Δh. This indicates that accompanying the flow is a loss of energy between the two points where the manometers contact the fluid. This energy loss is related to the viscous character of the fluid and its interaction with the sand; it is not related to the fact that the fluid is flowing "uphill."

1.6 SURFACE TENSION CONSIDERATIONS

While the physical effects of pressure on flow of a single fluid in a porous medium are rather straightforward to describe, the relation between the pressures in adjacent fluid phases separated by an interface is more complex. This situation arises when more than one fluid phase is present. Before discussing the impact of this situation within a porous medium, we consider a simple experiment wherein the behavior of a bead of water on a waxed surface is observed, such as a raindrop resting on the hood of a recently waxed car. The drop surface has the geometry of an oblate spheroid (squashed sphere), perhaps approaching spherical if it is small enough that gravitational effects are negligible. A circle of contact area exists between the spherical shape and the waxed surface with the radius of the contact area being somewhere between 0 and the radius of the droplet. If the same bead of water were placed on an unwaxed hood, it would spread over the surface as a thin film. Of interest is an explanation of the factors that influence the interactions between fluids and of fluids with surfaces.

The explanation of fluid-solid interactions lies in the molecular structure of the fluid comprising the drop and the way that structure relates to the molecules at the surface of the solid. In general, water molecules attract one another. At the boundary of the drop, however, the attractive forces of molecules within the drop are not balanced by the attractive forces of molecules outside the drop. The result is a modification of the structural arrangement of the molecules at the drop surface. Although the resulting molecular arrangement cannot be formally considered as a skin, it has characteristics often identified with a membrane under tension, such as the surface of a balloon filled with air.

The change in energy of a volume of static fluid due to an infinitesimal decrease in its volume is equal to $-pdV$. This expresses the fact that work must be done on the volume (i.e., it must be compressed by an external force per unit area) for its energy to increase. Similarly, a change in the surface area bounding a material is accompanied by a change in energy. Expansion of the surface requires that the attractive forces among the molecules be overcome. The change in energy will be designated γdA, where A is the surface area. For an interface between two different fluids, γ is referred to as the *interfacial tension*. In the case of an interface between a liquid and its own vapor, γ is called the *surface tension*. In many instances, these terms are used less precisely and interchangeably.

We emphasize that the change in energy of a surface is positive when the surface expands because work must be done to stretch that surface. On the other hand, the energy of a volume is increased by compressing that volume. The quantity, γ, has units of force per unit length, or energy per unit area. Mechanically, this can be understood by realizing that stretching a membrane may be accomplished by applying a force per unit length along the curve that bounds the membrane. Analogous to the fact that a force per unit area causes a change in volume, a force per unit length is needed to cause a change in area.

A droplet of water in air will attempt to minimize its surface area in response to the surface tension. The minimum surface area for a specified volume of fluid is a sphere. Thus one influence on the shape of a drop of water tends to cause the drop to be spherical. Gravity can cause deviation from the spherical shape, but in the immediate discussion, this force will be neglected. When the droplet is placed in contact with the solid, the force of interaction between the fluid and solid and the fact that the solid is rigid will influence the drop shape. The interface between the water and air tends to be spherical. If the water molecules are more attracted to each other than to the solid, as is the case for a waxed surface, the drop will minimize the area of contact with the solid. In the limit of no attraction to the solid, the droplet would be a sphere sitting on the solid. An increased attraction between the water and solid will cause the area of contact between the fluid and solid to increase. In the limit where the attraction to the solid is much stronger than the attraction between water molecules, the droplet will spread as a film over the solid. At intermediate levels of attraction, the surface area of contact between the fluid and solid will take on values intermediate between a complete sphere and a film. The forces acting on the surface of the droplet are the fluid pressures. At the curve at the edge of the water droplet where the solid, water, and air phases come together, the *common line*, interfacial tension and surface tension forces are operative. If the droplet surface or the common line contain significant mass, gravitational forces will also be operative.

To quantify the relationship between surface tension and the fluid pressures consider the simple geometry involved in the rise of water in a capillary tube (Figure 1.9a). If the water in the tube is in static equilibrium, then there must be a balance of forces acting on the interface between the air and the water phases. The forces that act on the interface are the pressure of the air acting on the concave side of the interface, the pressure of the water acting on the convex side, and the interfacial tension effects that act tangent to the surface. The fluid on the concave side, in this

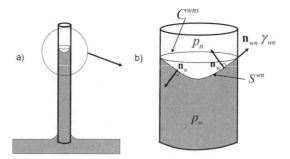

Figure 1.9: Definition sketch of the force balance between pressure and tensile forces at the static water-gas interface in a capillary tube: (a) the capillary tube; (b) forces acting on the interface.

case air, is referred to as the nonwetting fluid because it is less attracted to the solid than the other fluid. This pressure will be signified as p_n, where the subscript n indicates "nonwetting." The fluid on the convex side of the interface, in this case water, is referred to as the wetting fluid since it preferentially wets the the solid. The wetting phase pressure is indicated as p_w, where w indicates the "wetting" phase.[5] In addition to the pressure acting on the interface, at the edge of the interfacial surface forces are applied by the interfacial tension, γ_{wn}, and the surface tensions, γ_{ws} and γ_{ns}. These forces are exerted on the common line, the curve where the three phases come together. In addition, a lineal tension of the common line can contribute to a force balance. At equilibrium, the forces that act on the interface must be balanced, as must the forces that act on the common line.

To analyze the total force balance on the interface, integrals are formulated, each of which accounts for one of the forces summed over its geometric region of application. The forces acting on the wn interface include the fluid pressures that act on the surfaces and the surface tension along the common line. Analogously with a membrane, the pressures on the two sides will influence the shape of the interface while the tensions on the edge pin the location of the boundary or allow for slippage. The total force from the pressure is obtained as the integral of the pressure over the complete surface of the interface. The force from the surface tension is obtained as an integral over the common line.

The force on the wn interface due to the nonwetting air phase acts in a direction normal to the surface and in the direction pointing out from this phase. If this normal at any point on the surface is designated as $\mathbf{n}_n$, the total force due to the nonwetting phase may be calculated as the vector:

$$\mathbf{F}^n = \int_{S^{wn}} p_n \mathbf{n}_n \, dS \tag{1.25}$$

where S^{wn} is the surface of the interface. On the other side of the interface, the wetting phase exerts a force normal to the interfacial surface in the direction tangent

---

[5] Note that regardless of the chemical makeup of the two fluids in contact with the solid, the one that is more attracted to the solid is called the wetting phase. This can be confusing, for example, in the case of an oil and water mixture in contact with a plastic such that the oil is the "wetting" phase while water is referred to as "nonwetting."

to $\mathbf{n}_w$, the unit vector normal to the surface that points outward from the wetting phase. This force may be obtained as the integral:

$$\mathbf{F}^w = \int\limits_{S^{wn}} p_w \mathbf{n}_w \mathrm{d}S \qquad (1.26)$$

Finally, the force exerted by the interfacial tension along the bounding line is in a direction tangent to the unit vector $\mathbf{n}_{wn}$ that points outward from the interface in a direction that is normal to the bounding line and tangent to the surface. The total force exerted by the interfacial tension is therefore:

$$\mathbf{F}^{wn} = \int\limits_{C^{wns}} \gamma_{wn} \mathbf{n}_{wn} \mathrm{d}C \qquad (1.27)$$

where C^{wns} is the bounding curve where the wetting, nonwetting, and solid phases come together. Since the total force acting on the interface must be zero for the static case, the sum of the three forces in equations (1.25), (1.26), and (1.27) must be zero, or:

$$\int\limits_{S^{wn}} p_n \mathbf{n}_n \mathrm{d}S + \int\limits_{S^{wn}} p_w \mathbf{n}_w \mathrm{d}S + \int\limits_{C^{wns}} \gamma_{wn} \mathbf{n}_{wn} \mathrm{d}C = 0 \qquad (1.28)$$

The formulation of the total force in equation (1.28) has intuitive appeal on physical grounds. However, besides the total force balance for the interface, the forces at each point on the interface must also balance at equilibrium. Direct formulation of this balance is not intuitive. Nevertheless, application of mathematical theorems, that may require the user to leave intuition behind, to the physical equation (1.28) will lead to the desired result. The mathematical expression that is useful in this analysis converts the integral over the curve bounding the surface to an integral over the surface. This relation is the *divergence theorem* for a surface and has the general form [7]:

$$\int\limits_{C^{wns}} f\mathbf{n}_{wn} \mathrm{d}C = \int\limits_{S^{wn}} \nabla' f \, \mathrm{d}S - \int\limits_{S^{wn}} (\nabla' \cdot \mathbf{n}_n)\mathbf{n}_n f \, \mathrm{d}S \qquad (1.29)$$

where ∇' is the two-dimensional del operator acting in the surface. Application of this relation to the last term on the left side of equation (1.28) with f replaced by γ_{wn} yields:

$$\int\limits_{S^{wn}} p_n \mathbf{n}_n \mathrm{d}S + \int\limits_{S^{wn}} p_w \mathbf{n}_w \mathrm{d}S + \int\limits_{S^{wn}} \nabla' \gamma_{wn} \mathrm{d}S - \int\limits_{S^{wn}} (\nabla' \cdot \mathbf{n}_n)\mathbf{n}_n \gamma_{wn} \mathrm{d}S = 0$$

or, after collection of the terms in the integrands:

$$\int\limits_{S^{wn}} [p_n \mathbf{n}_n + p_w \mathbf{n}_w + \nabla' \gamma_{wn} - (\nabla' \cdot \mathbf{n}_n)\mathbf{n}_n \gamma_{wn}] \mathrm{d}S = 0 \qquad (1.30)$$

Since this relation must hold, regardless of the size or portion of the interfacial surface over which the integration is performed, the integrand in equation (1.30) must be zero at every point on the surface so that:

$$p_n \mathbf{n}_n + p_w \mathbf{n}_w + \nabla' \gamma_{wn} - (\nabla' \cdot \mathbf{n}_n) \mathbf{n}_n \gamma_{wn} = 0 \qquad (1.31)$$

This equation is a vector equation and has components in each of three orthogonal coordinate directions. The first, second, and fourth terms are vectors normal to the surface. The third term involves the gradient in the surface and thus is a vector tangent to the surface. Since equation (1.31) must be valid in any direction, the component of the equation in the directions tangent to the surface surface is obtained as:

$$\nabla' \gamma_{wn} = 0 \qquad (1.32)$$

Equation (1.32) indicates that at equilibrium, the surface tension of an interface between two fluids will be constant, independent of position, since its gradient is zero.

The balance of forces in the direction normal to the interface is obtained as the dot product of equation (1.31) with the unit vector normal to the surface, $\mathbf{n}_n$. Note that $\mathbf{n}_n = -\mathbf{n}_w$ such that $\mathbf{n}_n \cdot \mathbf{n}_n = 1$ and $\mathbf{n}_n \cdot \mathbf{n}_w = -1$. Thus the normal component of the force balance at any point on the interface is:

$$p_n - p_w - (\nabla' \cdot \mathbf{n}_n) \gamma_{wn} = 0 \qquad (1.33)$$

For the case where the interface is flat, the orientation of $\mathbf{n}_n$ does not change with position on the surface, and the divergence of this normal will be zero. Thus for a flat interface, equation (1.33) simplifies to the condition that the pressure across the interface will be continuous with $p_n = p_w$. This is the condition that is typically imposed on large-scale systems where the curvature is small, such as the surface of a bucket of water or at the top of a swimming pool. In general, the quantity $\nabla' \cdot \mathbf{n}_n$ is equal to the sum of the curvatures of the surface in any two orthogonal directions. If the *curvature of the surface* is denoted as $2/R_c$, where R_c is the *geometric mean*, the *radius of curvature*[6] is calculated as:

$$\nabla' \cdot \mathbf{n}_n = \frac{2}{R_c} = \left(\frac{1}{R_1} + \frac{1}{R_2} \right) \qquad (1.34)$$

where R_1 and R_2 are the radii of curvature of any two orthogonal curves on the surface and $\nabla' \cdot \mathbf{n}_n$ is called the *mean curvature*. Each of the radii of curvature is positive when the corresponding curve is concave on the n side and negative when the corresponding curve is concave on the w side. Then equation (1.33) can be rewritten in the form:

$$p_n - p_w = \frac{2}{R_c} \gamma_{wn} \qquad (1.35)$$

The *capillary pressure*, p_c, is now defined, in general, as the product of the mean curvature and the interfacial tension:

---

[6] For a spherically shaped surface, R_c is equal to the radius of the sphere; for a flat surface, R_c is infinite.

$$p_c = \frac{2}{R_c}\gamma_{wn} \tag{1.36}$$

Thus, equation (1.35) expresses the fact that at equilibrium:

$$p_n - p_w = p_c \tag{1.37}$$

Equation (1.35) is known as the *Laplace equation for capillary pressure*. Note that $p_n \neq p_w$ whenever $|R_c| \neq \infty$. Thus, when the surface tension is zero, the equilibrium pressure drop across an interface will be zero only if the interface is flat or at points on the interface where the radii of curvature are equal in magnitude and opposite in sign. Equation (1.35) is similar to that which describes the pressure difference across a physical membrane, such as for a balloon. If the surface tension of the membrane is inadequate to sustain the pressure difference, the balloon will burst.

Before illustrating the balance of forces on an interface between phases with an example calculation, we will develop the expression for the balance of forces on the common line. This expression provides information about the relationships among the interfacial tensions. Because the common line is located on the surface of the solid where the two fluid phases and the solid contact each other, the balance equation must account for the forces exerted by the interfaces. Consider the situation presented in Figure 1.10 where a drop of liquid is sitting on a solid surface, for example a drop of water on the hood of a car. The drop is a cap with spherical shape. A magnification of the region in the vicinity of the common line is provided as part (b) of the figure, and various unit vectors are illustrated there. The force exerted by

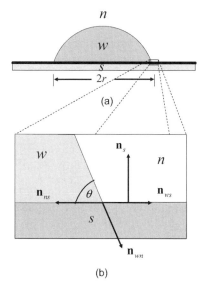

Figure 1.10: (a) Drop of water on a horizontal surface with angle of contact θ and circle of contact with radius r. (b) Enlargement of region where three phases meet with the unit normal vectors $\mathbf{n}_{wn}$, $\mathbf{n}_{ws}$, and $\mathbf{n}_{ns}$ indicated at a point on the common line.

each interface on the common line is tangent to the interface and exerts a force on the common line trying to displace it. The total force exerted on a segment of the common line by the interfaces is therefore given as the sum of the integrals of each of the three interface forces over the common line:

$$\mathbf{F}_{\text{interfaces}} = -\int_{C^{wns}} \gamma_{wn}\mathbf{n}_{wn}dC - \int_{C^{wns}} \gamma_{ws}\mathbf{n}_{ws}dC - \int_{C^{wns}} \gamma_{ns}\mathbf{n}_{ns}dC \qquad (1.38)$$

where each of the unit vectors is normal to the common line C, tangent to its corresponding interface, and points outward from the interface at the common line. In addition to these forces, the mechanical behavior of the solid and its surface roughness exert an attractive force on the common line with magnitude f_s normal to the smooth representation of the surface in the direction pointing into the solid. This force is $-f_s\mathbf{n}_s$. Finally, the common line has its own tension that acts along its axis, much like a rubber band or a spring. Thus forces are exerted at the ends of the common line segment by the line tension, γ_{wns}, in a direction tangent to the common line and pointing outward. These additional forces may be expressed as:

$$\mathbf{F}_{\text{line}} = -\int_{C^{wns}} f_s\mathbf{n}_s dC + (\gamma_{wns}\mathbf{n}_{wns})_1 + (\gamma_{wns}\mathbf{n}_{wns})_2 \qquad (1.39)$$

where the subscripts "1" and "2" refer to the two ends of the segment being studied and $\mathbf{n}_{wns}$ is the outward directed tangent vector at each end. If a curve being studied is a closed loop, the full circle of contact in the present case, the forces due to line tension will not appear as there are no ends. At equilibrium, the total force acting on the common line segment will be zero. Thus, addition of equations (1.38) and (1.39) yields:

$$-\int_{C^{wns}} \gamma_{wn}\mathbf{n}_{wn}dC - \int_{C^{wns}} \gamma_{ws}\mathbf{n}_{ws}dC - \int_{C^{wns}} \gamma_{ns}\mathbf{n}_{ns}dC$$
$$-\int_{C^{wns}} f_s\mathbf{n}_s dC + (\gamma_{wns}\mathbf{n}_{wns})_1 + (\gamma_{wns}\mathbf{n}_{wns})_2 = 0 \qquad (1.40)$$

The expression for the balance of forces at a point on the common line may be obtained by making use of the divergence theorem for a curve to relate the terms evaluated at the ends of the segment to integrals over the segment. This theorem has the form [7]:

$$(f\mathbf{n}_{wns})_1 + (f\mathbf{n}_{wns})_2 = \int_{C^{wns}} \nabla''f\, dC + \int_{C^{wns}} (\mathbf{n}_{wns}\cdot\nabla''\mathbf{n}_{wns})f\, dC \qquad (1.41)$$

where ∇'' is the one-dimensional del operator acting along the curve. Application of equation (1.41) to equation (1.40), with f replaced by γ_{wns}, to eliminate the terms at the ends of the curve then provides:

$$-\int_{C^{wns}} \gamma_{wn}\mathbf{n}_{wn}dC - \int_{C^{wns}} \gamma_{ws}\mathbf{n}_{ws}dC - \int_{C^{wns}} \gamma_{ns}\mathbf{n}_{ns}dC - \int_{C^{wns}} f_s\mathbf{n}_s dC$$
$$+ \int_{C^{wns}} \nabla''\gamma_{wns}dC + \int_{C^{wns}} (\mathbf{n}_{wns}\cdot\nabla''\mathbf{n}_{wns})\gamma_{wns}dC = 0 \qquad (1.42)$$

or, after collection of the integrands:

$$\int_{C^{wns}} [-\gamma_{wn}\mathbf{n}_{wn} - \gamma_{ws}\mathbf{n}_{ws} - \gamma_{ns}\mathbf{n}_{ns} - f_s\mathbf{n}_s + \nabla''\gamma_{wns}$$

$$+ (\mathbf{n}_{wns} \cdot \nabla''\mathbf{n}_{wns})\gamma_{wns}] dC = 0 \tag{1.43}$$

This equality must hold regardless of the length of common line segment over which the integration is performed. Therefore, the integrand itself must be zero or:

$$-\gamma_{wn}\mathbf{n}_{wn} - \gamma_{ws}\mathbf{n}_{ws} - \gamma_{ns}\mathbf{n}_{ns} - f_s\mathbf{n}_s + \nabla''\gamma_{wns} + (\mathbf{n}_{wns} \cdot \nabla''\mathbf{n}_{wns})\gamma_{wns} = 0 \tag{1.44}$$

The term in parentheses in this equation is the vector known as the *curvature* of the common line. It may be expressed in terms of two orthogonal vector components as:

$$\mathbf{n}_{wns} \cdot \nabla''\mathbf{n}_{wns} = \kappa_n \mathbf{n}_s + \kappa_g \mathbf{n}_{ws} \tag{1.45}$$

where κ_n is the *normal curvature* and κ_g is the *geodesic curvature* [4]. Substitution of this relation into equation (1.44) yields the force balance on the common line as:

$$-\gamma_{wn}\mathbf{n}_{wn} - \gamma_{ws}\mathbf{n}_{ws} - \gamma_{ns}\mathbf{n}_{ns} - f_s\mathbf{n}_s + \nabla''\gamma_{wns} + \kappa_n\gamma_{wns}\mathbf{n}_s + \kappa_g\gamma_{wns}\mathbf{n}_{ws} = 0 \tag{1.46}$$

This equilibrium balance of forces at a point on the common line is a vector equation that states the balance of forces in any coordinate direction. Important balance expressions may be obtained by calculating the dot product of equation (1.46) with each of the three orthogonal vectors $\mathbf{n}_{wns}$, $\mathbf{n}_s$, and $\mathbf{n}_{ws}$.

The dot product of $\mathbf{n}_{wns}$ with equation (1.46) yields:

$$\mathbf{n}_{wns} \cdot \nabla''\gamma_{wns} = 0 \tag{1.47}$$

where use has been made of the fact that all the unit vectors that appear explicitly in equation (1.46) are orthogonal to $\mathbf{n}_{wns}$. This equation expresses the equilibrium requirement that γ_{wns} be a constant along the common line in that its spatial derivative taken in the direction along the common line is zero.

The dot product of $\mathbf{n}_s$ with equation (1.46) gives the balance of forces in the direction normal to the solid surface:

$$-\gamma_{wn}\mathbf{n}_{wn} \cdot \mathbf{n}_s - f_s + \kappa_n\gamma_{wns} = 0 \tag{1.48}$$

The angle between the *wn* interface and the *ws* interface at the common line is called the *contact angle* and is designated as θ. Therefore:

$$\mathbf{n}_{wn} \cdot \mathbf{n}_s = -\sin\theta \tag{1.49}$$

and equation (1.48) may be written:

$$\gamma_{wn} \sin \theta - f_s + \kappa_n \gamma_{wns} = 0 \tag{1.50}$$

This equation indicates that the force exerted on the common line by the *wn* interface that tends to pull the common line away from the solid is balanced by an attractive force of the solid and a tendency of the curved common line to expand (or contract) due to the lineal tension. Note that if the common line has no curvature in the direction normal to the solid (such that $\kappa_n = 0$ as in Figure 1.10), the lineal tension will not result in a force in the direction normal to the solid.

Finally, the component of balance equation (1.46) in the direction tangent to the solid surface and normal to the common line (i.e., in direction $\mathbf{n}_{ws}$) is obtained from the dot product of equation (1.46) with $\mathbf{n}_{ws}$:

$$-\gamma_{wn} \cos \theta - \gamma_{ws} + \gamma_{ns} + \kappa_g \gamma_{wns} = 0 \tag{1.51}$$

where use has been made of the facts that $\mathbf{n}_{wn} \cdot \mathbf{n}_{ws} = \cos \theta$ and $\mathbf{n}_{ws} \cdot \mathbf{n}_{ws} = -\mathbf{n}_{ws} \cdot \mathbf{n}_{ns}$ $= 1$. This equation expresses the balance of forces in the direction tangent to the solid surface. When an imbalance exists, the common line will be displaced as one fluid spreads onto the surface. Equation (1.51) may be rearranged to the form:

$$\cos \theta = \frac{\gamma_{ns} - \gamma_{ws}}{\gamma_{wn}} + \frac{\kappa_g \gamma_{wns}}{\gamma_{wn}} \tag{1.52}$$

Typically, either from historical precedent or because the line tension and geodesic curvature are small, the last term in this expression is excluded so that the force balance becomes:

$$\cos \theta = \frac{\gamma_{ns} - \gamma_{ws}}{\gamma_{wn}} \tag{1.53}$$

Equation (1.53) is known as *Young's equation*.

The contact angle θ plays a very important role in multiphase flow. If the angle is less than 90 degrees, then the fluid *w* in Figure 1.10 is the wetting fluid. The fluid for which the wetting angle is greater than 90 degrees, the fluid *n* in the figure, is the nonwetting fluid. If $|\gamma_{ns} - \gamma_{ws}| > \gamma_{wn}$, then no equilibrium is possible as $|\cos \theta|$ would have to be greater than 1 for equation (1.53) to be satisfied. In this case, the wetting phase will spread until it completely coats the solid surface.[7]

In light of this theoretical analysis, we return to the question of why water will bead on a waxed car surface but not on an unwaxed surface. The explanation lies in the magnitude of the contact angle between the water and the painted surface as opposed to the contact angle between the water and the waxed surface. Water beads on the waxed solid surface because the contact angle between the water and the wax is greater than 90 degrees such that the air is the "wetting fluid" in this case. On the other hand, the contact angle between water and the unwaxed surface is less than 90 degrees such that the water tends to wet this surface. When the magnitude

---

[7] The wettability of various fluids can be influenced by additives that affect the surface tension of the fluids.

of the right side of equation (1.53) is greater than 1, as is the case for a poorly waxed car in need of a new paint job, the water tends to spread in sheets over the surface. In this comparison the *wettability* of the painted surface is said to be greater than that of the waxed surface.

Consider an experiment whereby a 0.10 cc droplet of water is placed on the hood of a car at a place where the hood is horizontal. Assume that the drop is small enough that its shape may be considered spherical. The contact angle between the water and solid phase is observed to be 100° such that the size of the droplet is more than half a sphere. The volume of the drop may be related to the radius of the *ws* circle of contact, *r*, and the contact angle, θ, according to:

$$V = \frac{\pi}{3}\left(\frac{r}{\sin\theta}\right)^3 (1 - \cos\theta)^2 (2 + \cos\theta) \tag{1.54}$$

For this system, where the surface is concave in the wetting phase, the mean radius of curvature is the negative of the radius of the drop cap:

$$R_c = -\frac{r}{\sin\theta} \tag{1.55}$$

If two variables from among V, r, θ, and R_c are observed, then the remaining variables may be calculated from the last two equations. If more than two of the variables are observed, the equations can serve as measures of the accuracy of the measurements. For the present case with $V = 0.1$ cc and $\theta = 100°$, rearrangement of equation (1.54) yields:

$$r = \left(\frac{3V}{\pi}\right)^{1/3} \left(\frac{1}{1 - \cos\theta}\right)^{2/3} \left(\frac{1}{2 + \cos\theta}\right)^{1/3} \sin\theta = 0.331\,\text{cm} \tag{1.56}$$

and equation (1.55) provides $R_c = -0.336$ cm. At room temperature, the surface tension of an air water interface[8] is approximately $\gamma_{wn} = 72.5$ dynes/cm [2]. Thus, from equation (1.36) the capillary pressure of the *wn* interface is:

$$p_n - p_w = p_c = \frac{2}{R_c}\gamma_{wn} = -4.32 \times 10^2 \text{ dynes}/\text{cm}^2 \tag{1.57}$$

where n refers to the air phase, w to the water phase, and the difference in the pressures is negative since the air preferentially "wets" the solid as indicated by the fact that the air-solid contact angle is less than 90°. For context, it may be helpful to note that this pressure is the same as the decrease experienced at the bottom of a glass of water if the depth of water is decreased by 0.44 cm. Although this magnitude of capillary pressure is small, it should seem reasonable that for flow in porous media where the curvatures can be orders of magnitude higher, capillary pressure is important.

---

[8] Surface tension for an air-water interface is approximately $\gamma_{wn} = 75.6 - 0.15\,T$ dynes/cm, where T is temperature in degrees Celsius.

With the information obtained thus far, we can also make use of equation (1.53) to determine the difference in interfacial tensions between the fluids and the solid:

$$\gamma_{ns} - \gamma_{ws} = \gamma_{wn}\cos\theta = 72.5\,\frac{\text{dynes}}{\text{cm}}\cos 100° = -12.6\,\frac{\text{dynes}}{\text{cm}} \qquad (1.58)$$

Although this result provides the magnitude by which $\gamma_{ws} > \gamma_{ns}$ for this system, no information is obtained about the values of each of these interfacial tensions.

Also, the result neglects the effects of common line tension. In fact, values of common line tensions involving a solid and a pair of fluids vary depending on the way the solid is prepared. The common line effects are considered to be small in porous media applications relative to other difficulties involving heterogeneity and scale. Nevertheless, one should be aware that this approximation is employed.

The *wn* interfacial tension exerts a force normal to the solid surface that is countered by the attraction of the solid to the common line. This attraction may be calculated using equation (1.50). Because the surface is flat and the force we are investigating acts normal to the surface on the curve, the normal curvature is zero. Therefore, the lineal tension does not affect the results for geometric reasons, and not by assumption. Equation (1.50) thus provides the attraction force per unit length of common line as:

$$f_s = \gamma_{wn}\sin\theta = 72.5\,\frac{\text{dynes}}{\text{cm}}\sin 100° = 71.4\,\frac{\text{dynes}}{\text{cm}} \qquad (1.59)$$

The preceding discussion and examples involve the effects of surface tension and capillary pressure at the microscale. The study of these phenomena in porous media columns or in the field is undertaken at the macroscale. At this scale, rather than looking at a particular interface or the points on an interface, the aggregate effects of interfacial processes within a representative region of the medium are accounted for. The term *capillarity* is commonly employed to refer to the fact that one fluid is preferentially drawn into a porous medium. This phenomenon is attributable to surface tension influences. Some aspects of this effect will now be examined briefly.

Generally, when water is brought into contact with dry soil, it is drawn into the soil. If the soil is above the water, the elevation to which the water will be drawn against gravity depends on the material that composes the soil and the pore sizes and geometry. To some degree, water will exist in soil above a saturated region and will be hydrodynamically connected to that region. The *water table* is the location in the subsurface where the water saturates the medium and where the pressure is equal to atmospheric. In fine-grained soil, water tends to be found at elevations higher above the water table than under similar conditions in more coarse soil. To demonstrate the reason this occurs, another simple experiment is considered.

The ends of two vertical clean glass tubes (*capillary tubes*) with inside diameters of 0.5 mm and 1.0 mm, respectively, are dipped into a beaker of water as depicted in Figure 1.11. The glass tubes are the solid phase, indicated as material s_g. After a short time, water is observed to have entered the tubes and risen above the level of

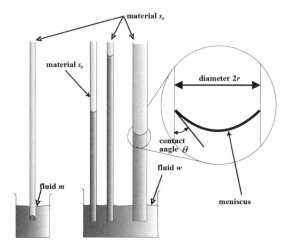

Figure 1.11: A suite of experiments demonstrating the relative wettability of glass and plastic in contact with air, water, and mercury.

the water in the beaker. Moreover, the water level in the 0.5 mm diameter tube is approximately twice as high as in the 1.0 mm tube.

This experiment demonstrates that the rise in the water level is roughly linearly inversely proportional to the radius of the tube. Also, at a horizontal observation level below the surface of the water in the beaker, the pressure in the water will be constant. Therefore if a mathematical plane is located such that it intersects a capillary tube, the water pressure in the tube at that position should be equal to the water pressure on the outside of the tube.

From this observation, the pressure of the water in the capillary tube at elevations above the water surface in the beaker can be calculated. Since the pressure difference between the air and the atmosphere across a flat surface will be zero according to equation (1.35), the water pressure at the water surface in the beaker is atmospheric. Therefore, the pressure in the capillary tube above this surface will be less than atmospheric. Pressure relative to atmospheric is called *gauge pressure*, and thus the pressure in the capillary tube is referred to as being *negative pressure*. Furthermore, since the water in the tube is static, the pressure decreases in direct proportion to elevation in the tube according to equation (1.21) in the form:

$$p_w = p_{atm} - \rho g Z \tag{1.60}$$

with a change in coordinate such that Z is the distance above the water level in the beaker and p_{atm} is atmospheric pressure. The higher the elevation above the water surface, the more negative will be the gauge pressure of the water. Since the density of air is very small, the pressure in the air may be considered to be constant in the study region (i.e., gravitational effects on the air pressure are negligible). At the top of the water column in the capillary tube, a jump in pressure takes place between the water and air phases. This jump is accounted for in equation (1.35). If the height of water in the capillary tube is designated as $Z = h$, combination of equations (1.35) and (1.60) gives:

$$p_w g h = \frac{2}{R_c} \gamma_{wn} \tag{1.61}$$

This equation indicates that *capillary rise* can occur in a capillary tube only if the surface tension between the two fluids is nonzero and the interface between the fluids is curved with the interface being convex (i.e., bulging outward) from the air side. If the interface is concave from (i.e., bulging into) the air side, the capillary rise will be negative such that the liquid level will decrease in a *capillary depression*. Equation (1.61) will be discussed further subsequently.

Now extend the experiment by dipping into the water a tube made of plastic material, denoted s_p, and with an inside diameter equal to that of the smaller glass tube, 0.5 mm (as indicated in Figure 1.11). Despite the equality of the tube diameters, the water does not rise as high in the plastic tube as in the glass tube. The interfacial tension of the air-water interface is unchanged, so based on equation (1.61), the lower rise in the plastic tube must be accompanied by a larger value of the radius of curvature of the interface, R_c, reflective of different interaction between the tube materials and the water.

As a final element of this experiment, dip a smaller diameter glass tube into a beaker of mercury (fluid m in Figure 1.11). In this case, the mercury does not penetrate the bottom of the tube and flow into the tube unless one pushes the tube deeper into the beaker. Indeed, the air displaces the mercury slightly as the mercury level in the tube is lower than that outside. In addition, the curved meniscus between the air and the fluid, that has thus far been convex from the air side, is now concave.

The explanation of the behavior of the water and mercury interactions with the tubes is found in the concept of *wettability*. In the air-water-solid systems, water preferentially wets the sides of the capillary tubes relative to air, and wets glass more enthusiastically than it wets plastic. The water is drawn up in the capillary tube because of its affinity for the sides of the tube and its ability to sustain a pressure jump across its interface with air.

The height of capillary rise is determined from the balance of forces as indicated in equation (1.61) at the meniscus between the water and the air. The contact angle between the meniscus and the tube wall, θ in Figure 1.11, is generally considered to be a property of the three phases. Although the roughness of the solid surface can impact the value of the contact angle, this effect will be neglected here. For the small capillary tubes under consideration, assume that the interface is essentially spherical such that the radius of curvature, R_c, is related to the tube radius, r, according to:

$$r = R_c \cos \theta \tag{1.62}$$

Substitution of this equation into equation (1.61) and rearrangement yields the height of the capillary rise as:

$$h = \frac{2\gamma_{wn} \cos \theta}{r \rho_w g} \tag{1.63}$$

Equation (1.63) demonstrates that the smaller the tube radius, the higher the meniscus will be drawn into the tube. As long as the tube is small enough that the approximation of a spherical interface applies, the capillary rise is linearly proportional to the inverse of the radius. Similarly, in naturally occurring fine-grained material, which would typically have smaller diameter pores than coarse-grained material, water is more easily held in the region above the water table. The observation that water does not rise as high in the plastic tube as in the glass tube suggests that the wettability of water relative to air on plastic is less than on glass. Finally, the observation that there is a depression rather than rise of mercury in the glass tube indicates that mercury does not wet glass relative to air.

While a capillary tube can, to a limited degree, be a surrogate for the channels in a porous medium, the analogy has some serious limitations. The pores in a porous medium are irregular in geometry and variable in shape. The cross-sectional characteristic length of a pore will vary over at least a couple of orders of magnitude. Pores are also connected to each other and allow flow to follow a complex, multi-directional path.

To obtain some insight into how a variable pore diameter might affect the behavior of multiphase fluids, consider yet another simple experiment. Figure 1.12 depicts a capillary tube with a small, but variable, diameter that is a simple surrogate for pores. The experiment begins by immersing the lower, flanged end of the capillary tube into a beaker of water. The tube is lowered very slowly until the water has just entered the tube up to point "1" in Figure 1.12b, the minimum constriction. This process is the *imbibition stage*, as water moves into the tube displacing the air that flows upward to leave the tube. If the capillary tube is lowered further into the water, for example by a distance δ, the water level in the capillary will increase only by an amount between 0 and δ. The reason the capillary rise is not sustained in this process is that the capillary forces are less effective because of the increased diameter of the tube. This difference will be minimized when the tube has been lowered such that the water just rises to point "2". If the tube is lowered further, the height of the water level in the capillary relative to the level in the beaker once again increases. This is due to the decrease in capillary diameter that gives rise to an increase in the

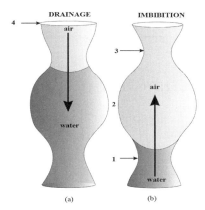

Figure 1.12: The constriction in the capillary tube tends to retain water at a higher elevation on drainage (*a*) and to preclude the water level from reaching a higher elevation on imbibition (*b*).

capillary force. Once location "3" is reached, the process continues in the same fashion as the tube is immersed further.

Let us now immerse the capillary tube to the location marked 4 in Figure 1.12a. Now slowly raise the capillary tube so the water moves downward in the capillary. In this, the *drainage stage*, the minimum cross section of the capillary tends to influence the behavior of the fluid by maintaining it at a level above the larger pore. Note that maintaining this higher level depends on the radius of the pore, the surface tension, and the wettability. The capillary forces are not influenced by the larger cross-section geometry of the capillary below the neck where the interface is established.

On the other hand, during imbibition the wider opening associated with the larger pore (or a large pore in a network in a natural soil) does not provide as attractive a region for water flow as does a small opening.[9] From another perspective, with all other factors being equal, the air-water interface tends to remain where the radius of the pore is the smallest. Thus water is precluded from advancing until the pressure is sufficient to overcome the impediment of the larger pore.

From this example it is evident that the physical behavior of the modified capillaries shown in Figure 1.12 is influenced by whether the water is draining or imbibing. In the macroscale study of porous media, the capillary pressure is related to the saturation. Thus it is not surprising that the equilibrium relationship between saturation and capillary pressure is not single-valued. The equilibrium relation depends on whether imbibition or drainage has occurred prior to examination of the state of the system. This situation is described by saying that the capillary pressure is a *hysteretic* function of saturation, that is, the relationship is dependent upon the history of the porous media system. Thus different equilibrium states (that is, distributions of fluid) may be obtained for the same external pressure conditions, depending upon the path taken to get to the new state. There are other reasons for hysteresis, and these will be considered in the chapter on constitutive relationships.

In summary, it is fair to say that the description of a porous media system composed of two or more fluids and a solid is complicated in comparison with the case of a single fluid and a solid by the presence of the interface between the fluids. The interactions between the fluids and between each of the fluids and the solids must be considered to obtain a complete description of the physical processes. Capillary effects allow for discontinuity of pressure across the fluid-fluid interfaces that is an important factor in modeling multiphase flow.

1.7 CONCEPT OF CONCENTRATION

One final basic concept, the *concentration* of a chemical species in a fluid phase, will be introduced here to conclude this chapter. The relatively short length of this section does not imply that the study of concentrations of species is simple, unim-

---

[9] Recall that water rises more readily in a small diameter capillary than one of large diameter.

portant, or straightforward. In fact, the study of the transport and reaction of chemicals within a phase and their transfer between phases is a challenging and timely problem that is seldom easy. As with saturation, porosity, and pressure, concentration is one of the primitive variables in mathematical simulation. Its evolution is described by a conservation equation, and properties of a fluid depend on the concentrations of its chemical constituents. Additionally, the concentration of a material is reflective of its amount in the total solution. Thus, issues of scale involving the size of the sample in which a concentration is measured are important. An average concentration of a contaminant in a groundwater system does not indicate whether hot pockets of concentration threaten water supply wells. In this section measures of the concentration of constituents dissolved in a fluid are explored. If the chemicals are dissolved in a water phase, they are referred to as the solutes in the water solvent, and the phase is referred to as an aqueous solution.

Concentration is generally expressed in terms of mass of solute per unit volume of solution. If, for example, one had a mass of 10 g of salt in a solution of one liter, the concentration could be expressed as a concentration of 10 g/1000 cc, or 0.01 g/cc. Alternatively, one could express the concentration as the number of grams of salt per given weight of solution. In the above example this would yield 10 g/1010 g, or 0.01. For dilute solutions, the difference between these two measures is insignificant.

In field situations, one normally works with concentration in terms of parts per million (ppm) defined as the dimensionless ratio:

$$\text{ppm} = \frac{\text{mass of solute}}{\text{mass of solution}} \times 10^6 \tag{1.64}$$

Alternatively, the concentration is sometimes expressed as parts per billion (ppb) where:

$$\text{ppb} = \frac{\text{mass of solute}}{\text{mass of solution}} \times 10^9 \tag{1.65}$$

For the case of ten grams of salt in a liter of water, the concentration is ten thousand parts per million or ten million parts per billion. Concentrations in terms of parts per billion are often used in studying organic contamination of water. Some chemicals are considered to be health hazards in concentrations on the order of 1 ppb. At such a low concentration, the dissolved chemical may have minimal impact on the flow properties of the phase. Nevertheless, the movement of the chemical relative to the phase is important to study.

Concentrations are also expressed in terms of moles of a constituent per volume. However, in this text, all references to concentration will be on a mass basis. When a contaminant adheres to a solid phase, its concentration relative to the solid may also be expressed as mass per mass of solid. Material attached to subsurface solids is immobile unless it dissolves into the fluids in contact with the solid. The scope of this text is primarily limited to subsurface fluid phases. However, it is not possible to completely isolate interacting components of environmental problems.

1.8 SUMMARY

The purpose of this chapter is to introduce the reader to the basic concepts that must be considered in studying flow in porous media. After introduction of several fundamental porous media properties, current methodology for quantification of porosity and grain-size distribution was presented.

Pressure has been described and defined for both saturated (single-phase flow) and unsaturated (multiphase, air-water) flow. Capillarity, a multiphase phenomenon, was introduced to demonstrate how negative gauge pressure can arise in water in air-water flow. The importance and behavior of the interface between fluid phases was presented along with the influence of surface tension, curvature, common lines, and hysteresis. Commonly used measures of solute concentration were briefly discussed.

1.9 EXERCISES

1. Indicate whether each of the following items exists at the microscale or the macroscale. For items that exist at both scales, propose a relation between the microscale and macroscale measures: (a) pressure; (b) velocity; (c) density; (d) porosity; (e) saturation; (f) capillary pressure; (g) chemical species concentration; (h) temperature.

2. Find the minimum pore diameter in a cubic packing of equal-sized spheres of radius R.

3. For the soil described by the data provided in Table 1.2, determine the coefficient of uniformity, C_u, the coefficient of curvature, C_c, and the possible soil type.

4. A porous medium is made up of grain particles with a density of 2.65 g/cc, water with a density of 1.0 g/cc, and air. A 1 cc sample taken from the medium has a mass of 2.05 g. If the porosity of the sample is 0.3, determine the water saturation and the water content.

5. A porous medium is constructed that is a collection of hollow spheres. A second porous medium is constructed using a similar collection of spheres except that these spheres are not hollow. Which medium has a higher porosity? Justify your answer.

6. The *void ratio* is defined as the volume of voids (i.e., the volume not occupied by solid particles) divided by the volume of solid particles in a sample of porous medium. Obtain a mathematical relation between the void ratio, e, and the porosity, ε.

7. Consider a 20 cm vertical column of water in a right circular cylinder. The pressure in the water is hydrostatic. Calculate the average pressure in the water obtained by averaging over the volume of the water. Compare this result with the average pressure in the water calculated by averaging over the surface that bounds the water. Can any general conclusions be drawn from this result?

8. Now consider a container that is a cylindrical column with radius R_1 for the bottom 10 cm and radius R_2 for the next 10 cm. Note that if $R_1 = R_2$, the system is the same as in the last problem. Calculate the average pressure of a 20 cm column of water in this cylinder based on averaging over the volume and then based on averaging over the surface that bounds the water. What are the implications of this result in relation to a porous media system in which the water content is not constant.

9. A capillary tube is constructed as an alternating sequence of water-wet and air-wet materials. Suppose that the air-wet material is the first (bottom) segment. Describe the behavior of the system as the tube is dipped into a beaker of water, pushed into the water, and withdrawn from the water. How would the behavior be different if the bottom segment were water-wet?

10. Consider a cone with the point on the bottom and with the angle at the tip designated as ϕ. Suppose the cylinder is filled to a depth of water b such that the radius of the common line circle is R. Provide expressions for the curvature of the common line, the normal curvature, and the geodesic curvature. Also provide a sketch that illustrates these quantities.

BIBLIOGRAPHY

[1] American Society for Testing and Materials, Standard Test Method for Particle-Size Analysis of Soils, Document Number ASTM D422-63, American Society for Testing Materials, Philadelphia, 8 pages, 2002.

[2] Brutsaert, W., *Hydrology: An Introduction*, Cambridge University Press, 618 pages, 2005.

[3] Bureau of Reclamation, United States Department of the Interior, *Earth Manual*, Part 1, Third Edition, U.S. Government Printing Office 329 pages, 1998.

[4] Do Carmo, M., *Differential Geometry of Curves and Surfaces*, Prentice Hall, 503 pages, 1976.

[5] Gavlak, R., D. Horneck, R.O. Miller, and J. Kotuby-Amacher, *Soil, Plant and Water Reference Methods for the Western Region*, WCC-103 Publication WREP-125, Second Edition, 199 pages, 2003.

[6] Graton, S.C., and H.S. Fraser, Systematic packing of spheres with particular reference to porosity and permeability, *J. Geol.* **43**(8), 785, 1935.

[7] Gray, W.G., A. Leijnse, R.L. Kolar, and C.A. Blain, *Mathematical Tools for Changing Spatial Scales in the Analysis of Physical Systems*, CRC Press, 232 pages, 1993.

[8] Greenkorn, R.A., *Flow Phenomena in Porous Media*, Marcel Dekker, Inc. 550 pages, 1983.

2

MASS CONSERVATION EQUATIONS

2.1 INTRODUCTION

Modeling of physical phenomena is most useful when it is based on a description of the underlying processes that produce an observed behavior rather than on correlations of observed data. If processes are described well, one may be able to predict how a system will react when subjected to a new or different stress. On the other hand, if a model is based on correlation of data, then when a system changes beyond the range of the data used in the correlation, the correlation will fail to describe it. For example, suppose one develops correlations for the fraction of rainwater that infiltrates into a field, as opposed to running off, as a function only of the rainfall intensity, duration, temperature, and the amount of rain in the preceding week. Such data could be collected over a period of years and used to predict contributions of the field to flood events. However, if one year the field is turned into a shopping mall, its response to a rainfall event will be radically different. Paving over a natural field greatly reduces infiltration and increases runoff. Therefore, the data and correlations developed for the field in the natural state are of no enduring value. Alternatively, if a model of the field is based on principles that reflect universal behavior, the model may also apply after the system has been modified dramatically, for example by covering soil with relatively impermeable asphalt.

One such principle is conservation of mass, which states that mass is neither created nor destroyed. Although the chemical composition of the mass in a region under study may undergo transformation due to chemical or biological reactions, the only way to alter the amount of mass in the system is by flow across the system boundaries. This principle applies to any region in space including the field discussed above, whether it is packed soil, newly plowed, overgrown with grass, or paved. Thus a model of a physical system that employs the principle of conservation of

Essentials of Multiphase Flow and Transport in Porous Media, by George F. Pinder and William G. Gray
Copyright © 2008 by John Wiley & Sons, Inc.

mass will have more generality than a mere correlation, and that part of the model will have applicability even if the properties of the system undergo considerable change.

In addition to the principle of conservation of mass, the principles of conservation of momentum and conservation of energy also must be satisfied. Problems can be posed in terms of these principles expressed mathematically as equations, but these equations alone are not capable of describing system behavior. The number of unknowns that arise is greater than the number of conservation equations, so the equations must be supplemented by correlations, or *constitutive relations*, that describe some system dynamics in terms of other system variables (e.g., mass density as a function of temperature and pressure, or stress as a function of strain). The need for constitutive relations certainly complicates efforts to model systems, but the situation is even more complex than might be realized at first glance.

The difficulties arise from the fact that depending on the temporal and spatial scales at which a system is to be described, the constitutive equations needed are different. For example, suppose one is interested in the weathering of a granite rock subjected to natural effects and uses a time scale of observation such that only the initial and final states are considered. The description of the process would be far different if a time scale of one second is of interest, where a fantastic model would indicate that the rate of weathering is zero, as opposed to a time scale of 10 millennia, where the changes due to physical and chemical effects would have to be taken into account. The difference in these time scales is ten orders of magnitude. Various investigators might consider problems with time scales ranging from less than the period of atomic motion (10^{-15} sec) to the age of the oldest rocks on earth (10^{16} sec). Very little imagination is required to conclude that constitutive relations describing processes of interest must be tailored to the the time scale of interest. Although the constitutive approximations will be scale dependent, the fundamental conservation principles will apply regardless of scale. Spatial scales of observation also may range over many orders of magnitude in that, for example, the radius of an electron is 10^{-15} m while the diameter of the earth is 10^7 m. Although earth dynamics could, in theory, be modeled by accounting for each electron, as well as other elemental particles, such an approach is infeasible from a practical perspective and unwarranted for the description of environmental processes of actual interest.

Faced with the need to enhance quantitative understanding of the environment, the modeler is always forced to work with average values of parameters. Still, a judicious decision must be made in the selection of the time and space scales that will provide useful information. For example, average precipitation rates differ depending on whether one is interested in a particular calendar day, a season, or an annual measure. Additionally, the average precipitation values for a city are different from those for an entire state. The forms of constitutive relations describing spatial variation of precipitation will vary depending on the distances between points of interest. Furthermore, the average values of precipitation employed in those correlations will also be different and based on corresponding temporal and spatial scales.

Depending on the problem to be studied, one must select the temporal and spatial scales that are appropriate. One does not plan a ski vacation based on the average annual snowfall in a state. Such time and space scales are too coarse to lead to a reasonable schedule for the vacation. Instead, a particular month and mountain have to be analyzed to provide a better prediction. Even with this more refined data, however, the avid skier can end up in poor ski conditions because of variations around the averages at a scale smaller than the averages considered. Such variations are called *subgridscale variations.* If these subscale variations could be modeled perfectly, in addition to a perfect model at the scale of observation, the skier would be able to plan for optimal conditions.

In summary, modeling makes use of conservation equations, correlations or constitutive relations, and variables and parameters averaged over appropriate time and space scales. Multiphase flow modeling in porous media also is developed considering these factors. The traditional approach to these models develops the equation of conservation of mass for each of the phases of interest and then expresses them at the appropriate scales. Although a full derivation of conservation of momentum would seem to be the next step in obtaining a description of porous media flow, this approach has not been followed for historical reasons. In 1856, Henry Darcy published his equation for steady flow of a single fluid phase in a column packed with a homogeneous sand [3]. The equation derived states that the steady volumetric flow rate per unit cross-sectional area in a column filled with a homogeneous sand is proportional to the head drop across the column. This correlation of experimental data was found to model flow in porous media well enough that its general form, with more complex coefficients of proportionality, has been extended to heterogeneous media, anisotropic solids, transient flow, and multiple fluid phases. The correlations, referred to as Darcy's law in all their incarnations, remain in near-universal use today in lieu of a derived momentum equation. The long history of collection of experimental data to support Darcian approaches and the fact that a full derivation of a general momentum equation is very complex have contributed to maintenance of the status quo. Innovative new experimental studies to support constitutive forms that arise in general momentum equation derivations are needed to advance the state of the art.

The study of multiphase flow that is the subject of this text will be based on the conservation of mass equation, Darcy's equation, and additional closure relations that complete the set of equations such that they can be solved. In particular, this chapter deals with the derivation of mass conservation equations that can be used in describing flow and chemical species transport in porous media.

The initial formulation of an equation of mass conservation will be at the most commonly employed *continuum scale,* referred to here as the *microscale.* Although real materials are composed of atoms and molecules that occupy only a small portion of space, with intermolecular regions being empty, the microscale views materials as continuous substances that occupy every point of a continuous region of space. The microscale averages the presence of molecules within a region such that they are considered to fill the space in some uniform fashion. The length scale of averaging is much greater than the mean free path in gases for molecular collisions, on the order of 10^{-7} m at standard pressure and temperature. This measure is used because the intermolecular distance in a solid and the mean free path for molecular collisions in a liquid are much smaller. The stipulation that the length scale be "much greater" than the mean free path ensures that enough molecules are included in the averaging region that the average value measured does not change with small changes in the size of the averaging region. The formulation of governing equations at the microscale has a rich tradition in fluid and solid mechanics. It has been possible to develop constitutive relations at this scale to supplement the conservation equations such that a wide array of problems of interest are well-described.

Modeling of flow in porous media at the microscale views the fluids and solids as being adjacent to each other and separated by sharp interfaces. Because the actual geometry of a natural porous solid phase is unknown, modeling of the evolution of the distribution of fluid phases within the pore space would be extremely difficult. The application of boundary conditions at interfaces of unknown location is an insurmountable challenge. For this reason, and also because a complete description at the microscale is more precise than what is required in many situations, multiphase flow is commonly modeled at the *macroscale*, a length scale that is much greater than a pore diameter but much less than the scale at which larger heterogeneities are observed. A commonly used estimate of the macroscale is on the order of ten to one hundred pore diameters. The averaging region is called a *representative elementary volume* (REV) and must produce averages that are insensitive to small changes in its size. At the macroscale, phases are considered to share a location and to each occupy a fraction of the averaging region. This is quite different from the microscale perspective with the phases adjacent to each other. In this chapter, methods will be developed that facilitate the systematic transformation of the microscale mass conservation equations to the macroscale. The assumptions that accompany this transformation will also be examined in some detail.

2.2 MICROSCALE MASS CONSERVATION

The equations of mass conservation can be applied to a chemical species within a solution or to the solution as a whole composed of various chemicals. In either case, at the microscale, the material under consideration is treated as a continuum. The following example, involving brewing of a cup of tea, provides an indication of the issues involved. This example is selected because the experimental setup is simple and can be easily reproduced. What is lacking in elegance and rigor is made up for by accessibility.

The experiment depicted in Figure 2.1, begins with a cup filled to a predetermined level with hot water. A tea bag containing some tea leaves is then immersed in the water. The tea bag is permeable to liquid, but the tea leaves are confined to the bag.

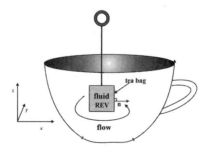

Figure 2.1: A permeable tea bag contains tea leaves. Although the leaves cannot leave the bag, hot water and tea dissolved in the water move easily through the bag surface.

A hot beverage is made when the liquid moves into the bag, dissolves some of the chemicals in the tea leaves and then carries those chemicals out of the bag and into the liquid in the cup. For the purposes of this discussion and the subsequent derivation, the fluid system will be idealized as composed of two chemicals, water and tea, although the "water" may initially contain chemical additives, such as fluoride, and the "tea" is actually a mixture of chemicals. The process of dissolving the tea can be influenced by stirring the liquid, if desired; and stirring will ensure that the distribution of tea in the water is uniform. However, whether or not the system is stirred, the equations of mass conservation must still apply. Mixing of water and tea occurs due to differences in the velocities of these species.

In this example, consider the fluid as residing in two different regions, within the tea bag (indicated as "bag") and outside the bag (indicated as "cup"). The total fluid in the tea cup is the sum of the fluid in the bag and the cup (indicated as "bag+cup"). Some general observations can be made about the fluid in the tea cup. In the absence of evaporation, pouring water into the cup, or drinking the liquid, the amount of bag+cup water is constant. Neither the amount of bag water nor the amount of cup water is necessarily constant, but their sum will be. The tea in bag+cup is constant, as long as no tea is consumed or spilled, even if some additional water is poured into the cup. Initially, some of the tea is in a solid phase. With time, the amount of tea in the fluid phase will increase, both in the bag and in the cup.

The processes of transfer of tea from the solid to the bag water, flow of the water and tea mixture between the bag and the cup, and mixing of portions of the solution containing high tea concentration with portions containing lower tea concentrations can be described by mass conservation equations. The equations express changes that occur over time and space due to dissolution of tea, changes of the mass fraction of tea in the solution, and the velocities at which the water and tea move. The equations are formulated here in terms of microscale variables integrated over a study region. Subsequently, they will be applied to the problem of brewing tea.

2.3   INTEGRAL FORMS OF MASS CONSERVATION

This derivation is concerned with the mass conservation equation for a chemical species in a mixture of other chemicals inside a volume of interest. For example, it could refer to the tea dissolved in water in the bag region, in the cup, or in the cup+bag depending on which volume is selected. Alternatively, the equation could refer to the water species. The derivation will involve microscale continuum quantities. The principle of mass conservation can be formulated for a chemical species i:

$$\begin{bmatrix} \text{rate of change of} \\ \text{mass of species } i \\ \text{in volume} \end{bmatrix} + \begin{bmatrix} \text{net rate of loss of mass} \\ \text{of species } i \text{ due to flow} \\ \text{across volume boundary} \end{bmatrix} - \begin{bmatrix} \text{rate of increase of species } i \\ \text{mass due to transformation} \\ \text{from other species} \end{bmatrix} = 0$$

$$(2.1)$$

The objective of the derivation is to convert this statement of mass conservation into an equation by obtaining mathematical analogs to each of the three terms.

To begin development of the expressions, denote the mass per volume of fluid at a microscale point in the volume as ρ and the fraction of the mass at this point that is species i as ω_i. The index i can take on values that correspond to each of the chemical constituents present. If there are N such different species, the sum of the mass fractions of all the species must be 1, or:

$$\sum_{i=1}^{N} \omega_i = 1 \tag{2.2}$$

Thus, the mass of species i per unit volume at a microscale point, also called the *concentration* of species i on a mass basis, is indicated as ρ_i and can be calculated as:

$$\rho_i = \rho \omega_i \tag{2.3}$$

The mass of species i at a point is therefore the concentration of i multiplied by the differential volume, $\rho_i\, dv$. Therefore, the total mass of dissolved i in a large study volume is the sum of the mass of i at all points in the volume. This is the integral over the volume, i.e.:

$$\text{mass of species } i \text{ in volume} = \int_V \rho_i dv \tag{2.4}$$

The first term in equation (2.1) is the rate of change of the mass of species as calculated in equation (2.4) such that:

$$\begin{bmatrix} \text{rate of change of} \\ \text{mass of species } i \\ \text{in volume} \end{bmatrix} = \frac{d}{dt} \int_V \rho_i dv \tag{2.5}$$

The second term in equation (2.1) is concerned with the flow of the chemical species across the boundaries of the volume of interest. This boundary is a closed surface. For example, if we are interested in the cup portion of the tea brewing situation, the boundary consists of the solid that is wet by the water–tea mixture, the surface of the mixture that is in contact with the air, and the tea bag membrane. The only way a chemical species could cross the boundary of the volume would be by flowing in a direction normal to the boundary. Flow tangent to the boundary remains in the volume. We designate the microscale velocity of species i at a point as $\mathbf{v}_i$ and the normal direction outward from the volume at a point on the surface as $\mathbf{n}$. Therefore the normal velocity of the chemical species i at a surface of the volume will be $\mathbf{v}_i \cdot \mathbf{n}$, where both the velocity and the normal are evaluated at the same point on the boundary. The value of the species velocity and the normal direction will change depending upon the point on the boundary being considered.

Now suppose that, with reference to the tea problem, one is walking with the tea cup, being careful not to spill. Since the cup is moving, all the fluid in the cup is moving. However, none of the water–tea mixture is leaving the cup. This highlights the fact that for mass to leave a volume at a point on the boundary of the volume under study, it must not only have a velocity normal to the surface, but that velocity must be different from the normal velocity of the surface itself. Designate the velocity of a microscale element of the surface of a volume as $\mathbf{w}$. The normal velocity of the surface at that point is then $\mathbf{w} \cdot \mathbf{n}$. The volume of species i leaving the cup on the element of surface ds is $(\mathbf{v}_i - \mathbf{w}) \cdot \mathbf{n}\, ds$. The mass per time leaving the volume is obtained by multiplying this expression by the mass of i per unit volume, ρ_i, and summing over all elements of the bounding surface, i.e., integrating over the complete surface of the volume to obtain:

$$\begin{bmatrix} \text{net rate of loss of mass} \\ \text{of species } i \text{ due to flow} \\ \text{across volume boundary} \end{bmatrix} = \int_S \rho_i(\mathbf{v}_i - \mathbf{w}) \cdot \mathbf{n}\, ds \qquad (2.6)$$

It is important to perform the integration specified in equation (2.6) over the entire surface of the volume being analyzed. For example, if one is interested in the cup region of the tea brewing problem, surface integration is over the complete boundary of the cup region. This surface includes the solid cup walls, the top surface of the fluid, and the boundary that separates the cup region from the bag region. The integral over the part of the surface where the fluid is in contact with the solid cup material is zero since no tea is moving through that surface or depositing on the surface. If no tea is being consumed or poured out of the cup, the integral over the surface of the fluid in contact with air will also be zero. Thus, the only way that tea can move out of the cup is by flowing into the tea bag through its surface. For the most part, the tea will be leached from the leaves into the fluid in the bag and then flow into the cup through the bag membrane that forms the final part of the cup boundary. Rather than this flow causing a loss of tea from the cup, there will be a gain of tea such that the term will be negative. This is accounted for by the fact that the normal vector $\mathbf{n}$ is oriented pointing out of the cup into the tea bag. Thus the flow term provided by equation (2.6) accounts for transfer across the entire boundary of a volume under study and correctly indicates whether the transfer results in gain or loss of material through the dot product of the velocity vector with the outwardly directed normal.

The third term in equation (2.1) accounts for the production of species i due to chemical or biological reactions within the fluid in the volume. Let the rate of mass production of i per unit volume per unit time at the microscale be designated as r_i. Multiplication of this quantity by the differential volume yields $r_i\, dv$, which is the rate of mass production per unit time in the differential volume. This quantity is summed over all points in the volume, i.e., integrated over the volume, to yield the desired expression:

$$\begin{bmatrix} \text{rate of increase of species } i \\ \text{mass due to transformation} \\ \text{from other species} \end{bmatrix} = \int_V r_i\, dv \qquad (2.7)$$

The chemical reaction term is one that, in general, must be approximated using a constitutive form. Typically, the approximations would provide the chemical reaction rate as a function of pressure, temperature, and the concentrations of the various chemical species in the system under study. For the tea brewing problem, the system is viewed as consisting of two species, tea and water, with no reaction occurring that transforms tea to water or vice-versa. Therefore the rate of change of tea mass in the cup due to chemical reaction is zero. However, in a general derivation, the reaction rate term must be carried along in its general form in anticipation of the potential study of systems where a chemical reaction is occurring.

Substitution of the terms developed in equations (2.5), (2.6), and (2.7) into equation (2.1) provides the equation of conservation of species i in a volume, V, with boundary, S:

$$\frac{d}{dt}\int_V \rho_i \, dv + \int_S \rho_i(\mathbf{v}_i - \mathbf{w})\cdot \mathbf{n} ds - \int_V r_i \, dv = 0 \tag{2.8}$$

This is the *integral form* of the *species conservation of mass equation* expressed in terms of microscale variables for species i in a volume. Note that the volume studied can be a physical space, such as the cup of tea, or a mathematical region, such as a cubic centimeter of space within the cup selected as being worthy of analysis. As long as the integrals are properly evaluated over the volume and the boundary of the volume, the analysis will be correct. The conservation of mass equation for species i may also be expressed in terms of the microscale solution density multiplied by the mass fraction of species i by substitution of equation (2.3) into equation (2.8) to obtain:

$$\frac{d}{dt}\int_V \rho\omega_i \, dv + \int_S \rho\omega_i(\mathbf{v}_i - \mathbf{w})\cdot \mathbf{n} ds - \int_V r_i \, dv = 0 \tag{2.9}$$

Equation (2.8) or equation (2.9) provides the mass balance equation for each species in a solution. Therefore, summation of the species balance equation over all the species will provide a conservation equation for the total mass. Summation of equation (2.9) over the N chemical species in the solution yields:

$$\frac{d}{dt}\int_V \rho\left(\sum_{i=1}^N \omega_i\right)dv + \int_S \rho\left[\left(\sum_{i=1}^N \omega_i\mathbf{v}_i\right) - \left(\sum_{i=1}^N \omega_i\right)\mathbf{w}\right]\cdot \mathbf{n} ds - \int_V\left(\sum_{i=1}^N r_i\right)dv = 0 \tag{2.10}$$

This equation is readily simplified making use of several definitions and observations. First, the sum of the mass fractions must be 1, as previously stated in equation (2.2). Thus equation (2.10) becomes:

$$\frac{d}{dt}\int_V \rho dv + \int_S \rho\left[\left(\sum_{i=1}^N \omega_i\mathbf{v}_i\right) - \mathbf{w}\right]\cdot \mathbf{n} ds - \int_V\left(\sum_{i=1}^N r_i\right)dv = 0 \tag{2.11}$$

Second, since the chemical reaction terms account for the rate of mass production of species at a point, they must sum to zero even if the terms are not individually zero. If the rates of mass production did not sum to zero, this would indicate that mass is being created or destroyed when species are transformed from one chemical to another, which is not physical in the absence of nuclear reactions. The sum of the reactions over all species is therefore zero whether or not each of the individual reactions is zero. This reduces the total mass balance equation further to:

$$\frac{d}{dt}\int_V \rho \, dv + \int_S \rho \left[\left(\sum_{i=1}^N \omega_i \mathbf{v}_i\right) - \mathbf{w}\right] \cdot \mathbf{n} \, ds = 0 \tag{2.12}$$

Finally, define the *barycentric velocity*, **v**, of the fluid as a mass fraction weighted velocity summed over all species:

$$\mathbf{v} = \sum_{i=1}^N \omega_i \mathbf{v}_i \tag{2.13}$$

If this identity is employed in equation (2.12), the total mass conservation equation for the fluid in a volume in terms of microscale variables is:

$$\frac{d}{dt}\int_V \rho \, dv + \int_S \rho(\mathbf{v} - \mathbf{w}) \cdot \mathbf{n} \, ds = 0 \tag{2.14}$$

This is the conservation equation used if one is interested in modeling the total amount of fluid in a volume without being concerned about the composition of that fluid.

Now that the barycentric velocity has been defined, it is useful to make a few comments about the velocity of the chemical species or, in particular, the velocity of a chemical species relative to this velocity. Modify equation (2.9) by adding and subtracting the barycentric velocity in the second integral and then regrouping terms to obtain:

$$\frac{d}{dt}\int_V \rho\omega_i \, dv + \int_S \rho\omega_i(\mathbf{v} - \mathbf{w}) \cdot \mathbf{n} \, ds + \int_S \rho\omega_i(\mathbf{v}_i - \mathbf{v}) \cdot \mathbf{n} \, ds - \int_V r_i \, dv = 0 \tag{2.15}$$

In this form of the equation, the second term accounts for species i that leaves the volume across its boundary due to flow moving at the barycentric velocity. This part of the flow is referred to as *convection*. The next integral accounts for the flow of species i across the boundary of the volume due to its velocity relative to the mean velocity of flow. When the fluid is not being stirred and not flowing relative to a volume of fixed shape and size, it is quiescent and the relative movement of constituent i is spreading due to random collisions between molecules of different types. This process of spreading is referred to as *diffusion*.

If one observes a tea bag sitting in a cup of quiescent liquid, tea can be seen to move slowly into the cup from the bag. Typically, more tea will exist at the bottom of the cup than towards the top. Also, more will be close to the bag than away from it. To distribute the tea uniformly through the cup by diffusion, one can wait a long time, much longer than it takes the hot water to cool. As an alternative to the patient approach, one may stir the fluid. Stirring causes the water and tea to mix more completely to the point where the velocities of the water and tea will quickly become equal and their concentrations will be uniform. Thus a brewed cup of tea can be produced within a time frame that the liquid is hot, while assuring the quality of the concoction is uniform.

This process of mixing that makes use of mechanical intervention or induces turbulence to eliminate concentration gradients is called *dispersion*. Both diffusion and dispersion are measures of flow relative to the barycentric velocity. Thus they can be designated for the species i as $\mathbf{j}_i$, where [2]:

$$\mathbf{j}_i = \rho \omega_i (\mathbf{v}_i - \mathbf{v}) \tag{2.16}$$

Substitution of this expression into equation (2.15) yields:

$$\frac{d}{dt} \int_V \rho \omega_i \, dv + \int_S \rho \omega_i (\mathbf{v} - \mathbf{w}) \cdot \mathbf{n} \, ds + \int_S \mathbf{j}_i \cdot \mathbf{n} \, ds - \int_V r_i \, dv = 0 \tag{2.17}$$

For this equation to be employed in the solution of a species transport problem, a constitutive relation will have to be introduced for the diffusive/dispersive transport term, $\mathbf{j}_i$. An observation that will prove useful in hypothesizing a constitutive equation is to note that species i will have a velocity different from that of the mixture only if its concentration is not uniform. This observation will be exploited subsequently.

The primary results of this section are two integral equations that express mass conservation. The first is the integral expression for the conservation of a chemical species measured as a microscale property in a system whose length scale is much greater than the microscale. The most commonly used form is given as equation (2.17), which generally must be supplemented with constitutive expressions for the chemical reaction rate and the diffusion or dispersion vector. Use of equation (2.17) in a modeling exercise does not produce information about the distribution of the chemical species of interest within the study region. This equation only provides representative values obtained as integrals over the system. The second equation that is of particular importance is equation (2.14), which expresses total mass conservation for a material that occupies a volume of interest. This equation also is not capable of providing information about the distribution of density or velocity within a system as it makes use of average values or totals obtained by integrating over the system. Additionally, equation (2.14) does not distinguish among species that may be present in the fluid. This has the advantage of not requiring constitutive expressions for chemical reactions and diffusion/dispersion to proceed with calculations, but it has a disadvantage if the composition of the mixture is of interest.

Note that if a system is composed of N chemical constituents, a species equation similar to equation (2.17) can be written for each of the constituents. The resulting N equations can be summed to obtain equation (2.14), making the total number of equations $N + 1$. However, only N of these equations are independent in that given any N of the equations, the remaining equation in the set of $N + 1$ equations may be derived. Typically, equation (2.14) is solved in conjunction with $N - 1$ of the species mass balance equations.

2.4 INTEGRAL THEOREMS

In some instances, it is desirable to employ a mass conservation equation that provides the ability to model the variation of concentrations, density, and velocities at the microscale within a volume. Such an equation can be obtained from the corresponding integral conservation form by employing two classical mathematical theorems that relate integrals of a function over a volume to the integral of the function over the surface of the volume. These integral theorems will only be stated here, not derived, as they can be found in basic calculus texts.

2.4.1 Divergence Theorem

The first integral theorem is known as the *divergence theorem*. In fact the one- and two-dimensional forms of this theorem were employed in the last chapter when assessing a balance of forces on an interface or common line. The form of interest here relates the integral over a volume of the divergence of a vector to the integral over the volume surface of the normal component of the vector according to:

$$\int_V \nabla \cdot \mathbf{F} \, dv = \int_S \mathbf{F} \cdot \mathbf{n} \, ds \tag{2.18}$$

where V is the volume of interest, S is the boundary surface of the volume, $\mathbf{F}$ is a vector function with continuous first spatial derivatives within the volume, and $\mathbf{n}$ is the unit vector normal to the surface oriented to be positive pointing outward from the volume. With regard to the derivation of mass conservation equations, $\mathbf{F}$ may be any of the flux vectors that appear in the integral equations. For example, for $\mathbf{F} = \rho \omega_i \mathbf{v}_i$, the divergence theorem provides:

$$\int_V \nabla \cdot (\rho \omega_i \mathbf{v}_i) \, dv = \int_S \rho \omega_i \mathbf{v}_i \cdot \mathbf{n} \, ds \tag{2.19}$$

Physically, the divergence of a vector can be thought of as the extent to which the vector behaves as a source or sink. If we consider equation (2.19), it says that the net increase of mass per time, as expressed on the left side of the equation as the integral of the divergence over the volume, is equal to the net flux over the boundary of the volume. This interpretation applies to a volume that is fixed in space.

2.4.2 Transport Theorem

The second theorem that will be employed is the *transport theorem*. This equation states that a rate of change of the integral of a function over a volume is related to the change in the value of the function in the volume and any change in the size of the volume due to the movement of its boundaries. For a scalar quantity f with a continuous first derivative with respect to time, the transport theorem is:

$$\int_V \frac{\partial f}{\partial t} \, dv = \frac{d}{dt} \int_V f \, dv - \int_S f \mathbf{w} \cdot \mathbf{n} \, ds \tag{2.20}$$

where V is the volume of interest, S is the boundary surface of the volume, f is a function with a continuous first derivative in time within the volume, $\mathbf{w}$ is the velocity of the boundary of the volume, and $\mathbf{n}$ is the unit vector normal to the surface oriented to be positive outward from the volume. If, for example, this theorem is to be applied to the integral of the partial time derivative of the mass of species i per unit volume, $f = \rho \omega_i$, the transport theorem states:

$$\int_V \frac{\partial(\rho \omega_i)}{\partial t} dv = \frac{d}{dt} \int_V \rho \omega_i dv - \int_S \rho \omega_i \mathbf{w} \cdot \mathbf{n} \, ds \qquad (2.21)$$

This equation states that the sum over the volume of the rate of change of the density of species i at all points in the volume is equal to the rate of change of the mass of species i in the volume minus the rate of gain of mass of species i due to the expansion of the volume into regions that contain additional amounts of species i.

2.5   POINT FORMS OF MASS CONSERVATION

The divergence and transport theorems can be applied to the equation of conservation for the total fluid as given in equation (2.14) to obtain an expression for mass conservation that is valid at all points in the fluid. If $f = \rho$ and $\mathbf{F} = \rho \mathbf{v}$, the surface integrals can be eliminated from equation (2.14) to obtain:

$$\int_V \left[\frac{\partial \rho}{\partial t} + \nabla \cdot (\rho \mathbf{v}) \right] dv = 0 \qquad (2.22)$$

Inherent in equation (2.22) are the constraints that ρ is continuous in time and $\rho \mathbf{v}$ is continuous in space. Neither of these conditions is necessary for the more general equation (2.14) to be valid, but the constraints are applied by virtue of the conditions necessary for the divergence and transport theorems to be applicable.

Equation (2.22) expresses the mathematical fact that the average value of the integrand is zero, although the integrand itself is not necessarily zero. However, from a physical perspective, note that this equation applies to any subregion of the volume that could be selected for analysis. Therefore equation (2.22) must apply for integration over V as well as if the integral were taken over $V - \delta V$, where δV is a small portion of the region with characteristic length on the order of the microscale. In other words, the volume of integration is arbitrary such that equation (2.22) applies for integration over any volume of size ranging from the microscale to the full extent of the system of interest. For this to be the case, the integrand itself must be zero at each microscale point such that the point equation for mass conservation is:

$$\frac{\partial \rho}{\partial t} + \nabla \cdot (\rho \mathbf{v}) = 0 \qquad (2.23)$$

This equation describes the conservation of total mass at a microscale point with the first term accounting for the rate of change of mass at the point while the second term is the net outward flow of mass per unit time from the point.

The divergence and transport theorems may also be applied to the species mass balance equation (2.17) as an important step toward obtaining the point form of this equation. The transport theorem is applied to the first term to provide the partial time derivative while the divergence theorem is applied twice to eliminate the surface integrals. The result of these manipulations is:

$$\int_V \left[\frac{\partial(\rho\omega_i)}{\partial t} + \nabla\cdot(\rho\omega_i\mathbf{v}) + \nabla\cdot\mathbf{j}_i - r_i \right] dv = 0 \tag{2.24}$$

By the argument that this equation must apply for an arbitrary volume, the integrand itself must be zero so the microscale point species concentration equation is obtained as:

$$\frac{\partial(\rho\omega_i)}{\partial t} + \nabla\cdot(\rho\omega_i\mathbf{v}) + \nabla\cdot\mathbf{j}_i - r_i = 0 \tag{2.25}$$

Often the second and third terms in this equation are grouped together for simplicity of interpretation in the form:

$$\frac{\partial(\rho\omega_i)}{\partial t} + \nabla\cdot(\rho\omega_i\mathbf{v} + \mathbf{j}_i) - r_i = 0 \tag{2.26}$$

The first term is the rate of change of mass of species i at a point, the divergence term is the net loss of species i due to flow from the point, and r_i is the rate at which species i is formed due to chemical transformation from other species.

Recall that equation (2.16) provided the decomposition of the flow term with

$$\rho\omega_i\mathbf{v}_i = \rho\omega_i\mathbf{v} + \mathbf{j}_i \tag{2.27}$$

Thus the net outward flux of the chemical species with its velocity $\mathbf{v}_i$ is equal to the flux due to the barycentric velocity of the fluid $\mathbf{v}$ plus a dispersion or diffusion term that accounts for the fact that when there are gradients in concentration, a species may flow at a velocity different from the average. Equations (2.26) and (2.27) can be combined to obtain the microscale species conservation equation without dispersion indicated explicitly but with the velocity being that of constituent i:

$$\frac{\partial(\rho\omega_i)}{\partial t} + \nabla\cdot(\rho\omega_i\mathbf{v}_i) - r_i = 0 \tag{2.28}$$

Equations of the form (2.26) and (2.28) may be written for each of the species in a mixture regardless of their concentrations. As was done with the integral form of the species balance equation, the point form may be summed over all species to provide an expression for the conservation of total mass at the point. Summation of equation (2.26) over the N species that comprise the solution yields:

$$\frac{\partial}{\partial t}\left[\rho\left(\sum_{i=1}^{N}\omega_i\right)\right] + \nabla\cdot\left[\rho\left(\sum_{i=1}^{N}\omega_i\right)\mathbf{v}\right] + \nabla\cdot\left(\sum_{i=1}^{N}\mathbf{j}_i\right) - \sum_{i=1}^{N}r_i = 0 \tag{2.29}$$

Equation (2.2) indicates that the sum of the mass fractions of all species present must equal 1. Additionally, since mass cannot be created by the reaction term, although the species can be converted from one to another by reaction, the sum of the rate of mass generation terms, r_i, must be zero. With these conditions employed, equation (2.29) becomes:

$$\frac{\partial \rho}{\partial t} + \nabla \cdot (\rho \mathbf{v}) + \nabla \cdot \left(\sum_{i=1}^{N} \mathbf{j}_i \right) = 0 \qquad (2.30)$$

Finally, the sum of the dispersion terms is the sum of the deviations of the species velocities from the barycentric flow velocity of the fluid. Inherent in the definition of a mean velocity is the fact that the sum of deviations must be zero. Therefore:

$$\sum_{i=1}^{N} \mathbf{j}_i = 0 \qquad (2.31)$$

Application of this last condition to equation (2.30) produces the simplified form:

$$\frac{\partial \rho}{\partial t} + \nabla \cdot (\rho \mathbf{v}) = 0 \qquad (2.32)$$

which was derived previously by considering the total fluid system as equation (2.23).

The derivation of the point forms of the microscale equations has been shown to follow directly from the integral forms written in terms of microscale variables. It should be no surprise that the sum of mass conservation equations for all the species present at a microscale point gives rise to the microscale mass conservation equation for the fluid. This summation result applies whether one considers the integral or the point forms. The point form of the species conservation equation offers the opportunity to calculate the distribution of the species within the fluid at the microscale while the integral form expresses the relation for the entire system. The integral form applies for any system, while the reduction to the point form requires that the first time and space derivatives of the quantities under consideration exist.

2.6 THE MACROSCALE PERSPECTIVE

In the preceding section, the integral and point conservation of mass equations were derived in terms of microscale variables. The properties of the fluid or solid at a microscale "point" are obtained as averages over a large number of molecules comprising the substance of interest in the vicinity of the point. For the study of porous media, the same philosophy applies to the calculation of "point" values, but the region of averaging is much larger than that used to obtain microscale values. The scale of study is referred to as the *macroscale*, and its characteristic length is on the order of tens to hundreds of pore diameters.

There are at least two possible approaches to the calculation of macroscale variables. The first is to determine macroscale quantities directly from molecular behavior. This approach is similar to that followed in obtaining microscale variables with the difference between microscale and macroscale average quantities attributable to the much larger size of the averaging region for the macroscale. The second approach involves starting with the variables obtained at the microscale and averaging them over a volume of interest to obtain macroscale average values. This second approach is employed here. However, it should be noted that if changes of scale are carried out in a mathematically consistent fashion, different methods used to effect the change should lead to the same results. In fact this observation suggests the further conclusion that study of a problem using theoretical, experimental, and computational methods should lead to the same conclusions if all methods are perfect. The differences in results obtained using these methods provide fertile ground for research studies about the systems under study and the methods themselves.

In the previous section, the microscale mass conservation equations for a single phase, possibly composed of a number of chemical species, were obtained. Phases were considered to be adjacent to each other and separated by sharp interfaces. In this section, the concepts are extended to the macroscale case wherein phases are viewed as overlapping with each occupying a fraction of the volume of the macroscale "point." The averaging of the microscale quantities is performed over a region known as the *representative elementary volume*, or *REV*. This concept is discussed in the next subsection. Subsequently, macroscale theorems analogous to the divergence and transport theorems for microscale variables are introduced. Then, after average quantities are defined, the mass conservation equations are obtained expressed in terms of macroscale variables.

2.6.1 The Representative Elementary Volume

The transformation of microscale equations to the macroscale requires that the microscale quantities be averaged over a macroscale volume. To understand this process, consider a sphere that may be located with its centroid at any point in the region of interest. The characteristic length of the sphere, i.e., its diameter, must be much larger than the microscale but much less than the dimension of the system under study. Envision that an average value of some microscale property is obtained by computation of an average for the sphere. This macroscale value is then associated with the location of the centroid of the sphere. The sphere can be moved around through the system and averages can be calculated to be associated with every centroid point. The spheres may overlap. It should be clear that average values of a property can be calculated and associated with every point in the system except those closer to the boundary than the radius of the sphere. This procedure provides averages that vary in space and time. The goal of a systematic transformation from the microscale to the macroscale is to develop equations that are expressed in terms of the macroscale field. In particular, for mass conservation, the macroscale density and velocity are variables of importance in the governing equations.

For averaging of microscale variables to the macroscale, the volume of averaging is considered to be independent of the location of its centroid, of its orientation, and of time. For convenience, a sphere of constant size will be used as the averaging volume. The size of the averaging volume is selected such that average values calculated are invariant with respect to small changes in the size of the averaging volume. There has been much debate over whether or not such an averaging volume exists for a porous medium, but we will simply note that in many cases such a volume can be identified. This volume is the REV. As long as it exists, its precise size is unimportant.

We note that the equations to be developed here do not require that the average values be independent of the size of the REV. However, if the averages did depend on the size of the REV, then quantities describing the system would be difficult to measure in an unambiguous way.

2.6.2 Global and Local Coordinate Systems

One of the most useful concepts in the development of equations for porous media, or for any system, at the macroscale is the introduction of a *local coordinate system* whose origin is located at the centroid of the macroscale volume or REV. As a preliminary exercise to the formal introduction of this coordinate system, consider the following scenario.

The weekend has arrived and a student at the University of Vermont decides to leave Burlington to go off into the mountains to contemplate porous media flow. The student extracts a Vermont Vacation atlas from under the front seat of her car, where it has been conveniently stored, along with some loose change and a variety of crumbs of unknown origin, and begins to plan the trip. The area around Camels Hump mountain is of interest, particularly the town of Duxbury, and the location map in the front of the atlas provides direction to the page depicting a detailed map of the Camels Hump area.

The location map consists of the entire state of Vermont subdivided into twenty-nine rectangular areas with a number associated with each rectangle. The number on the rectangle corresponds to the page number where the detailed map of the area indicated is found. The Camels Hump area is indicated as being on page 18. The town of Duxbury is clearly presented on the detailed map on this page at a scale much larger than that provided on the index map. Page 18 provides good information for travel in the Camels Hump area, but to determine the route to this area, it is necessary to refer back to the index map as well as maps on other pages that show the intermediate roads of travel from Burlington to Duxbury.

The location of Duxbury can be found in two ways. One is to locate it directly on the index map at the front of the atlas. The second is to locate it on page 18 and then locate rectangle 18 on the index map. The former approach uses one coordinate system, the latter uses two. The two-coordinate system includes a local coordinate system associated with page 18 and a second global coordinate system that locates the local coordinate with respect to the global coordinates.

Each of the twenty-nine pages can be located using the single global coordinate system, but each page has its own local coordinate system. Note that it is possible to set up the atlas such that each of the twenty-nine pages is overlapping. This creates no additional complication. However, one would then have options for locating Duxbury using the global coordinate system to get to the appropriate page and then the coordinate system of the page selected to find this town. The relative location of the town on the two pages will be different (i.e., the town will have different local coordinates) because of the different global coordinates (i.e., page numbers providing windows of regions of Vermont).

In the development of the macroscale mass conservation equation, a global coordinate system will locate the centroid of the REV while a local coordinate system will be set up relative to the REV centroid. Thus all microscale points in the domain of study can be identified by using a global coordinate system directly. This location will be indicated as point $\mathbf{r}$. Alternatively, one can use the global coordinate system to locate the centroid of an REV, denoted as $\mathbf{x}$, and then find the point of interest by using a local coordinate system, ξ, relative to that centroid.

Associated with each REV in the porous medium is a *local coordinate system ξ*. The origin of this coordinate system is at the centroid of the REV. Any point in space can be located using this coordinate system. For example the vector ξ_1 in Figure 2.2 locates a point uniquely relative to the center of the REV_1, $\mathbf{x}_1$. Besides being the centroid of REV_1, $\mathbf{x}_1$ is also the location of the origin of the local ξ coordinate system. In comparison to the mapping exercise, the local coordinate system is analogous to the letters and numbers along the perimeter of the local map that facilitate finding a specific location.

When the location of the origin of a local coordinate system is known relative to the location of the origin of the global coordinate system, any point in space may be described relative to the origin of the global coordinate system. For example, for point "2" in the figure, this fact is expressed as:

$$\mathbf{r}_2 = \mathbf{x}_2 + \xi_2 \qquad (2.33)$$

where these vectors are found in Figure 2.2. Thus the point "2" is located by the global vector $\mathbf{r}_2$ or by the equivalent sum of the two vectors $\mathbf{x}_2$ and ξ_2. Note also that point "2" may be located as the sum of $\mathbf{x}_1$ and a local vector with respect to the origin of REV_1 even though the local vector will extend to a point outside REV_1.

In summary, one can locate a point in the porous medium by either using global coordinates directly, or by using global coordinates to locate the origin of a local

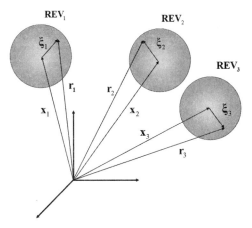

Figure 2.2: Use of the equivalent position vectors $\mathbf{r}$ and $\mathbf{x} + \xi$ to locate a point within an REV. The vector $\mathbf{x}$ locates the centroid of the REV while ξ is the position relative to the centroid [4].

coordinate system and then locating the desired point in the local coordinate system.

Once an averaging volume for a physical system has been selected, it is useful to discuss a coordinate system that can be used to describe the averaging volume and its properties. Consider an averaging volume such as depicted in Figure 2.2. The global coordinate system is indicated as being the $\mathbf{r}$ system such that the location of any point in space is indicated as $\mathbf{r}$. Let the centroid of an REV be indicated as $\mathbf{x}$. Thus if a coordinate system, ξ, is established with its origin at the centroid of the REV, any point in space can be located either directly using the main coordinates or relative to the centroid of the REV since:

$$\mathbf{r} = \mathbf{x} + \xi \tag{2.34}$$

The attractive feature of this coordinate specification is that since the REV's are all spheres of the same size, the space occupied by an REV of radius R, regardless of the location of its centroid, is given by:

$$0 < |\xi| \le R \tag{2.35}$$

Note that if one moves an increment of space $d\mathbf{r}$ in the global coordinate system, this is equivalent to an incremental change in the location of the REV centroid plus a change in the local system coordinate obtained as the differential of equation (2.34):

$$d\mathbf{r} = d\mathbf{x} + d\xi \tag{2.36}$$

According to this expression, if the REV being considered does not change such that the location of the centroid is fixed, $\mathbf{x}$ will be constant and the change in global position $\mathbf{r}$ is equivalent to a change in the local coordinate ξ such that:

$$d\mathbf{r} = d\xi \quad \text{with } \mathbf{x} \text{ constant} \tag{2.37}$$

Alternatively, movement of an increment in space $d\mathbf{r}$ can be made by holding the local coordinate (i.e., the position relative to the centroid) constant but moving the location of the centroid. The incremental motion satisfies the relationship:

$$d\mathbf{r} = d\mathbf{x} \quad \text{with } \xi \text{ constant} \tag{2.38}$$

A final case is to examine a fixed position in space such that $d\mathbf{r} = 0$. Although the actual global position does not change, the position can be viewed by changing the REV centroid as long as an appropriate change is made in the local coordinate relative to that centroid. If infinitesimal changes in these two coordinates are made while holding the absolute position constant, then, based on equation (2.36):

$$0 = d\mathbf{x} + d\xi \quad \text{with } \mathbf{r} \text{ constant} \tag{2.39}$$

Of additional importance is the behavior of spatial derivatives taken with respect to the coordinate system $\mathbf{r}$, the REV centroid coordinate $\mathbf{x}$, and the coordinate system relative to the centroid ξ. Consider a function of interest, such as the microscale species mass fraction, ω_i, that is a function of space and time. In terms of a global coordinate system in space, the functional dependence of the mass fraction is expressed as:

$$\omega_i = \omega_i(t, \mathbf{r}) \tag{2.40}$$

However, equation (2.34) provides the relation between the $\mathbf{r}$ coordinate system and the specification of position using the local coordinates relative to the centroid of the averaging volume. Therefore:

$$\omega_i = \omega_i(t, \mathbf{r}) = \omega_i(t, \mathbf{x} + \xi) \tag{2.41}$$

It is important to observe that the functional dependence of ω_i on $\mathbf{x}$ and ξ is of the particular form involving the sum of these two variables, a less general form than dependence on the two variables $\mathbf{x}$ and ξ independently. For this special case, spatial derivatives of ω_i taken with respect to the $\mathbf{r}$ coordinates is equivalent to the spatial derivative taken with respect to the centroid coordinate, $\mathbf{x}$, while holding the local coordinate, ξ, constant and also equivalent to the spatial derivative taken with respect to local coordinate, ξ, while holding the centroid position, $\mathbf{x}$, constant. In other words, since the function ω_i satisfies the functional dependence on spatial coordinates as given in equation (2.41):

$$\nabla_{\mathbf{r}}\omega_i = \nabla_{\mathbf{x}}\omega_i = \nabla_{\xi}\omega_i \tag{2.42}$$

where $\nabla_{\mathbf{x}}$ is taken holding the ξ coordinates fixed and ∇_{ξ} is taken while holding the centroid $\mathbf{x}$ fixed. For a vector, such as velocity $\mathbf{v}$, that is a function of $\mathbf{r}$ or $\mathbf{x} + \xi$, the divergence obeys a similar relation with:

$$\nabla_{\mathbf{r}}\cdot\mathbf{v} = \nabla_{\mathbf{x}}\cdot\mathbf{v} = \nabla_{\xi}\cdot\mathbf{v} \tag{2.43}$$

2.6.3 Macroscopic Variables

To this point, a formal definition of a *macroscopic variable* has not been provided. Macroscopic variables are consistent with a larger length scale on the order of tens to hundreds of pore diameters and are the quantities observed in most porous media experiments. To model macroscale processes, it is convenient to define macroscale variables in terms of their microscale counterparts and then introduce theorems that facilitate the transfer of the mass conservation equation forms from the microscale to the macroscale.

Earlier the concept of porosity in a porous medium was introduced. This idea may be generalized to the concept of *volume fraction*, the fraction of space in an REV occupied by a phase. Consider an REV with centroid located at position $\mathbf{x}$. Assume that contained in the REV are a number of fluid and solid phases separated by distinct interfaces. Arbitrarily designate one of the fluid phases as the α phase, where α may refer to a wetting phase, a nonwetting phase, or a solid. A microscale element of volume is indicated as dv_{ξ}, where the subscript ξ indicates that this is a microscale element in the local coordinate system. The volume of the α phase is therefore the sum of these elements over the volume of α phase in the REV according to:

$$\delta V^\alpha(t, \mathbf{x}) = \int\limits_{\delta V^\alpha(t, \mathbf{x})} dv_\xi \qquad (2.44)$$

The volume of the α phase in an REV may be a function of time since the mix of phases within the REV may change during dynamic processes. The value of δV^α may be a function of the location of the REV since materials can be distributed differently at different positions. Although the volume occupied by a phase in an REV can change, the total volume of the REV does not vary with time or position. Note also that since integration is performed over the microscale local coordinate system, the microscale distribution of the phase is not provided by knowledge of δV^α. The volume of α phase is a function of the location of the centroid of the REV rather than the distribution within the REV and is therefore a macroscale quantity.

A conceptually and mathematically helpful way to consider the distribution of phases within an REV is to make use of the *phase-distribution function*. This function will be denoted as $\gamma_\alpha(t, \mathbf{r})$, where the subscript α denotes the phase associated with with the phase distribution function. For a system composed of wetting phase, w, nonwetting phase, n, and a solid, s, α can be w, n, or s, depending on which phase is under discussion. The phase distribution function is defined such that, for example, $\gamma_w(t, \mathbf{r})$ will equal 1 at any microscale point within the w phase and will equal 0 at microscale points in all other phases. The distribution function is a microscale function whose value is determined at every point in space except at an interface where its value changes from 1 to 0. Since γ_w (or more generally γ_α, where α refers to any phase of interest), is a function of global position $\mathbf{r}$, it may also be expressed as a function of the centroid coordinate plus a local coordinate such that:

$$\gamma_\alpha(t, \mathbf{r}) = \gamma_\alpha(t, \mathbf{x} + \boldsymbol{\xi}) \qquad (2.45)$$

subject to:

$$\gamma_\alpha(t, \mathbf{x} + \boldsymbol{\xi}) = \begin{cases} 1 & \text{at microscale points in the } \alpha \text{ phase} \\ 0 & \text{at microscale points in all other phase} \end{cases} \qquad (2.46)$$

The remaining prickly issue is to designate the way in which a distribution function transitions from a value of "1" to a value of "0" at an interface. We will consider the function to undergo this change of value smoothly, such that the derivative is defined, but nevertheless over an infinitesimal distance such that for purposes of integrating this function over the REV, the distance is negligible. Although a more rigorous statement of this situation can be provided, for the purposes here, this statement is adequate.

One of the uses of the distribution function is to move the functional dependence on the macroscale coordinate $\mathbf{x}$ from the limits of the integration in equation (2.44) to the integrand. This is accomplished by noting that integration over the α phase within δV^α is equivalent to integration over δV if the integrand is zero except within the α phase. The distribution function provides this assurance such that equation (2.44) can be extended further to read:

$$\delta V^\alpha(t, \mathbf{x}) = \int\limits_{\delta V^\alpha(t,\mathbf{x})} dv_\xi = \int\limits_{\delta V} \gamma_\alpha(t, \mathbf{x} + \boldsymbol{\xi}) dv_\xi \qquad (2.47)$$

It is redundant, but nevertheless worthwhile, to emphasize that this integration moves the scale of observation from the microscale of the integrand to the macroscale of the evaluated integral whereby a function takes on a value associated with the REV that has its centroid at $\mathbf{x}$. After the integration is performed, information about microscale behavior is lost in favor of a description on the average. Integrations over a phase with in an REV are facilitated by the use of the distribution function to change the scale of description from the *microscopic* to the *macroscopic* level of observation.

The expression given in equation (2.47) is useful for defining the *volume fraction* of a phase α, ε^α, as the fraction of the REV space occupied by phase α according to:

$$\varepsilon^\alpha(t, \mathbf{x}) = \frac{\delta V^\alpha(t, \mathbf{x})}{\delta V} = \frac{1}{\delta V} \int\limits_{\delta V} \gamma_\alpha(t, \mathbf{x} + \boldsymbol{\xi}) dv_\xi \qquad (2.48)$$

The sum of the volume fractions of all phases present in the REV must be 1. Thus, for a system composed of a solid, s, a wetting fluid phase, w, and a nonwetting fluid phase, n, the following volume fractions exist, respectively:

$$\varepsilon^s(t, \mathbf{x}) = \frac{1}{\delta V} \int\limits_{\delta V} \gamma_s(t, \mathbf{x} + \boldsymbol{\xi}) dv_\xi \qquad (2.49)$$

$$\varepsilon^w(t, \mathbf{x}) = \frac{1}{\delta V} \int\limits_{\delta V} \gamma_w(t, \mathbf{x} + \boldsymbol{\xi}) dv_\xi \qquad (2.50)$$

$$\varepsilon^n(t, \mathbf{x}) = \frac{1}{\delta V} \int\limits_{\delta V} \gamma_n(t, \mathbf{x} + \boldsymbol{\xi}) dv_\xi \qquad (2.51)$$

and they satisfy the condition:

$$1 = \varepsilon^s + \varepsilon^w + \varepsilon^n \qquad (2.52)$$

Also, the porosity, ε is the fraction of the REV not occupied by solid. Therefore, it can be calculated from the wetting and nonwetting fluid fractions according to:

$$\varepsilon = \varepsilon^w + \varepsilon^n \qquad (2.53)$$

or from the solid fraction as:

$$\varepsilon = 1 - \varepsilon^s \qquad (2.54)$$

These relations describe the fraction of the REV occupied by each phase, but the microscale information that would describe in detail how the phases are distributed through the REV is not provided by these macroscale quantities.

While the definition of volume fraction is straightforward, the integration of other microscale variables to obtain their macroscale forms requires some additional discussion. Several criteria are adopted for the calculation of macroscale variables to assure there is consistency between the definitions across scales and that the macroscale quantities correspond to some physical reality. These criteria are as follows:

1. The macroscopic quantities shall exactly account for the total amount of the corresponding microscopic quantity.
2. Only an additive form of a thermo-mechanical quantity may appear in the argument of an averaging operator.
3. The primitive concept of a physical quantity, as first introduced into the microscale conservation equations, must be preserved by proper definition of the macroscopic quantities.
4. The average value of a microscopic quantity must be clearly related to its measurement in the field.

The first of these criteria simply says that a quantity may be neither created nor destroyed through the averaging process. Thus, for example, the total mass in a system obtained based on microscale observations must be the same amount obtained if macroscale observations are employed.

The second criterion says that the microscale quantities that are summed, or integrated, to obtain macroscale quantities must be extensive variables, variables with values proportional to the quantity of material being considered. Mass, energy, and momentum, for example, are *extensive quantities* (see [1]) and can be added to obtain their total amounts in a system. Mass density, pressure, and temperature, on the other hand, are examples of *intensive variables*, properties of a system that do not add to form any meaningful quantity.

The third criterion states that one cannot create new physical quantities through the averaging process. It would not be acceptable, for example, to average the amount of mass present at the microscopic level and redefine this macroscopic average quantity as physically different from mass defined at the microscopic level. Similarly velocities at the macroscale must be related to their microscale counterparts. For intensive variables, such as temperature, assuring that this criterion is satisfied requires that the macroscale temperature be defined in some consistent manner such that it relates to the microscale temperature.

The fourth criterion recognizes the importance of defining quantities that are measurable. For example, the definition of the macroscale mass density must not be just a mathematical construct for convenience but must also be a quantity that is generally measurable at the macroscopic level.

2.6.4 Definitions of Macroscale Quantities

With the above definitions and criteria in hand, we can now proceed to define the macroscale quantities needed for the derivation of the macroscale mass conservation equations. In the microscale development, the only microscale quantities that appear are mass density (ρ), mass fraction (ω_i), species velocity ($\mathbf{v}_i$), barycentric phase velocity ($\mathbf{v}$), dispersion or diffusion vector ($\mathbf{j}_i$), and the chemical reaction rate (r_i). The subscript i denotes the chemical species. Use of i as a subscript, rather than a superscript, indicates that the quantity it adorns is a microscale variable. By criterion 3 above, these microscale variables must be systematically transformed to the macroscale. That task will be accomplished here.

The mass density is the first item to be considered. To satisfy criterion 1, first multiply the microscale mass density, the intensive quantity $\rho(t, \mathbf{x} + \boldsymbol{\xi})$, by the microscale element of volume to obtain the mass at the microscale point, $\rho\, dv_\xi$. The mass of the α phase within the REV is then equal to the sum of the mass at all the points within the α phase, or, in other words, the integral of the density over the α phase volume within the REV:

$$\text{mass in } \alpha \text{ phase} = \int\limits_{\delta V^\alpha(t,\, \mathbf{x})} \rho(t, \mathbf{x}+\boldsymbol{\xi})dv_\xi \tag{2.55}$$

This integral over δV^α may be expressed as an integral over the entire REV if the distribution function is introduced into the integrand:

$$\text{mass in } \alpha \text{ phase} = \int\limits_{\delta V} \rho(t, \mathbf{x}+\boldsymbol{\xi})\gamma_\alpha(t, \mathbf{x}+\boldsymbol{\xi})dv_\xi \tag{2.56}$$

The change from equation (2.55) to equation (2.56) moves all dependence on the macroscale coordinate $\mathbf{x}$ into the integrand, and the region of integration does not now depend on the location of the REV being considered.

The macroscale density of the α phase at a point is now defined as the mass of α phase divided by the volume of the α phase within the REV and is indicated as ρ^α, with the superscript indicating a macroscale quantity such that:

$$\rho^\alpha(t, \mathbf{x}) = \frac{1}{\delta V^\alpha(t, \mathbf{x})} \int\limits_{\delta V} \rho(t, \mathbf{x}+\boldsymbol{\xi})\gamma_\alpha(t, \mathbf{x}+\boldsymbol{\xi})dv_\xi \tag{2.57}$$

The average defined in this manner is often referred to as the *intrinsic phase average* which has the property that when the microscale quantity being averaged is constant in a phase, the macroscale average is equal to the microscale value. Another form of the average that is helpful to employ is obtained by multiplying equation (2.57) by $\delta V^\alpha/\delta V$ and then making use of equation (2.48) to obtain:

$$\varepsilon^\alpha(t, \mathbf{x})\rho^\alpha(t, \mathbf{x}) = \frac{1}{\delta V} \int\limits_{\delta V} \rho(t, \mathbf{x}+\boldsymbol{\xi})\gamma_\alpha(t, \mathbf{x}+\boldsymbol{\xi})dv_\xi \tag{2.58}$$

The preceding equation converts the microscale mass per volume to the macroscale mass of α phase per REV. When only the α phase is present in the REV, such that $\varepsilon^\alpha = 1$, and the REV length scale is the same as the micro length scale, the microscale and macroscale densities will be equal. This is a reassuring observation, although not of particular utility in modeling porous media flow, that points to the consistency between the microscale and macroscale definitions.

The next quantity to be considered at the macroscale is the mass fraction. The mass fraction is an intensive quantity in that the addition of the mass fraction at a number of points is meaningless. Multiplication of the microscale mass fraction by density and the differential element of volume, however, is equal to the amount of mass of the chemical species at the point. This is an extensive quantity and can be added. Therefore:

$$\text{mass of species } i \text{ in } \alpha \text{ phase} = \int_{\delta V^\alpha(t,\mathbf{x})} \rho(t, \mathbf{x}+\boldsymbol{\xi})\omega_i(t, \mathbf{x}+\boldsymbol{\xi})dv_\xi \qquad (2.59)$$

In this equation, the density is actually a weighting function for averaging of the mass fraction. The distribution function for the α phase may be used to change the integral such that it is over the entire REV:

$$\text{mass of species } i \text{ in } \alpha \text{ phase} = \int_{\delta V} \rho(t, \mathbf{x}+\boldsymbol{\xi})\omega_i(t, \mathbf{x}+\boldsymbol{\xi})\gamma_\alpha(t, \mathbf{x}+\boldsymbol{\xi})dv_\xi \qquad (2.60)$$

The macroscale quantity obtained after division by δV^α provides a definition of the macroscale mass fraction as follows:

$$\rho^\alpha(t, \mathbf{x})\omega^{i\alpha}(t, \mathbf{x}) = \frac{1}{\delta V^\alpha(t, \mathbf{x})}\int_{\delta V} \rho(t, \mathbf{x}+\boldsymbol{\xi})\omega_i(t, \mathbf{x}+\boldsymbol{\xi})\gamma_\alpha(t, \mathbf{x}+\boldsymbol{\xi})dv_\xi \qquad (2.61)$$

or:

$$\varepsilon^\alpha(t, \mathbf{x})\rho^\alpha(t, \mathbf{x})\omega^{i\alpha}(t, \mathbf{x}) = \frac{1}{\delta V}\int_{\delta V} \rho(t, \mathbf{x}+\boldsymbol{\xi})\omega_i(t, \mathbf{x}+\boldsymbol{\xi})\gamma_\alpha(t, \mathbf{x}+\boldsymbol{\xi})dv_\xi \qquad (2.62)$$

Although the preceding equation provides a formal definition of the macroscale mass fraction, it is important to ensure that this quantity is meaningful and maintains the same character as its microscale counterpart according to criterion 3. To gain insight into the properties of the macroscale mass fraction, begin by summing both sides of equation (2.62) over the species present. If there are N species, then:

$$\varepsilon^\alpha(t, \mathbf{x})\rho^\alpha(t, \mathbf{x})\left[\sum_{i=1}^{N} \omega^{i\alpha}(t, \mathbf{x})\right] = \frac{1}{\delta V}\int_{\delta V} \rho(t, \mathbf{x}+\boldsymbol{\xi})\left[\sum_{i=1}^{N} \omega_i(t, \mathbf{x}+\boldsymbol{\xi})\right]\gamma_\alpha(t, \mathbf{x}+\boldsymbol{\xi})dv_\xi \qquad (2.63)$$

where rearrangement is facilitated by the fact that ε^α, ρ^α, ρ, and γ_α can be removed from the summation since they do not depend on the species index. By definition, the sum of the microscale species mass fraction is 1 so that equation (2.63) simplifies to:

$$\varepsilon^\alpha(t, \mathbf{x})\rho^\alpha(t, \mathbf{x})\left[\sum_{i=1}^{N} \omega^{i\alpha}(t, \mathbf{x})\right] = \frac{1}{\delta V}\int_{\delta V} \rho(t, \mathbf{x}+\boldsymbol{\xi})\gamma_\alpha(t, \mathbf{x}+\boldsymbol{\xi})dv_\xi \qquad (2.64)$$

The right side of this equation is known from equation (2.58) to be equal to $\varepsilon^\alpha\rho^\alpha$. Therefore, division of both sides of equation (2.64) by $\varepsilon^\alpha\rho^\alpha$ yields:

$$\sum_{i=1}^{N} \omega^{i\alpha}(t, \mathbf{x}) = 1 \tag{2.65}$$

The following facts argue that the mass weighted macroscale mass fraction is a useful quantity: (i) it equals the microscale value when the microscale value is constant in the α phase in the REV; (ii) it is calculated from an extensive quantity as required in criterion 1 above (see the end of Subsection 2.6.3); (iii) it is the value typically measured when collecting a sample; (iv) it satisfies the constraint that the sum of fractions of a whole is equal to 1. Indeed, the macroscale definition of the mass fraction is consistent with the microscale definition and thus satisfies criterion 3 as well.

The macroscopic value of the velocity of a chemical species is calculated from averaging of the species momentum. Velocity is an intensive variable. Multiplication of the species velocity by the mass per volume, that is, the species density, gives the momentum per volume. Then integration over the volume of the phase in the REV gives the total momentum of the species in the phase under consideration in the REV. The microscale point momentum of a chemical species is $\rho\omega_i\mathbf{v}_i dv_\xi$. Summation of this quantity over the α phase proceeds by integration, as in the preceding cases, to obtain:

$$\text{momentum of species } i \text{ in } \alpha \text{ phase} = \int_{\delta V} \rho\omega_i\mathbf{v}_i\gamma_\alpha dv_\xi \tag{2.66}$$

For economy of space, the dependence of each of the four quantities in the integral on t and $\mathbf{x} + \boldsymbol{\xi}$ is not listed explicitly but is understood. Division of equation (2.66) by δV provides the macroscale expression for the momentum of species i in the α phase per unit of system volume:

$$\varepsilon^\alpha(t, \mathbf{x})\rho^\alpha(t, \mathbf{x})\omega^{i\alpha}(t, \mathbf{x})\mathbf{v}^{i\alpha}(t, \mathbf{x}) = \frac{1}{\delta V} \int_{\delta V} \rho\omega_i\mathbf{v}_i\gamma_\alpha dv_\xi \tag{2.67}$$

The only quantity on the left side of equation (2.67) that has not been defined previously is the macroscale species velocity. Therefore, this equation provides the definition of the macroscale species average velocity. In the interest of compact notation, the explicit dependence of the macroscale quantities on t and $\mathbf{x}$ will not be indicated in subsequent manipulations but will be understood. Consider the summation of equation (2.67) over all the species in the system. If there are N of these species, then:

$$\varepsilon^\alpha\rho^\alpha\left[\sum_{i=1}^{N}\omega^{i\alpha}\mathbf{v}^{i\alpha}\right] = \frac{1}{\delta V}\int_{\delta V}\rho\left[\sum_{i=1}^{N}\omega_i\mathbf{v}_i\right]\gamma_\alpha dv_\xi \tag{2.68}$$

Substitution of equation (2.13), the definition of the microscale barycentric velocity, into the right side of this equation provides the relation:

$$\varepsilon^\alpha\rho^\alpha\left[\sum_{i=1}^{N}\omega^{i\alpha}\mathbf{v}^{i\alpha}\right] = \frac{1}{\delta V}\int_{\delta V}\rho\mathbf{v}\gamma_\alpha dv_\xi \tag{2.69}$$

which suggests the definition of the barycentric macroscale phase velocity for the α phase as:

$$\mathbf{v}^\alpha = \sum_{i=1}^{N} \omega^{i\alpha} \mathbf{v}^{i\alpha} \tag{2.70}$$

Simple combination of the last two equations also provides the direct definition of the macroscale intrinsic average of the phase velocity:

$$\varepsilon^\alpha \rho^\alpha \mathbf{v}^\alpha = \frac{1}{\delta V} \int_{\delta V} \rho \mathbf{v} \gamma_\alpha \mathrm{d} v_\xi \tag{2.71}$$

The average velocities obtained are consistent with the requirements of the criteria listed in the previous subsection. Because the phase velocity is a sum of the species velocities weighted with respect to the macroscale mass fractions, it is a barycentric velocity.

The macroscale dispersion vector is defined by analogy with the microscale dispersion vector in equation (2.16) as:

$$\mathbf{j}^{i\alpha} = \rho^\alpha \omega^{i\alpha} \left(\mathbf{v}^{i\alpha} - \mathbf{v}^\alpha \right) \tag{2.72}$$

This definition satisfies the necessary condition that summation of the dispersion vector over all species is zero since it accounts for deviations of species velocities from the average.

Some additional insight into this equation may be gained by multiplying by the volume fraction and expanding the right side such that:

$$\varepsilon^\alpha \mathbf{j}^{i\alpha} = \varepsilon^\alpha \rho^\alpha \omega^{i\alpha} \mathbf{v}^{i\alpha} - \varepsilon^\alpha \rho^\alpha \omega^{i\alpha} \mathbf{v}^\alpha \tag{2.73}$$

Now substitute equations (2.67) and (2.71) into this equation to replace two of the products of macroscale quantities with integral expressions such that:

$$\varepsilon^\alpha \mathbf{j}^{i\alpha} = \frac{1}{\delta V} \int_{\delta V} \rho \omega_i \mathbf{v}_i \gamma_\alpha \mathrm{d} v_\xi - \mathbf{v}^\alpha \frac{1}{\delta V} \int_{\delta V} \rho \omega_i \gamma_\alpha \mathrm{d} v_\xi \tag{2.74}$$

At this point, it is useful to recall that the macroscale velocity $\mathbf{v}^\alpha$ does not depend on the microscale coordinate ξ. Therefore, it may be moved inside the integral that involves integration over the ξ coordinates without error. Thus equation (2.74) may be written:

$$\varepsilon^\alpha \mathbf{j}^{i\alpha} = \frac{1}{\delta V} \int_{\delta V} \rho \omega_i \left(\mathbf{v}_i - \mathbf{v}^\alpha \right) \gamma_\alpha \mathrm{d} v_\xi \tag{2.75}$$

The microscale dispersion involves $\mathbf{v}_i - \mathbf{v}$ and this term can be introduced into equation (2.75) by adding and subtracting $\mathbf{v}$ in the term in parentheses. Then splitting the resulting integral into two parts yields:

$$\varepsilon^\alpha \mathbf{j}^{i\alpha} = \frac{1}{\delta V} \int_{\delta V} \rho \omega_i (\mathbf{v}_i - \mathbf{v}) \gamma_\alpha \mathrm{d} v_\xi + \frac{1}{\delta V} \int_{\delta V} \rho \omega_i (\mathbf{v} - \mathbf{v}^\alpha) \gamma_\alpha \mathrm{d} v_\xi \qquad (2.76)$$

The integrand in the first integral is the microscale dispersion times the distribution function, that is, $\mathbf{j}_i \gamma_\alpha$. Add and subtract the macroscale mass fraction $\omega^{i\alpha}$ to ω^i in the second integral. This leaves the value of the term unchanged but facilitates rearrangement of the form of the equation such that equation (2.76) becomes:

$$\varepsilon^\alpha \mathbf{j}^{i\alpha} = \frac{1}{\delta V} \int_{\delta V} \mathbf{j}_i \gamma_\alpha \mathrm{d} v_\xi + \frac{1}{\delta V} \int_{\delta V} \rho \omega^{i\alpha} (\mathbf{v} - \mathbf{v}^\alpha) \gamma_\alpha \mathrm{d} v_\xi$$
$$+ \frac{1}{\delta V} \int_{\delta V} \rho (\omega_i - \omega^{i\alpha})(\mathbf{v} - \mathbf{v}^\alpha) \gamma_\alpha \mathrm{d} v_\xi \qquad (2.77)$$

The macroscale variable $\omega^{i\alpha}$ may be moved outside the second integral since it does not depend on ξ, the microscale variable of integration. Then equation (2.71) indicates that the remaining integration of the second term is zero. The remaining expression provides the macroscale dispersion vector for species i consistent with definition (2.72) in terms of integrals from the microscale as:

$$\varepsilon^\alpha \mathbf{j}^{i\alpha} = \frac{1}{\delta V} \int_{\delta V} \mathbf{j}_i \gamma_\alpha \mathrm{d} v_\xi + \frac{1}{\delta V} \int_{\delta V} \rho (\omega_i - \omega^{i\alpha})(\mathbf{v} - \mathbf{v}^\alpha) \gamma_\alpha \mathrm{d} v_\xi \qquad (2.78)$$

This expression provides insight into the physical processes that contribute to macroscale dispersion. First, as might be expected, the microscale dispersion is converted to the macroscale by the first integral in equation (2.78). The second term in equation (2.78) indicates that macrodispersion is enhanced further by deviations of the microscale mass fractions and velocity from their macroscale averages. The presence of this extra term is consistent with criterion 3 above in that the deviation quantities contribute to mixing within the REV consistent with the physical effect of dispersion.

The last quantity that appears in the microscale mass conservation equation that will be transformed from the microscale to the macroscale is the rate of species production per unit volume, r_i. The microscale rate of production of species i is $r_i \mathrm{d} v_\xi$, which can be integrated (summed) over the α phase within the REV to obtain:

$$\text{rate of mass production of species } i \text{ in } \alpha \text{ phase} = \int_{\delta V} r_i \gamma_\alpha \mathrm{d} v_\xi \qquad (2.79)$$

Division of this expression by the REV leaves the volume fraction multiplied by the intrinsic macroscale rate of production of species i per volume of α phase:

$$\varepsilon^\alpha r^{i\alpha} = \frac{1}{\delta V} \int_{\delta V} r_i \gamma_\alpha \mathrm{d} v_\xi \qquad (2.80)$$

2.6.5   Summary of Macroscale Quantities

To facilitate easy access for subsequent use, it seems prudent to summarize the macroscale variables that have been developed. These are tabulated here for a chemical species i in phase α. All the macroscale variables are functions of time and of the macroscale coordinate that locates the centroid of the averaging volume. For reference, the identities are expressed in terms of integrals over the fixed REV making use of the distribution function and as integrals over the volume of α phase within the REV.

Volume Fraction, ε^α

$$\varepsilon^\alpha = \frac{1}{\delta V}\int_{\delta V}\gamma_\alpha dv_\xi = \frac{1}{\delta V}\int_{\delta V^\alpha}dv_\xi \tag{2.81}$$

Intrinsic Mass Density, ρ^α

$$\varepsilon^\alpha \rho^\alpha = \frac{1}{\delta V}\int_{\delta V}\rho\gamma_\alpha dv_\xi = \frac{1}{\delta V}\int_{\delta V^\alpha}\rho dv_\xi \tag{2.82}$$

Intrinsic Mass Fraction, $\omega^{i\alpha}$

$$\varepsilon^\alpha \rho^\alpha \omega^{i\alpha} = \frac{1}{\delta V}\int_{\delta V}\rho\omega_i\gamma_\alpha dv_\xi = \frac{1}{\delta V}\int_{\delta V^\alpha}\rho\omega_i dv_\xi \tag{2.83}$$

Barycentric Intrinsic Velocity of Species i, $\mathbf{v}^{i\alpha}$

$$\varepsilon^\alpha \rho^\alpha \varepsilon^{i\alpha}\mathbf{v}^{i\alpha} = \frac{1}{\delta V}\int_{\delta V}\rho\omega_i\mathbf{v}_i\gamma_\alpha dv_\xi = \frac{1}{\delta V}\int_{\delta V^\alpha}\rho\omega_i\mathbf{v}_i dv_\xi \tag{2.84}$$

Barycentric Intrinsic Velocity of Phase α, $\mathbf{v}^\alpha$

$$\varepsilon^\alpha \rho^\alpha \mathbf{v}^\alpha = \frac{1}{\delta V}\int_{\delta V}\rho\mathbf{v}\gamma_\alpha dv_\xi = \frac{1}{\delta V}\int_{\delta V^\alpha}\rho\mathbf{v} dv_\xi \tag{2.85}$$

Dispersion Vector for Species i, $\mathbf{j}^{i\alpha}$

$$\varepsilon^\alpha \mathbf{j}^{i\alpha} = \varepsilon^\alpha \rho^\alpha \omega^{i\alpha}\left(\mathbf{v}^{i\alpha} - \mathbf{v}^\alpha\right) = \frac{1}{\delta V}\int_{\delta V^\alpha}\mathbf{j}_i dv_\xi + \frac{1}{\delta V}\int_{\delta V^\alpha}\rho(\omega_i - \omega^{i\alpha})(\mathbf{v} - \mathbf{v}^\alpha)dv_\xi \tag{2.86}$$

Chemical Rate of Mass Production for Species i, $r^{i\alpha}$

$$\varepsilon^\alpha r^{i\alpha} = \frac{1}{\delta V}\int_{\delta V}r_i\gamma_\alpha dv_\xi = \frac{1}{\delta V}\int_{\delta V^\alpha}r_i dv_\xi \tag{2.87}$$

All of these definitions satisfy the four criteria for definition of macroscale properties enumerated previously at the end of Subsection 2.6.3. Furthermore, the macroscale quantities are easily seen to be averages of microscale counterparts that will be equal to their microscale counterparts for cases where the microscale value is constant within the REV or when $\varepsilon^\alpha = 1$ and the scale of the REV is equal to the microscale. Physical understanding of the relations between microscale and macroscale quantities will prove helpful in navigating the mathematical development that follows.

2.7 THE AVERAGING THEOREMS

The macroscale perspective was introduced in the last section, and the definitions of variables were also provided. These concepts are necessary for transforming the mass balance equations in terms of microscale variables to mass balance equations in terms of macroscale variables. This transformation is accomplished by averaging the point microscale conservation equation over an REV. Besides averaging quantities, this procedure also involves averaging of the time and space derivatives of those quantities. Therefore, additional tools needed to facilitate the transformation to the macroscale are two theorems that convert the average of a derivative to a derivative of the average. These theorems are extensions of the divergence and transport theorems considered previously; they will be developed here from those microscale formulas.

The averaging theorems, like the divergence and transport theorems, are relations between averages over volumes and averages over surfaces of the volume. The volume to be treated here is that occupied by a phase within an REV of volume δV. If the phase of interest is denoted as the α phase, its volume within the REV is δV^α. The fact that δV is a constant independent of time or location while δV^α depends on both time and the macroscale coordinate is important in the development here.

The boundary of the α phase is composed of two different types of surfaces. The first is the interface between the α phase and other phases. These surfaces are in the interior of δV and will be denoted as $S^{\alpha\beta}$, where the order of the indices is irrelevant and $S^{\alpha\beta}$ is the interface between the α and β phases where $\alpha \neq \beta$. The second part of the boundary of the α phase is on the exterior of the REV. On this mathematical, rather than physical, surface, the α phase is in contact with α phase. Thus, this surface is designated as $\delta S^{\alpha\alpha}$. It is across this boundary that α phase material enters and leaves the REV. This boundary is different from the interior boundaries between phases because it cannot move in the normal direction since the REV does not deform with time and because it is a mathematical surface rather than one associated with a boundary between phases. In fact this part of the boundary is completely within the α phase and therefore physical properties do not undergo any discontinuity across this boundary.

For example, consider the two-phase system as depicted in Figure 2.3 composed of a solid, s, and a fluid phase, w. The REV is indicated as the spherical region, δV. The w and s phases within δV occupy volumes of size δV^w and δV^s, respectively. Therefore $\delta V = \delta V^w + \delta V^s$. The quantity S^{ws} is the total interfacial area within δV between the w and s phases internal to the REV. A portion of this area is highlighted in the figure with respect to one of the solid grains within the REV. This quantity is typically large relative to the external surface of the REV because of the complex

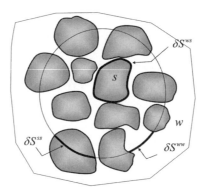

Figure 2.3: An REV, δV, for a system composed of w and s phases. The interface between phases, S^{ws} and the surfaces that make up the boundary of the REV, δS^{ss} and δS^{ww} are also indicated.

interfacial geometry that exists within the REV. This area within the REV may move due to movement of the phases. The area δS^{ww} is the portion of the external surface of the REV that encounters the w phase on both sides. Similarly, the area δS^{ss} is the portion of the REV surface that intersects the solid phase. Portions of both of these surfaces are indicated in the figure. For the two-phase system with a spherical REV indicated, the external surface, δS, satisfies the relation $\delta S = \delta S^{ww} + \delta S^{ss}$. Coincidental concurrence of part of the ws interface with the boundary of the REV is neglected. Because an REV is fixed in space and does not deform, δS is a constant. However, as one moves from REV to REV, the fraction of this surface composed of δS^{ss} interface as opposed to δS^{ww} interface will change due to the heterogeneity of the system. Note also that the velocities of the δS^{ss} and δS^{ww} interfaces associated with a particular REV are zero because an REV is fixed in space. However, the relative sizes of these interfaces can change due to movement of the materials. If a system consisting of more than two phases is of interest, the boundary of a phase consists of intersections of the phase with the boundary of the REV plus the boundary of the phase with all other phases at interfaces within the REV.

With these considerations, we now proceed to extend the divergence theorem and the transport theorem to their macroscale counterparts.

2.7.1 Spatial Averaging Theorem

The spatial averaging theorem is developed to allow the average of the divergence of a microscale quantity to be related to the divergence of a macroscale (averaged) quantity. It is based on the divergence theorem given in equation (2.18) and rewritten here for the case when the volume of concern is the α phase within an REV, δV^α:

$$\int_{\delta V^\alpha} \nabla \cdot \mathbf{F} \, dv_\xi = \int_{\delta S^{\alpha\alpha}} \mathbf{F} \cdot \mathbf{n}_\alpha \, ds_\xi + \sum_{\beta \neq \alpha} \int_{S^{\alpha\beta}} \mathbf{F}|_\alpha \cdot \mathbf{n}_\alpha \, ds_\xi \tag{2.88}$$

where $\mathbf{F}$ is a microscale vector that is a function of t and $\mathbf{x} + \boldsymbol{\xi}$ and continuous in the α phase, $\mathbf{n}_\alpha$ is a unit vector on the boundary of the α phase positive outward from the α phase in the REV, the summation is over all interfaces between the α

phase and other phases, designated as β phases, within the REV, and, as noted earlier, $\delta S^{\alpha\alpha}$ is the portion of the external surface of the REV that intersects the α phase. The notation $\mathbf{F}|_\alpha$ is employed to indicate that the microscale vector $\mathbf{F}$ is evaluated in the α phase at the interface.

Now we make use of a trick! Recall that the phase distribution function γ_α is nonzero only in the α phase, where it is equal to 1. Therefore, if $\mathbf{F}$ is multiplied by γ_α in the first integral on the right side of equation (2.88), the integration may be performed over the entire external boundary of the REV, denoted δS. The fact that γ_α is zero on parts of δS other than $\delta S^{\alpha\alpha}$ preserves the equality written as:

$$\int_{\delta V^\alpha} \nabla \cdot \mathbf{F} dv_\xi = \int_{\delta S} (\mathbf{F}\gamma_\alpha) \cdot \mathbf{n}_\alpha ds_\xi + \sum_{\beta \neq \alpha} \int_{S^{\alpha\beta}} \mathbf{F}|_\alpha \cdot \mathbf{n}_\alpha ds_\xi \qquad (2.89)$$

This seemingly minor change is important because it converts the first integral on the right-hand-side of equation (2.88) from an integration over portions of a surface to an integration over a closed surface in equation (2.89). Therefore, the divergence theorem may be applied to this term to convert the integral over the external closed boundary of the REV to an integral over the REV:

$$\int_{\delta V^\alpha} \nabla \cdot \mathbf{F} dv_\xi = \int_{\delta V} \nabla_\xi \cdot (\mathbf{F}\gamma_\alpha) dv_\xi + \sum_{\beta \neq \alpha} \int_{S^{\alpha\beta}} \mathbf{F}|_\alpha \cdot \mathbf{n}_\alpha ds_\xi \qquad (2.90)$$

Since the quantities in parentheses in this equation are both functions of $\mathbf{x} + \boldsymbol{\xi}$ rather than of $\mathbf{x}$ and $\boldsymbol{\xi}$ independently, identity (2.43) may be employed to replace the divergence operator with respect to $\boldsymbol{\xi}$ coordinates with the divergence operator in $\mathbf{x}$ coordinates:

$$\int_{\delta V^\alpha} \nabla \cdot \mathbf{F} dv_\xi = \int_{\delta V} \nabla_\mathbf{x} \cdot (\mathbf{F}\gamma_\alpha) dv_\xi + \sum_{\beta \neq \alpha} \int_{S^{\alpha\beta}} \mathbf{F}|_\alpha \cdot \mathbf{n}_\alpha ds_\xi \qquad (2.91)$$

With the equation in this form, since the divergence is taken with respect to the macroscale coordinate and the integration is over a volume that is independent of this coordinate, the divergence operator may be moved outside the integral. Also, multiply by $1/\delta V$, which is a constant and may be moved inside the divergence operator, to obtain the form:

$$\frac{1}{\delta V}\int_{\delta V^\alpha} \nabla \cdot \mathbf{F} dv_\xi = \nabla_\mathbf{x} \cdot \left[\frac{1}{\delta V}\int_{\delta V} (\mathbf{F}\gamma_\alpha) dv_\xi \right] + \sum_{\beta \neq \alpha} \frac{1}{\delta V}\int_{S^{\alpha\beta}} \mathbf{F}|_\alpha \cdot \mathbf{n}_\alpha ds_\xi \qquad (2.92)$$

The presence of γ_α in the integrand may be removed by restricting the region of integration to being over the α phase. We thus obtain the *spatial averaging theorem*:

$$\frac{1}{\delta V}\int_{\delta V^\alpha} \nabla \cdot \mathbf{F} dv_\xi = \nabla_\mathbf{x} \cdot \left[\frac{1}{\delta V}\int_{\delta V^\alpha} \mathbf{F} dv_\xi \right] + \sum_{\beta \neq \alpha} \frac{1}{\delta V}\int_{S^{\alpha\beta}} \mathbf{F}|_\alpha \cdot \mathbf{n}_\alpha ds_\xi \qquad (2.93)$$

With $\mathbf{F}$ replaced by a microscale vector of interest, such as the mass flux $\rho\omega_i \mathbf{v}_i$ from the species conservation equation, the definitions of averaged quantities may be employed. The term on the left side of equation (2.93) is an average of the divergence and it is equal to the first term on the right side, the divergence of an average (a macroscale quantity), plus a term that must be evaluated on the interface between the α phase and the other phases present. This last term will be discussed further in the context of the averaged mass conservation equation.

The trick of using the distribution function γ_α facilitated the derivation from equation (2.88) to equation (2.93) yet it appears in neither the initial nor the final form. Thus γ_α may be thought of as a "mathematical catalyst" that enables a transformation but is not, itself, transformed.

2.7.2 Temporal Averaging Theorem

The temporal averaging theorem relates the average of the partial time derivative of a microscale quantity to the partial time derivative of an averaged macroscale quantity. It follows directly from the transport theorem equation (2.20). As a starting point for the derivation, the transport theorem is rewritten here for the α phase volume within an REV, δV^α:

$$\int_{\delta V^\alpha} \frac{\partial f}{\partial t} dv_\xi = \frac{d}{dt} \int_{\delta V^\alpha} f dv_\xi - \int_{\delta S^{\alpha\alpha}} f\mathbf{w}\cdot\mathbf{n}_\alpha ds_\xi - \sum_{\beta \neq \alpha} \int_{S^{\alpha\beta}} f|_\alpha \mathbf{w}\cdot\mathbf{n}_\alpha ds_\xi \qquad (2.94)$$

where $f|_\alpha$ is the microscale quantity f evaluated in the α phase at the interface. Two important observations simplify this equation. First, the location of the averaging volume is fixed in macroscale space. Therefore the time derivative in the first integral on the right side, indicating that the volume is being followed, reduces to a partial time derivative. Second, the normal velocity of the surface of the REV is zero so that $\mathbf{w} \cdot \mathbf{n}_\alpha = 0$ on $\delta S^{\alpha\alpha}$; and the second integral on the right side must be zero. The identity (2.94) therefore simplifies to:

$$\int_{\delta V^\alpha} \frac{\partial f}{\partial t} dv_\xi = \frac{\partial}{\partial t} \int_{\delta V^\alpha} f dv_\xi - \sum_{\beta \neq \alpha} \int_{S^{\alpha\beta}} f|_\alpha \mathbf{w}\cdot\mathbf{n}_\alpha ds_\xi \qquad (2.95)$$

To express the terms in this equation as averages, divide by the constant size of the averaging volume, δV, to obtain the *temporal averaging theorem*:

$$\frac{1}{\delta V} \int_{\delta V^\alpha} \frac{\partial f}{\partial t} dv_\xi = \frac{\partial}{\partial t}\left[\frac{1}{\delta V} \int_{\delta V^\alpha} f dv_\xi \right] - \sum_{\beta \neq \alpha} \frac{1}{\delta V} \int_{S^{\alpha\beta}} f|_\alpha \mathbf{w}\cdot\mathbf{n}_\alpha ds_\xi \qquad (2.96)$$

This equation is even more similar to the transport theorem than the spatial averaging theorem is to the divergence theorem. Physically, this equation states that the average rate of change of a microscale function multiplied by the volume of interest identified with phase α is equal to the rate of change of the product of the macroscale average value of the function with the α phase volume minus a term that accounts for changes due to expansion or contraction of the volume of α phase due to movement of its boundary within the REV.

The derivation of the spatial and temporal averaging theorem completes the arsenal of tools needed to transform a mass conservation equation for a phase from the microscale to the macroscale. The application of these theorems will be explored in the next section.

2.8 MACROSCALE MASS CONSERVATION

The derivation of the macroscale mass conservation equations proceeds by integrating the corresponding point mass conservation equation over the phase of interest in an REV, applying the temporal and spatial averaging theorems, then expressing the integrals that remain in terms of average quantities. The heavy spade work has been done in the preceding sections; what remains is to apply the equations and interpret the terms that arise. Although the derivation of the microscale conservation equations proceeded from the integral forms to the point forms, the derivation of the macroscale equations first integrates from the microscale to the macroscale and then can be integrated from the macroscale to the integral form for the entire system.

2.8.1 Macroscale Point Forms

In this subsection, we seek the form of a macroscale mass conservation equation that allows for solution of a field of macroscale quantities within the system domain. This is analogous to the microscale point conservation equation which retains partial derivatives in both time and space. The starting point for this derivation will be the species conservation equation (2.28), although any of the forms of the microscale species conservation equations is an equally good starting point for development of the macroscale conservation equation. We are seeking the mass conservation equation for species i in the α phase. Integrate equation (2.28) over the α phase portion of the REV and divide the integral by the REV size to obtain:

$$\frac{1}{\delta V} \int_{\delta V^\alpha} \left[\frac{\partial(\rho\omega_i)}{\partial t} + \nabla \cdot (\rho\omega_i \mathbf{v}_i) - r_i \right] \mathrm{d}v_\xi = 0 \qquad (2.97)$$

To most easily treat each of the three terms that appear in the integral, separate the single integral into three parts:

$$\frac{1}{\delta V} \int_{\delta V^\alpha} \frac{\partial(\rho\omega_i)}{\partial t} \mathrm{d}v_\xi + \frac{1}{\delta V} \int_{\delta V^\alpha} \nabla \cdot (\rho\omega_i \mathbf{v}_i) \mathrm{d}v_\xi - \frac{1}{\delta V} \int_{\delta V^\alpha} r_i \mathrm{d}v_\xi = 0 \qquad (2.98)$$

Now each of these three integrals will be examined in turn with the integrals of derivatives being converted to derivatives of integrals using the averaging theorems with the definitions of averages also being employed.

The order of integration and differentiation for the first term in equation (2.98) is exchanged using temporal averaging theorem (2.96) with f replaced by $\rho\omega_i$ to obtain:

$$\frac{1}{\delta V} \int_{\delta V^\alpha} \frac{\partial(\rho\omega_i)}{\partial t} \mathrm{d}v_\xi = \frac{\partial}{\partial t} \left[\frac{1}{\delta V} \int_{\delta V^\alpha} \rho\omega_i \mathrm{d}v_\xi \right] - \sum_{\beta \neq \alpha} \frac{1}{\delta V} \int_{S^{\alpha\beta}} (\rho\omega_i)|_\alpha \mathbf{w} \cdot \mathbf{n}_\alpha \mathrm{d}s_\xi \qquad (2.99)$$

The integral appearing in brackets has been defined in terms of macroscopic quantities in equation (2.83). Substitution of this expression into the equation yields:

$$\frac{1}{\delta V} \int_{\delta V^\alpha} \frac{\partial(\rho\omega_i)}{\partial t} dv_\xi = \frac{\partial(\varepsilon^\alpha \rho^\alpha \omega^{i\alpha})}{\partial t} - \sum_{\beta \neq \alpha} \frac{1}{\delta V} \int_{S^{\alpha\beta}} (\rho\omega_i)|_\alpha \mathbf{w} \cdot \mathbf{n}_\alpha ds_\xi \qquad (2.100)$$

The order of differentiation and integration in the second term in equation (2.98) is reversed using the spatial averaging theorem for the divergence, equation (2.93) with $\mathbf{F}$ replaced by $\rho\omega_i\mathbf{v}_i$:

$$\frac{1}{\delta V} \int_{\delta V^\alpha} \nabla \cdot (\rho\omega_i\mathbf{v}_i) dv_\xi = \nabla_{\mathbf{x}} \cdot \left[\frac{1}{\delta V} \int_{\delta V^\alpha} \rho\omega_i\mathbf{v}_i dv_\xi \right] + \sum_{\beta \neq \alpha} \frac{1}{\delta V} \int_{S^{\alpha\beta}} (\rho\omega_i\mathbf{v}_i)|_\alpha \cdot \mathbf{n}_\alpha ds_\xi \qquad (2.101)$$

The quantity in brackets in this case has been defined previously in terms of macroscale quantities in equation (2.84). The macroscale quantities are substituted in and the subscript $\mathbf{x}$ on the divergence operator is dropped for convenience so that equation (2.101) simplifies to:

$$\frac{1}{\delta V} \int_{\delta V^\alpha} \nabla \cdot (\rho\omega_i\mathbf{v}_i) dv_\xi = \nabla \cdot (\varepsilon^\alpha \rho^\alpha \omega^{i\alpha} \mathbf{v}^{i\alpha}) + \sum_{\beta \neq \alpha} \frac{1}{\delta V} \int_{S^{\alpha\beta}} (\rho\omega_i\mathbf{v}_i)|_\alpha \cdot \mathbf{n}_\alpha ds_\xi \qquad (2.102)$$

Equations (2.100) and (2.102) are now substituted into equation (2.98). Additionally, the definition of the macroscale reaction rate term as given by equation (2.87) is employed so that the macroscale species mass conservation equation becomes:

$$\frac{\partial(\varepsilon^\alpha \rho^\alpha \omega^{i\alpha})}{\partial t} + \nabla \cdot (\varepsilon^\alpha \rho^\alpha \omega^{i\alpha} \mathbf{v}^{i\alpha}) - \varepsilon^\alpha r^{i\alpha}$$

$$+ \sum_{\beta \neq \alpha} \frac{1}{\delta V} \int_{S^{\alpha\beta}} (\rho\omega_i)|_\alpha (\mathbf{v}_i|_\alpha - \mathbf{w}) \cdot \mathbf{n}_\alpha ds_\xi = 0 \qquad (2.103)$$

Examination of this equation reveals great similarity with the form of microscale equation (2.28) if one accounts for the volume fraction and notes that the quantities are macroscale rather than microscale. The main difference in structure between the two equations is the integral over the interface between the phases that appears in equation (2.103). A corresponding term does not appear in the microscale equation because there is no internal structure at a microscale point, and there are no interfaces between phases. This integral will now be discussed in additional detail.

The part of the integral in equation (2.103) that warrants the most attention is the term $(\mathbf{v}_i|_\alpha - \mathbf{w}) \cdot \mathbf{n}_\alpha$. This term is equal to the difference at the phase interface between the velocity of species i and the velocity of the interface in the normal direction. If this term is zero, the normal velocities of the chemical species and the interface will be equal such that no species i will enter or leave the α phase due to transfer across the interface. This can be understood, for example, by considering a can of soda. If the can is sealed and tossed around, although the fluid inside has a nonzero velocity, no liquid or gas will leave the can, even if the can becomes dented, because the normal velocity of the boundary of the can is equal to the normal velocity of the fluid at the boundary. If, however, tossing the can around causes it to be punctured, fluid will leave because it will achieve a normal velocity greater than that of the boundary surface composed of aluminum plus holes in the aluminum. The term $(\mathbf{v}_i|_\alpha - \mathbf{w}) \cdot \mathbf{n}_\alpha$ is the *relative microscale normal velocity* of species i with respect to the $\alpha\beta$ interface. When this is zero at every point on the interface, there is no transfer across the interface; at points where it is positive, species i is leaving the α phase and entering the β phase; where it is negative, species i is being transferred to the α phase from the β phase. In fact $(\mathbf{v}_i|_\alpha - \mathbf{w}) \cdot \mathbf{n}_\alpha ds_\xi$ is the volumetric rate of transfer at a point; and multiplication of this by the mass of species i per volume, $\rho\omega_i$, gives the rate of transfer of mass. Thus the sum over all interfaces of the integral over the various interfaces accounts for the net transfer of mass of species i out of the α phase.

At the macroscale, it is common to account for the presence of wells that pump out or inject a fluid into a groundwater aquifer or a petroleum reservoir. This process can be considered to be transfer between the fluid in the porous medium, for example in the α phase, to the "well" phase. The flow occurs across the "interface" between the α phase and the well. At the macroscale, a well is considered to operate at a macroscale point, and the volume of species i that leaves the system at that point is $(\mathbf{v}_i|_\alpha - \mathbf{w}) \cdot \mathbf{n}_\alpha ds_\xi$. Typically, at the well the surface area is very small but the velocity is high enough that the amount of fluid entering or leaving the system is large.

Based on this discussion, special notation is adopted to account for the mass transfer of species i between phases, with units of mass of species i per unit volume per unit time. The exchange term is broken into two main parts: one that accounts for mass transfer between phases within the porous medium and a second that accounts for transfer at wells. The notation employed is as follows:

$$\sum_{\beta \neq \alpha} \frac{1}{\delta V} \int_{S^{\alpha\beta}} (\rho\omega_i)|_\alpha (\mathbf{v}_i|_\alpha - \mathbf{w}) \cdot \mathbf{n}_\alpha ds_\xi = -\sum_{\beta \neq \alpha} e_{\alpha\beta}^{i\alpha} - \sum_{W=1}^{N_W} \rho_W^\alpha \omega_W^{i\alpha} Q_W^\alpha \delta(\mathbf{x} - \mathbf{x}^W) \qquad (2.104)$$

where $e_{\alpha\beta}^{i\alpha}$ is the transfer of species i between the α and β phases in the porous medium that is positive when species i is being transferred into the α phase from the adjacent β phase; N_W is the number of wells in the system; Q_W^α is the volume per unit time being pumped into the α phase at location $\mathbf{x}^W$; $\delta(\mathbf{x} - \mathbf{x}^W)$ is the Dirac delta function associated with location $\mathbf{x}^W$. The Dirac delta $\delta(\mathbf{x} - \mathbf{x}^W)$ is zero at all points except $\mathbf{x} = \mathbf{x}^W$ and has the properties that, if V_∞ is the total macrospace domain of the system under study and dV is a macrospace element of volume:

$$\int_{V_\infty} \delta(\mathbf{x} - \mathbf{x}^W) dV = 1 \qquad (2.105)$$

and

$$\int_{V_\infty} \rho_W^\alpha \omega_W^{i\alpha} Q_W^\alpha \delta(\mathbf{x} - \mathbf{x}^W) dV = \rho_W^\alpha \omega_W^{i\alpha} Q_W^\alpha \qquad (2.106)$$

Note that if the macroscale point under consideration is not a location where a well exists then the summation over the well terms is zero such that equation (2.104) reduces to:

$$\sum_{\beta \neq \alpha} \frac{1}{\delta V} \int_{S^{\alpha\beta}} (\rho \omega_i)|_\alpha (\mathbf{v}_i|_\alpha - \mathbf{w}) \cdot \mathbf{n}_\alpha \mathrm{d}s_\xi = -\sum_{\beta \neq \alpha} e^{i\alpha}_{\alpha\beta} \qquad (2.107)$$

This form is also employed if one chooses to account for wells through boundary conditions when solving the point species balance equation. Here, equation (2.104) is selected.

One final point worth mentioning is that when Q^α_W is negative such that α phase is being pumped out of the system at well W, the density and mass fraction are equal to the values in the system at point $\mathbf{x}^W$ such that:

$$\rho^\alpha_W \omega^{i\alpha}_W = \rho^\alpha(t, \mathbf{x}^W) \omega^{i\alpha}(t, \mathbf{x}^W)$$

However, if Q^α_W is positive, indicating that material is being pumped into the α phase, the values of ρ^α_W and $\omega^{i\alpha}_W$ for the fluid being injected must be specified.

Substitution of equation (2.104) into equation (2.103) provides the macroscale point conservation equation for chemical species i in the α phase:

$$\frac{\partial(\varepsilon^\alpha \rho^\alpha \omega^{i\alpha})}{\partial t} + \nabla \cdot (\varepsilon^\alpha \rho^\alpha \omega^{i\alpha} \mathbf{v}^{i\alpha}) = \varepsilon^\alpha r^{i\alpha} + \sum_{\beta \neq \alpha} e^{i\alpha}_{\alpha\beta} + \sum_{W=1}^{N_W} \rho^\alpha_W \omega^{i\alpha}_W Q^\alpha_W \delta(\mathbf{x} - \mathbf{x}^W) \quad (2.108)$$

This equation has units of mass of species i in the α phase per unit volume of REV per unit time. The first term is the rate of accumulation at a macroscale point. The second term is the net loss due to flow. The three terms on the right side are sources due to chemical reaction, phase change, and well injection, respectively. Equation (2.108) is a form of the species conservation equation that can be re-expressed in other equivalent forms and also summed over the chemical species and over the phases. Some of these alternative forms will be presented next.

The velocity of species i can be removed from equation (2.108) by expressing it in terms of the phase velocity and a dispersion term using equation (2.86) such that:

$$\frac{\partial(\varepsilon^\alpha \rho^\alpha \omega^{i\alpha})}{\partial t} + \nabla \cdot (\varepsilon^\alpha \rho^\alpha \omega^{i\alpha} \mathbf{v}^\alpha + \varepsilon^\alpha \mathbf{j}^{i\alpha})$$

$$= \varepsilon^\alpha r^{i\alpha} + \sum_{\beta \neq \alpha} e^{i\alpha}_{\alpha\beta} + \sum_{W=1}^{N_W} \rho^\alpha_W \omega^{i\alpha}_W Q^\alpha_W \delta(\mathbf{x} - \mathbf{x}^W) \qquad (2.109)$$

This is the form that is commonly employed in simulating flow, and constitutive equations are employed for the dispersion and mass transfer terms.

The mass conservation equation for the α phase as a whole may be obtained by summing over the species balance equation for that phase. Summation of equation (2.108) over all N species in the α phase gives:

$$\sum_{i=1}^{N} \frac{\partial\left(\varepsilon^{\alpha} \rho^{\alpha} \omega^{i\alpha}\right)}{\partial t} + \sum_{i=1}^{N} \nabla \cdot \left(\varepsilon^{\alpha} \rho^{\alpha} \omega^{i\alpha} \mathbf{v}^{i\alpha}\right)$$

$$= \sum_{i=1}^{N} \varepsilon^{\alpha} r^{i\alpha} + \sum_{i=1}^{N} \sum_{\beta \neq \alpha}^{N} e_{\alpha\beta}^{i\alpha} + \sum_{i=1}^{N} \sum_{W=1}^{N_W} \rho_W^{\alpha} \omega_W^{i\alpha} Q_W^{\alpha} \delta(\mathbf{x} - \mathbf{x}^W) \qquad (2.110)$$

The following five identities are applied to this equation:

$$\sum_{i=1}^{N} \frac{\partial\left(\varepsilon^{\alpha} \rho^{\alpha} \omega^{i\alpha}\right)}{\partial t} = \frac{\partial\left(\varepsilon^{\alpha} \rho^{\alpha}\right)}{\partial t} \quad \text{since} \quad \sum_{i=1}^{N} \omega^{i\alpha} = 1 \qquad (2.111)$$

$$\sum_{i=1}^{N} \nabla \cdot \left(\varepsilon^{\alpha} \rho^{\alpha} \omega^{i\alpha} \mathbf{v}^{i\alpha}\right) = \nabla \cdot \left(\varepsilon^{\alpha} \rho^{\alpha} \mathbf{v}^{\alpha}\right) \quad \text{since} \quad \sum_{i=1}^{N} \omega^{i\alpha} \mathbf{v}^{i\alpha} = \mathbf{v}\alpha \qquad (2.112)$$

$$\sum_{i=1}^{N} \varepsilon^{\alpha} r^{i\alpha} = 0 \quad \text{since there is no net mass production by chemical reactions} \qquad (2.113)$$

$$\sum_{i=1}^{N} e_{\alpha\beta}^{i\alpha} = e_{\alpha\beta}^{\alpha} \quad \text{where } e_{\alpha\beta}^{\alpha} \text{ is total mass transfer to } \alpha \text{ from } \beta \text{ phase} \qquad (2.114)$$

$$\sum_{i=1}^{N} \rho_W^{\alpha} \omega_W^{i\alpha} Q_W^{\alpha} \delta(\mathbf{x} - \mathbf{x}^W) = \sum_{i=1}^{N} \rho_W^{\alpha} Q_W^{\alpha} \delta(\mathbf{x} - \mathbf{x}^W) \quad \text{since} \quad \sum_{i=1}^{N} \omega_W^{i\alpha} = 1 \qquad (2.115)$$

With these conditions employed, equation (2.110) simplifies to the mass balance equation for the α phase:

$$\frac{\partial\left(\varepsilon^{\alpha} \rho^{\alpha}\right)}{\partial t} + \nabla \cdot \left(\varepsilon^{\alpha} \rho^{\alpha} \mathbf{v}^{\alpha}\right) = \sum_{\beta \neq \alpha} e_{\alpha\beta}^{\alpha} + \sum_{W=1}^{N_W} \rho_W^{\alpha} Q_W^{\alpha} \delta(\mathbf{x} - \mathbf{x}^W) \qquad (2.116)$$

This equation contains terms similar to those in the species equation (2.108) except for a reaction term. This is consistent since the mass of a phase cannot be generated or destroyed by chemical or biological reactions. For a multispecies phase, equation (2.116) is commonly used as one of the mass conservation equations along with N – 1 of the conservation equations for the N species. Solution of the point equations of mass conservation provides the macroscale field of mass fractions throughout the study region.

2.8.2 Integral Forms

Besides existing as macroscale point forms, the mass conservation equations may be integrated over the entire system of interest. Here, the integral of the macroscale species mass conservation equation (2.108) will be developed. Then the integral of the macroscale equation for the conservation of total mass of the α phase, equation (2.116), will be employed. The results will be compared with the integral forms based on the microscale quantities.

The integral of the species conservation equation (2.108) over a volume with length scale much greater than the macroscale length scale of the REV is obtained as:

$$\int_V \frac{\partial(\varepsilon^\alpha \rho^\alpha \omega^{i\alpha})}{\partial t} dV + \int_V \nabla \cdot (\varepsilon^\alpha \rho^\alpha \omega^{i\alpha} \mathbf{v}^{i\alpha}) dV - \int_V \varepsilon^\alpha r^{i\alpha} dV$$

$$- \int_V \sum_{\beta \neq \alpha} e^{i\alpha}_{\alpha\beta} dV - \int_V \sum_{W=1}^{N_W} \rho^\alpha_W \omega^{i\alpha}_W Q^\alpha_W \delta(\mathbf{x} - \mathbf{x}^W) dV = 0 \qquad (2.117)$$

At the macroscale, the system is viewed as containing all phases at all points. The interfaces between phases in the interior of the system are not boundaries. Therefore the transport and divergence theorems, equations (2.20) and (2.18), respectively, may be applied to the first two integrals in equation (2.117). This will give rise to surface integrals over the external boundary of the system, S, with outwardly directed unit normal $\mathbf{n}^{\text{ext}}$. The result is:

$$\frac{d}{dt} \int_V \varepsilon^\alpha \rho^\alpha \omega^{i\alpha} dV + \int_S \varepsilon^\alpha \rho^\alpha \omega^{i\alpha} (\mathbf{v}^{i\alpha} - \mathbf{w}) \cdot \mathbf{n}^{\text{ext}} dS - \int_V \varepsilon^\alpha r^{i\alpha} dV$$

$$- \int_V \sum_{\beta \neq \alpha} e^{i\alpha}_{\alpha\beta} dV - \int_V \sum_{W=1}^{N_W} \rho^\alpha_W \omega^{i\alpha}_W Q^\alpha_W \delta(\mathbf{x} - \mathbf{x}^W) dV = 0 \qquad (2.118)$$

An alternative form of this equation is obtained by making use of the dispersion vector as given in equation (2.86). This form is typically employed with the species velocity replaced by the phase velocity plus the velocity of the species relative to the phase:

$$\frac{d}{dt} \int_V \varepsilon^\alpha \rho^\alpha \omega^{i\alpha} dV + \int_S \varepsilon^\alpha \rho^\alpha \omega^{i\alpha} (\mathbf{v}^\alpha - \mathbf{w}) \cdot \mathbf{n}^{\text{ext}} dS + \int_S \varepsilon^\alpha \mathbf{j}^{i\alpha} \cdot \mathbf{n}^{\text{ext}} dS$$

$$- \int_V \varepsilon^\alpha r^{i\alpha} dV - \int_V \sum_{\beta \neq \alpha} e^{i\alpha}_{\alpha\beta} dV - \int_V \sum_{W=1}^{N_W} \rho^\alpha_W \omega^{i\alpha}_w Q^\alpha_W \delta(\mathbf{x} - \mathbf{x}^W) dV = 0 \qquad (2.119)$$

Equation (2.118) may be compared to equation (2.9) that involves the integration of microscale quantities. Equation (2.9) will now be written in notation that indicates we are concerned with the α phase such that integration is only over that phase. The boundary of the α phase as viewed from the microscale perspective includes the internal interfaces with other phases and that portion of the external boundary of the system that cuts through the α phase. The global equation in terms of microscale variables is therefore:

$$\frac{d}{dt} \int_{V^\alpha} \rho \omega_i dv + \int_{S^{\alpha\alpha}} \rho \omega_i (\mathbf{v}_i - \mathbf{w}) \cdot \mathbf{n}_\alpha ds - \int_{V^\alpha} r_i dv - \sum_{\beta \neq \alpha} \int_{S^{\alpha\beta}} (\rho \omega_i)|_\alpha (\mathbf{w} - \mathbf{v}_i|_\alpha) \cdot \mathbf{n}_\alpha ds$$

$$- \sum_W \int_{S^W} (\rho_W \omega_{iW})|_\alpha (\mathbf{w} - \mathbf{v}_{iW}|_\alpha) \cdot \mathbf{n}_\alpha ds = 0 \qquad (2.120)$$

This equation has been written to facilitate term-by-term comparison with equation (2.118), the species mass balance equation written in terms of the macroscale properties. The main difference is that the macroscale version is somewhat "smeared" in that corresponding integrals use a volume fraction in the integrand along with integration over the entire domain whereas integration over the α phase based on the microscale variables sharply defines the microscale surface and volumes under study.

Summation of equations (2.118) and (2.120) over the N species comprising the phase is equivalent to integration of the total mass equations over the volume. The results are, respectively:

$$\frac{d}{dt}\int_V \varepsilon^\alpha \rho^\alpha dV + \int_S \varepsilon^\alpha \rho^\alpha (\mathbf{v}^\alpha - \mathbf{w}) \cdot \mathbf{n}^{ext} dS - \int_V \sum_{\beta \neq \alpha} e^\alpha_{\alpha\beta} dV$$
$$- \int_V \sum_{W=1}^{N_W} \rho^\alpha_W Q^\alpha_W \delta(\mathbf{x} - \mathbf{x}^W) dV = 0 \qquad (2.121)$$

and:

$$\frac{d}{dt}\int_{V^\alpha} \rho dv + \int_{S^{\alpha\alpha}} \rho(\mathbf{v} - \mathbf{w}) \cdot \mathbf{n}_\alpha ds - \sum_{\beta \neq \alpha} \int_{S^{\alpha\beta}} (\rho)|_\alpha (\mathbf{w} - \mathbf{v}|_\alpha) \cdot \mathbf{n}_\alpha ds$$
$$- \sum_W \int_{S^W} (\rho_W)|_\alpha (\mathbf{w} - \mathbf{v}_W|_\alpha) \cdot \mathbf{n}_\alpha ds = 0 \qquad (2.122)$$

Comparison of these equations reveals that, as with the species forms, the differences relate to scale effects while the equations describe the total system comparably.

2.9 APPLICATIONS

The derivation of the mass conservation equations in the last section has involved quite a bit of mathematics. This mathematics has carried us from a physically based statement of the conservation of mass in equation (2.1), to the mathematical expression of mass conservation as given in equation (2.9), through an examination of microscale and macroscale integral theorems, and finally to the macroscale species conservation equation at a point given by equation (2.110), the macroscale conservation equation for a phase at a point given in equation (2.116), and the integral forms of these equations over a system given respectively as equations (2.119) and (2.121). At the end of this mathematical journey, it is useful to look at the equations we have obtained and to relate them to physical problems. This exercise will also be enlightening in indicating where supplemental information is needed so that these equations can be applied to problems of interest.

Our focus is on the macroscale description of multiphase flow in porous media. We will therefore restrict the applications considered here to the macroscale. For our purposes, we have started from the physical description of the problem, passed through the microscale, and arrived at the macroscale. Although we will be describing problems only from a macroscale perspective, it will be helpful to keep in mind the microscale roots of our equations and the relations between microscale and macroscale quantities as given by equations (2.81) through (2.87) and the interface transfer terms as defined in equation (2.104). With a firm grasp on these concepts, we proceed to a relatively simple application that will add further insight.

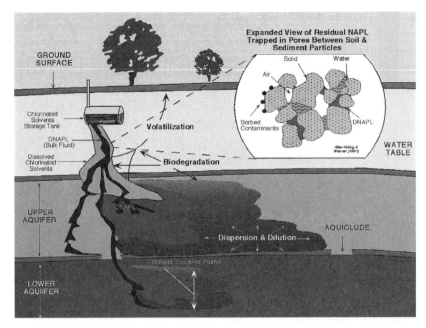

Figure 2.4: A subsurface region that contains air, water, and a contaminant phase. A slightly soluble liquid with a density greater than that of water enters the subsurface and moves vertically downwards through the water table to contaminate both an upper and a lower aquifer (from EPA [5]).

Consider the system depicted in Figure 2.4. Suppose that in what follows we are concerned with the subsurface region consisting of the unsaturated zone above the water table and the saturated zone above some confining geologic formation. The system contains four phases: air(a), water(w), organic fluid(n), and solid (s). Because of interaction among the phases, we will assume that each phase is a mixture of chemical constituents. We will address here the effectiveness of integral and point models based on the mass conservation equation in simulating this system.

2.9.1   Integral Analysis

Using equation (2.119), we can analyze a chemical constituent, i, in phase α. For purposes of discussion here, suppose that the organic chemical that has been spilled is benzene which makes up the bulk of phase n but is a constituent dissolved in the water phase designated as component i. In this integral analysis, let the volume studied, V, be the unsaturated plus the saturated zone indicated in Figure 2.4. The bounding surface of this region, S, is the land surface, the confining bed at the bottom, the interface with the river, and the vertical surfaces that define the lateral extent of the system. For the water phase, therefore, the integral equation (2.119) can be expressed:

$$\frac{\mathrm{d}M^{iw}}{\mathrm{d}t} + F^{iw} = 0 \qquad (2.123)$$

where M^{iw} is the mass of benzene species in the water phase defined as:

$$M^{iw} = \int_V \varepsilon^w \rho^w \omega^{iw} dV \tag{2.124}$$

and F^{iw} accounts for the remaining terms in equation (2.119):

$$F^{iw} = \int_S \varepsilon^w \rho^w \omega^{iw} (\mathbf{v}^w - \mathbf{w}) \cdot \mathbf{n}^{ext} dS + \int_S \varepsilon^w \mathbf{j}^{iw} \cdot \mathbf{n}^{ext} dS - \int_V \varepsilon^w r^{iw} dV$$
$$- \int_V \sum_{\beta=a,n,s} e^{iw}_{w\beta} dV - \int_V \sum_{W=1}^{N_W} \rho^w_W \omega^{iw}_W Q^w_W \delta(\mathbf{x} - \mathbf{x}^W) dV \tag{2.125}$$

Expression of the integral conservation equation in this form indicates several points:

- The integral species balance equation is used to solve for the total mass of the species in the system under consideration as a function of time, $M^{iw}(t)$. It is not able to provide information concerning the distribution of the species within the system. It provides information as to how the amount of species i within phase w changes with time.
- Solution of equation (2.123) requires an initial condition that specifies the initial total amount of species i in phase w. We will designate this as M_0^{iw}.
- Equation (2.123) cannot be solved unless F^{iw} can be expressed as a function only of M^{iw} and t or unless some additional equations, for example involving other variables of importance, are available.

The first and third item present particular limitations on the integral analysis. In fact, they imply that the integral equation can be solved only if it is supplemented by experimental data or with relations capable of expressing the integrals that comprise F^{iw} in terms of time and total mass of benzene in the w phase. For example, if water is being pumped from the system at various wells, the value of ω^{iw}_W must be known at each well. Because this is unknown, it could be measured and inserted in the equation. However, the integral model could not be used for predictions without knowledge of the concentrations that will result at the wells. Only if the concentration is uniform in the aquifer such that $\omega^{iw}_W = \omega^{iw}$ will the concentration be known. This would be the case if the water were uncontaminated, but otherwise would not be encountered in a physical situation. Furthermore, expressions are needed in terms of M^{iw} and t for the fluxes at the boundary of the system, for the reaction rates, and for the exchange of benzene between phases if the equation is to be solved. In essence, the integral equation for conservation of a chemical species in the subsurface is not useful for modeling the amount of that species because the species is not distributed uniformly in the region modeled.

The mass balance equation is useful for modeling a phase that does not exchange mass with other phases. This equation is the sum of a species equation over all the species in the phase such that the chemical reactions sum to zero and with the exchange terms, $e^\alpha_{\alpha\beta}$, between phases set to zero. The differential equation to be solved for the w phase is based on equation (2.121) with $\alpha = w$:

$$\frac{dM^w}{dt} + F^w = 0 \tag{2.126}$$

where M^w is the mass of the water phase defined as:

$$M^w = \int_V \varepsilon^w \rho^w dV \tag{2.127}$$

and F^w accounts for the flow in and out of the system:

$$F^w = \int_S \varepsilon^w \rho^w (\mathbf{v}^w - \mathbf{w}) \cdot \mathbf{n}^{ext} dS - \int_V \sum_{W=1}^{N_W} \rho_W^w Q_W^w \delta(\mathbf{x} - \mathbf{x}^W) dV \tag{2.128}$$

Equation (2.126) can be used to model the amount of water in an aquifer that is subject to withdrawal of water through wells, recharge from rainfall and rivers, and leakage exchange with other geologic units. Its primary utility arises in long term resource assessment. The initial condition and the fluxes must be specified. Of course, if a subsurface region is specified as being saturated, changes in the mass of water stored would require that the storage space be modified, through consolidation of the solid phase, or that the density of the water be changed in reaction to changes in pressure and/or temperature. These changes either have to be specified through monitoring of the system or obtained as part of the model through other equations. Thus, it should be clear that modeling of a multiphase system requires information about the behavior of the other phase as well. The interaction between phases, indeed, is what makes multiphase systems so much more complex than single-phase systems.

2.9.2 Point Analysis

Equation (2.109) provides a good starting point for the analysis of the movement of chemical constituent i in phase α from the local macroscale perspective. To make the discussion more concrete, let us again frame this for the case of benzene dissolved in a water phase. The overall system consists of the solid phase, s, a benzene organic phase, n, a vapor phase, a, and the water phase, w. With reference to Figure 2.4, we are considering the subsurface region composed of both saturated and unsaturated zones. The conditions at a point are reflected by the macroscale properties of the REV associated with that point. The species equation for benzene in the water phase based on equation (2.109) is:

$$\frac{\partial(\varepsilon^w \rho^w \omega^{iw})}{\partial t} + \nabla \cdot (\varepsilon^w \rho^w \omega^{iw} \mathbf{v}^w) + \nabla \cdot (\varepsilon^w \mathbf{j}^{iw})$$
$$= \varepsilon^w r^{iw} + \sum_{\beta=a,n,s} e_{w\beta}^{iw} + \sum_{W=1}^{N_W} \rho_W^w \omega_W^{iw} Q_W^w \delta(\mathbf{x} - \mathbf{x}^W) \tag{2.129}$$

It is desired to use this equation to solve for the evolving distribution of the benzene mass fraction, ω^{iw}, as a function of time and position. In theory, this can be accomplished if the initial distribution of benzene in the water phase is known, if some appropriate boundary conditions are specified (where "appropriateness" depends on the mathematical character of the equation), and if the other quantities appearing in the equation are known as functions of ω^{jw}, $\mathbf{x}$, and t. This final stipulation carries with it the following difficulties:

- The volume fraction of water phase, ε^w, will vary with time and position even if the solid phase is homogeneous because the volume fraction of the benzene and air phases will be functions of time and position. This observation implies that solution of equation (2.129) cannot be accomplished in isolation without considering the movement of the other phases.

- The density of the water phase, ρ^w, is a function of pressure in the water phase, temperature, and the composition of that phase. In cases where the temperature is uniform and the constituents dissolved in the water are present in small concentrations, it may be possible to consider the density as depending only on pressure. However, in general, this simplification is not possible and thus knowledge of the distribution of all chemicals is needed.

- The macroscale velocity of the water phase has to be somehow specified if equation (2.129) is to be solved. Note that since $\mathbf{v}^w$ is a vector, it has three components at every point. In general, this velocity will be highly variable. For a very simple problem, such as steady flow through a column of homogeneous porous medium saturated with water, it may be possible to specify the flow field based on the rate at which water flows out of the system. However, in general additional conditions that describe the velocity field are needed. Although, from physical considerations, we might anticipate that an appropriate macroscale equation can be derived based on the principle of conservation of momentum, in fact such an equation is complex in form. As will be discussed in the next chapter, the description of the flow field is typically based on a correlation extracted from experimental observations.

- Specification of the diffusion/dispersion vector $\mathbf{j}^{iw}$ as a function of the mass fraction, time, and position is also needed. This has proven to be a difficult task as the correlation most commonly used typically involves a coefficient whose magnitude depends on the scale at which the problem is studied.

- Any reactions involving benzene will depend on the presence of other chemical species and their distribution in the system. Thus additional information is needed to specify the chemical reaction field within the system.

- Knowledge is needed of the interphase exchange of benzene. This will also be a function of time, position, and the properties of the phases present at a point. Specification of the exchange term $e_{w\beta}^{iw}$ is therefore complicated.

- Finally, knowledge of the pumping rates and of the composition of water phase being injected into the system is needed. Fortunately, the equation provides the composition if water is being extracted, but the individual well pumping rates must be specified.

A masochist could, at least in theory, set up a monitoring program to measure all the variables needed to supplement equation (2.129) and provide an opportunity to solve it for the distribution of benzene in water. However, this could only be done in conjunction with an active experiment. Use of the model for predictive purposes, where values of quantities that would need to be measured are not known, is impossible. Furthermore, as an alternative to setting up a complex monitoring program to support an indirect method of providing the concentration distribution as the solution to a differential equation, it certainly would be more sensible and cost-effective to simply measure the concentrations! At this point, we are unable to use the mass conservation equation for any constructive purpose. We desperately need additional equations or conditions for the variables. The approaches to obtaining that information are provided in the next chapter.

Before heading to the next chapter, perhaps discouraged by the volume of supplementary information needed to employ point species balance equations in any meaningful way, it is encouraging to note that if we are interested in only the behavior of a phase and not in the distribution of species in the phase, then the mass balance equation used is equation (2.116). This equation is the sum of the species equations within a phase or the equation for a phase composed of a single species. If, in addition, we assume that the exchange of chemical species with adjacent phases is negligible for purposes of modeling the flow of a w phase, the phase balance equation becomes:

$$\frac{\partial(\varepsilon^w\rho^w)}{\partial t} + \nabla\cdot(\varepsilon^w\rho^w\mathbf{v}^w) = \sum_{W=1}^{N_W}\rho_W^w Q_W^w\delta(\mathbf{x}-\mathbf{x}^W) \qquad (2.130)$$

Equation (2.130) can be used to model the distribution of mass of w phase per REV, the product $\varepsilon^w\rho^w$, if the velocity field and pumping rates are specified. When used to model groundwater with only a single fluid phase, the quantities $\mathbf{v}^w$, ρ^w, and ε^w are typically expressed as functions of pressure or, more generally, stress. Then if the velocity can be determined as a function of the pressure, the equation will be solvable (with the pumping rates specified). When employed with more than one fluid phase, equation (2.130) is often expressed in terms of saturation as:

$$\frac{\partial(\varepsilon s^w\rho^w)}{\partial t} + \nabla\cdot(\varepsilon s^w\rho^w\mathbf{v}^w) = \sum_{W=1}^{N_W}\rho_W^w Q_W^w\delta(\mathbf{x}-\mathbf{x}^W) \qquad (2.131)$$

where $\varepsilon^w = \varepsilon s^w$ and ε is the porosity. In cases when the changes in porosity are small in comparison to the changes in saturation that occur, equation (2.131) is solved for saturation, provided the velocity is specified in sufficient detail.

The difficulties that arise in modeling a multiphase system using a point macroscale perspective are different from those that appear when using the integral, or system, equation. The possibility of solving the equation to obtain a distribution of mass fractions or saturations throughout a system of interest is attractive, but the task of specifying enough unknowns to make the equation solvable appears to be daunting. Nevertheless, we know that if we make approximations to terms that appear in the mass conservation equations in a way such that the equality specified by the equation is violated, the information we glean from the analysis will not be correct. Subsequently, we will search for relations that allow us to use the mass conservation equations in a predictive mode not simply as regulations that must not be violated.

2.10 SUMMARY

In this chapter, we have derived species and phase mass balance equations at the microscale and then applied averaging theorems to convert them to the macroscale. These equations are suitable for use in modeling multiphase flow in porous media. Unfortunately, the number of unknowns that appear in the equations is greater than the number of equations. Therefore additional conditions are needed for the problem to be completely specified. In multiphase flow modeling, the classical approach is to obtain the additional conditions from examination of correlations of experimental data and from insightful approximations. In the ensuing chapter, we will make some inroads into this problem by providing approximations for the parameters that have proven, in hindsight through their utility in applications, to be useful.

2.11 EXERCISES

1. Show that if a mixture contains only chemical species, equation (2.9) is identical to equation (2.14).

2. Instead of writing an equation of conservation of mass of a chemical species, it is possible to work with an equation that expresses the conservation of moles. If c is the number of moles per volume and x_i is the mole fraction, develop the molar conservation equation for species i that is analogous to mass conservation equation (2.9) in the form:

$$\frac{d}{dt}\int_V cx_i dv + \int_S cx_i(\mathbf{v}_i - \mathbf{w})\cdot\mathbf{n}\,ds - \int_V R_i dv = 0 \qquad (2.132)$$

Explain the meaning of each term and obtain the relation between r_i and R_i.

3. Start with equation (2.132) and show that the integral form of the molar conservation equation for a solution is:

$$\frac{d}{dt}\int_V c\,dv + \int_S c(\mathbf{v}_c - \mathbf{w})\cdot\mathbf{n}\,ds - \int_V \sum_{i=1}^N R_i dv = 0 \qquad (2.133)$$

where $\mathbf{v}_c$ is the molar average velocity of the mixture.

4. Show that a dispersion vector defined as:

$$\mathbf{J}_i = cx_i(\mathbf{v}_i - \mathbf{v}_c) \qquad (2.134)$$

may be employed to obtain a point molar conservation equation in the form:

$$\frac{\partial(cx_i)}{\partial t} + \nabla\cdot(cx_i\mathbf{v}_c + \mathbf{J}_i) - R_i = 0 \qquad (2.135)$$

5. The notation D/Dt is used to indicate a *material derivative*, a time derivative that follows the function it acts on. Thus, for example, Df/Dt assesses the time rate

of change of function f associated with a fixed part of material. At the microscale, this derivative is defined as:

$$\frac{Df}{Dt} = \frac{\partial f}{\partial t} + \mathbf{v} \cdot \nabla f \tag{2.136}$$

This expression indicates that the change in property f of material moving at velocity $\mathbf{v}$ is caused by both the change f at a point fixed in space and the changes in f within the field. For example, a particle can be heated both by being situated at a point where the temperature is increasing and by moving into regions of higher temperature.

Show that if a fluid behaves such that the densities of fluid particles (i.e., fluid associated with a microscale point) do not change with time as they flow through a system of interest, the fluid may be said to be *incompressible* and its mass conservation equation can be expressed as:

$$\nabla \cdot \mathbf{v} = 0 \tag{2.137}$$

6. Explain the difference between a constant density fluid and an incompressible fluid.

7. Justify the fact that if a volume does not deform or move through space, the transport theorem for that volume may be written:

$$\int_V \frac{\partial f}{\partial t} dv = \frac{\partial}{\partial t} \int_V f \, dv \tag{2.138}$$

8. A special form of the transport theorem is known as the *Reynolds transport theorem*. This theorem is applied to a volume that distorts such that no fluid crosses its boundary. Show that the Reynolds transport theorem may be expressed:

$$\int_V \frac{\partial f}{\partial t} dv = \frac{D}{Dt} \int_V f \, dv - \int_S f\mathbf{v} \cdot \mathbf{n} ds \tag{2.139}$$

9. If a closed volume is studied that allows no fluid to enter or leave, justify the statement of conservation of mass as:

$$\frac{D}{Dt} \int_V \rho \, dv = 0 \tag{2.140}$$

Make use of the Reynolds transport theorem, equation (2.139), to obtain the point mass conservation equation. Explain differences and similarities between the form you derive and equation (2.23).

10. Derive the macroscale point equation of conservation of moles of species i starting from equation (2.135). Define all macroscale average values explicitly.

11. Use the spatial averaging theorem, equation (2.93), to show that:

$$\nabla \varepsilon^\alpha = -\sum_{\beta \neq \alpha} \frac{1}{\delta V} \int_{S^{\alpha\beta}} \mathbf{n}_\alpha ds_\xi \tag{2.141}$$

12. Use the temporal averaging theorem, equation (2.96), to show that:

$$\frac{\partial \varepsilon^\alpha}{\partial t} = \sum_{\beta \neq \alpha} \frac{1}{\delta V} \int_{S^{\alpha\beta}} \mathbf{w} \cdot \mathbf{n}_\alpha ds_\xi \tag{2.142}$$

13. Show that summation of species conservation equation (2.109) over all chemical species i present in a solution leads to the mass balance equation for the α phase identical to equation (2.116).

14. Discuss the term-by-term correspondence of equations (2.121) and (2.122).

15. In equation (2.107), the term that accounts for mass transfer of species i into the α phase from the β phase is denoted as $e_{\alpha\beta}^{i\alpha}$ and is defined:

$$e_{\alpha\beta}^{i\alpha} = -\frac{1}{\delta V} \int_{S^{\alpha\beta}} (\rho\omega_i)|_\alpha (\mathbf{v}_i|_\alpha - \mathbf{w}) \cdot \mathbf{n}_\alpha ds\xi \tag{2.143}$$

Explain/prove the fact that:

$$e_{\alpha\beta}^{i\alpha} + e_{\alpha\beta}^{i\beta} = 0 \tag{2.144}$$

for any two adjacent phases α and β.

BIBLIOGRAPHY

[1] Bear, J., and Y. Bachmat, *Introduction to Modeling of Transport Phenomena in Porous Media*, Kluwer Academic Publishers, 553 pages, 1990.

[2] Bird, R.B., W.E. Stewart, and E.N. Lightfoot, *Transport Phenomena* Second Edition, John Wiley and Sons, New York, 920 pages, 2002.

[3] Darcy, H. (1856), Determination of the laws of flow of water through sand, in *Physical Hydrology*, (R.A. Freeze and W. Back, eds.), Hutchinson Ross, Stroudsburg, PA, 1983.

[4] Gray, W.G., A. Leijnse, R.L. Kolar, and C.A. Blain, *Mathematical Tools for Changing Spatial Scales in the Analysis of Physical Systems*, CRC Press, Boca Raton, FL, 232 pages, 1993.

[5] U.S. EPA, ' EPA/600/F-98/022, 1999.

3

FLOW EQUATIONS

3.1 INTRODUCTION

The mass conservation equations derived in the previous chapter describe the transport of chemical components between and within phases as the phases flow and intermingle within a porous medium. However, mass conservation equations alone are insufficient to describe the system behavior. They can be solved for the distribution of chemical species only if additional information is provided concerning the velocity field, dispersion vector, chemical reaction rate expressions, and interphase transport relations.

Two general approaches can be employed to specify this additional information. One involves working with species-based conservation equations for mass, momentum, and energy and obtaining relations for all the supplementary information needed (e.g., values for coefficients that arise and some relations among variables) such that the equations can be solved for species concentrations, velocity fields, and temperature distributions. For the case where the phase is composed of a single species, or we are interested only in the behavior of the phase as a whole, the species-specific terms will drop out, and the conservation equations will model the density of the phase as a whole. This method of working with an array of conservation equations related to each species may be appealing because of its generality and because nothing is neglected. However, it is a brute force approach that requires a very large amount of measured information and insight about mechanisms that influence behavior for the equations to be solvable. This method also suffers because it includes the overhead cost of solving for possible complexities, even when they are insignificant, at the cost of obliterating opportunities for clever equation development. The second approach is to simplify the general equations in light of various special cases of interest. Then effort can be expended in providing information that

Essentials of Multiphase Flow and Transport in Porous Media, by George F. Pinder and William G. Gray
Copyright © 2008 by John Wiley & Sons, Inc.

makes the conservation equations for those special cases solvable. As experience is gained in dealing with the special cases, confidence can be built to tackle the more general forms insightfully.

Both approaches start with the same equation sets. As long as all important processes are modeled with the same accuracy, they will yield the same results. The first approach is analogous to using a sledgehammer to drive a wedge or a thumb tack. The second is analogous to matching the size of the hammer to the task. It requires having more hammers in the tool box, but matches the effort needed to complete the job with the job at hand. The second approach will be employed in this and subsequent chapters.

In the present chapter, the focus is on the equations that describe the behavior of a phase as a unit and the mechanical interactions between the phases present. Analyses involving individual chemical species that comprise the phase are postponed to the next chapter. Although the phases may be composed of a mixture of chemical species, we will here consider the behavior of each phase without concern for the distribution of the chemicals that comprise the phase. Phases may exchange mass. Chemical reactions that might occur within the phase will alter the composition of the phase but do not alter the total amount of mass in the phase. Although the chemical composition typically alters the property of a phase, we will not consider this effect since we are not modeling compositional changes. Furthermore, chemical diffusion and dispersion are not modeled since those are processes that involve chemical species within a phase.

For a phase designated as α the *mass balance equation* is equation (2.116):

$$\frac{\partial(\varepsilon^\alpha \rho^\alpha)}{\partial t} + \nabla \cdot (\varepsilon^\alpha \rho^\alpha \mathbf{v}^\alpha) = \sum_{\beta \neq \alpha} e^\alpha_{\alpha\beta} + \sum_{W=1}^{N_W} \rho^\alpha_W Q^\alpha_W \delta(\mathbf{x} - \mathbf{x}^W) \qquad (3.1)$$

The difficulty we encounter in trying to solve this equation is that, while we have one equation for each phase, the number of unknowns in each phase mass conservation equation is more than one. For solution of an equation set, we require the same number of equations as unknowns, as well as conditions at the boundary of the system, and initial conditions for transient problems. If the well pumping rates, Q^α_W are specified at each well, ρ^α_W is specified for injection processes, and $e^\alpha_{\alpha\beta}$, the mass exchange term between α and β phases, is negligible, the left side of equation (3.1) still contains five unknowns for each phase (ρ^α, ε^α, and three components of the vector $\mathbf{v}^\alpha$). Four additional conditions are therefore needed if the equation is to be solved for any phase.

One set of sources of these additional conditions is the momentum and energy conservation equations. Although these are attractive because they express fundamental conditions that must not be violated, they are problematic because they also introduce new variables (e.g., stress tensor, heat conduction vector, interphase momentum and energy exchange terms, and internal energy) that must be accounted for. The number of variables grows faster than the number of equations. Although approaches to overcoming this problem are an active topic of research, a theoretically derived momentum equation is traditionally not employed in solving porous media problems, and the energy equation is only invoked when heat transfer processes are important.

To overcome the deficit of information, the mass conservation equation is supplemented instead with *constitutive equations*. A constitutive equation is a relation among system variables that is based on experimental observation or mathematical conjecture other than a conservation equation. Constitutive relations vary from material to material, are only approximations of material behavior under certain conditions, and are subject to revision and improvement as new insights are gained. Thus, the character of a constitutive equation is quite different from the fundamental inviolacy of a conservation equation that must apply exactly to any material. Constitutive equations include relationships such as equations of state for an equilibrium system (e.g., the ideal gas law), dynamic expressions for diffusive fluxes in terms of gradients of system variables (e.g., mass diffusion related to gradients in mass fractions via Fick's law or heat conduction related to temperature gradients using Fourier's law), and chemical reaction rate and equilibrium expressions. We will invoke constitutive equations in this chapter to obtain the additional conditions needed to specify mathematically the behavior of solid and fluid phases in a porous medium.

Three additional observations are warranted at this point. As noted above, the mass conservation equation for a phase contains five unknowns and nominally requires four additional conditions to be applied if it is to be solved. However, in practice additional variables to those that appear in equation (3.1) are introduced so that more than four conditions are needed. For example, the pressure will be seen to be a very useful variable, but its introduction into the problem requires that it be related to the variables that appear in the mass conservation equation; additionally, some parameters that appear in the new equation must be specified. Second, the solution of the mass balance equation for one phase cannot be achieved independently of some information about the behavior of the other phases in the system. This coupling is particularly important when the phases exchange significant amounts of mass and is more complex the more phases that are present. Third, all supplementary relations must be expressed in terms of macroscale variables. Caution must be exercised in applying an expression derived in terms of microscale variables at the macroscale because systems behave differently when viewed from the perspectives of different scales. However, in some instances this kind of direct extension may be employed.

In this chapter the additional conditions used to model the phases in a porous medium system are presented. Historically, the first condition has been based on the experiments of Henry Darcy. These experiments and their interpretation will be considered in the next section following the presentation in [22].

3.2 DARCY'S EXPERIMENTS

Flow in porous media is important in a wide range of applications including geological formations (e.g., groundwater aquifers, petroleum reservoirs), near surface soils (e.g., infiltration of water, pollutant leakage from landfills), biological materials (e.g., the spinal disk) and engineered materials (e.g., composites, mortar, catalysts, woven materials). Quantitative descriptions of flow and transport processes in porous media date from the experimental studies of Henry Darcy in 1856 [12, 13]. Darcy's studies involved packing a vertical column with sand as depicted in Figure

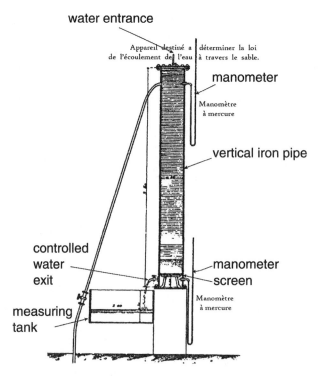

water entrance

manometer

vertical iron pipe

controlled
water
exit

manometer
screen

measuring
tank

Figure 3.1: Apparatus intended to determine the law of the water flow through sand [12].

3.1. U-shaped tubes, called *manometers*, tapped into fluid reservoirs at each end of the column and were open to the atmosphere. Consider that the lower manometer has been cut off in the figure such that it actually extends upward to the same elevation as the upper manometer. In the same way that the water level in a straw inserted into a cup containing a beverage will be independent of how far the straw is inserted into the drink, the water levels in the two manometers will be equal when there is no flow in the column. These water levels are referred to as *hydraulic head*, or simply head, which is the height above some reference datum to which water will rise in a manometer. Darcy's experiments involved observing the difference in the head in the manometers for the case when water completely fills the pore space and is flowing through the packed column at a steady flow rate. The column was packed with sand to four different heights. The sand used in each packing differed primarily by the degree of washing employed.

The idea was to maintain a constant water level in the reservoir at the top of the column that would cause water to flow downward through the column. The head measured at the top reservoir is denoted as h_2 while that for the bottom reservoir is h_1. No flow occurs when $h_2 = h_1$; flow is downward when $h_2 > h_1$; and flow would be upward for $h_2 < h_1$.

Despite difficulties in obtaining a constant water source, the experiments demonstrated that the volumetric flow rate down through the porous medium, Q, is proportional to the head difference across the sand column, $h_2 - h_1$, and the cross-sectional flow area, A, and is inversely proportional to the packed height of the column, L, such that:

$$Q = KA \left| \frac{h_2 - h_1}{L} \right| \tag{3.2}$$

In this algebraic expression, K is referred to as *hydraulic conductivity* and it is a function of both the porous medium and the fluid properties. Darcy found that K was essentially constant for a particular packing that he employed and did not change when the flow rate in the column changed.

It is important to note that Darcy's experiments actually provide no direct information about any properties within the packed column. All data was collected at locations external to the packed column, and Darcy's algebraic expression provides effective information for the column as a whole. In equation (3.2), the hydraulic conductivity is characteristic of the column and provides no indication of the degree of homogeneity of the packing in the column or of any dependence of K on position within the column. Neither the volumetric flow rate, Q, nor the volumetric flow rate per area, Q/A, indicates the speed of the water flowing within the pores since they consider the total column cross section, not just the fraction of the column actually available for flow due to the solid. The area, A, is a property of the column that was packed and does not relate to the effective cross-sectional area of flow. The length parameter, L, is the distance between the sampling points of the manometers and does not indicate the travel distance for fluid moving through the porous medium, which is influenced by the tortuous path created by the medium. Finally, as noted earlier, even the head values, h_1 and h_2, are obtained from measurements taken in reservoirs outside of the column.

In reporting his experimental results, Darcy was careful not to overstate their utility and implications. However, the need for scientists to model flow and transport in porous media systems has led to general acceptance of equations similar to Darcy's expression (3.2), with the state of the art reported in references such as [4, 5, 10, 14, 16]. These extended expressions are used to account for multiple fluid phases, inhomogeneous and anisotropic systems, and changes of h with position within the porous medium.

Despite their utility, these extended equations cannot be justified based exclusively on Darcy's experimental data. Indeed, if examined carefully, even the simplest differential form of Darcy's equation for single-phase flow in a porous medium is highly restricted. Subsequent extensions to multiphase flow are even more problematic (e.g., [20]).

This situation has arisen because systematic procedures for the derivation of equations for flow in porous media, such as averaging theory, began to develop approximately 100 years after Darcy's experiments. The need to model porous media systems in the intervening years produced heuristic equations with variables and parameters that are not precisely related to measurements. Evolving theoretical understanding, beyond the scope of the present work, is beginning to make it possible to relate experimental measurements to equation variables, thus facilitating transfer of data between scales. Here we will make note of the Darcy equation and will develop some extensions of this equation that are used as constitutive expressions that supplement equation (3.1) such that the porous medium flow is fully described.

3.3 FLUID PROPERTIES

The formulation of equations that govern the fluid flow can be accomplished by fitting curves through data to obtain parameters, as was done in analyzing the Darcy experiment. However, if one first identifies some properties of the materials being studied that can impact the system behavior, the equations obtained will be more robust in that the values of parameters obtained can be adjusted based on the system and not just arbitrarily based on the system behavior. The important concept to understand here is that the fluid properties must be quantifiable in terms of thermo-mechanical variables that we can measure, and about which we have, or can obtain, information. For the fluid, the important properties are the mass density (ρ), composition (ω_i), pressure (p), dynamic viscosity (μ), and temperature (T). These are indicated here as microscale quantities because that is the scale at which most handbooks provide values and relations among these variables. However, since our interest is in porous media flows, we note that these quantities must be transferred to the macroscale.

The *macroscale density* ρ^α is the mass of α-phase fluid per volume of α phase and has previously been defined in equation (2.82). The macroscale mass fraction of species i in the α phase, $\omega^{i\alpha}$, was defined in equation (2.83). Pressure, with dimensions $[M/(Lt^2)]$, can be considered to be either a force per area or an energy per volume. The former suggests that a definition of macroscale pressure might best be obtained by averaging over a surface within an averaging volume; the latter suggests defining average pressure based on an intrinsic phase average. Here, we will make the assumption that the difference between these two definitions is negligible and choose the macroscale pressure as the volume average:

$$\varepsilon^\alpha p^\alpha = \frac{1}{\delta V} \int_{\delta V} p\gamma_\alpha \mathrm{d}v_\xi = \frac{1}{\delta V} \int_{\delta V^\alpha} p\,\mathrm{d}v_\xi \tag{3.3}$$

The microscale dynamic viscosity μ with dimensions $[M/(Lt)]$ is a coefficient of proportionality between the shear stress and the rate of strain. We will designate an appropriate macroscale version of this coefficient as some sort of average of μ over the α phase within the REV and designate it as μ^α. However, it is important to note that the macroscale dynamic viscosity defined in this manner is not the proportionality coefficent between macroscale stress and macroscale rate of strain defined in terms of the macroscale velocity. Commonly, the fact that a macroscale measure of the dynamic viscosity is needed for a macroscale equation is overlooked, and the microscale value is used directly. However, because equations should be formulated at a single scale, we will explicitly recognize that we are using a macroscale coefficient. Of course, when μ is constant within an REV, $\mu^\alpha = \mu$.

The fact that equality of temperature of two bodies is used as a measure of equilibrium complicates the definition of macroscale temperature for the situation when the temperature is not uniform within the averaging volume. The question of whether to use a volume average temperature or some weighted average temperature, where the weighting function might be a quantity such as entropy or heat capacity, could be important if the temperature variation within an averaging volume is great. In the present discussion, we are not going to consider energy transport, so we can avoid further consideration of this issue using the galling statement that the definition of average temperature is beyond the scope of this book. We will designate the macroscale temperature as T^α, but we will not provide a precise definition of this quantity. In fact, in all the developed equations in this book, we will consider the temperature in a system to be independent of space and time. We will point out instances where nonconstant temperature complicates the analysis. When the temperature within an REV is constant, the definition of temperature will not change with scale such that $T = T^\alpha$.

Although we have identified ρ^α, p^α, $\omega^{j\alpha}$, and μ^α as fluid properties of interest, we cannot specify each of them independently of the others. Equations that relate these parameters are known as *equations of state*, and these form the topic of the next section.

3.4 EQUATIONS OF STATE FOR FLUIDS

Equations of state, or state constitutive equations, provide relations among the properties of a material. Although these state equations apply strictly only at equilibrium, they are generally considered to hold in most dynamic situations when the deviations from equilibrium or the rates of change of properties are "small" (where smallness is defined, using the circular argument, as pertaining to the situation where the equilibrium relations can be employed with accuracy).

Equations of state for fluid properties are typically developed experimentally using well-mixed systems such that the values of the parameters being studied do not vary with position. Then the relations are expressed in terms of microscale variables. When working with porous media, the fact that we are dealing with macroscale quantities complicates the specification of the state equations.

The fact that a microscale quantity is not necessarily uniform in an REV can introduce errors into the state equation if one simply assumes that the form developed in terms of microscale variables is appropriate when written in terms of macroscale variables. For cases where the gradients in the variables within an REV are small, this issue is not important. For our purposes, we will develop constitutive equations in terms of macroscale variables recognizing that lack of correspondence between microscale and macroscale quantities could be a source of error in the analysis of a system.

We now turn to the equations of state that are employed for study of a fluid phase.

3.4.1 Mass Fraction

If a fluid phase α is composed of N chemical species, only $N - 1$ of them can be specified independently because the sum of the mass fractions must be 1. If, for convenience, we designate the Nth species as the one that is dependent, the equation of state or constitutive formula is:

$$\omega^{N\alpha} = 1 - \sum_{i=1}^{N-1} \omega^{i\alpha} \tag{3.4}$$

Similar equations may be written for each phase with one species in each phase selected to be the dependent species. This selection need not be the same for each phase, and the number of species in the different phase can be different.

We see that equation (3.4) states that if we know the mass fractions of species 1 through $N - 1$, we know the mass fraction of species N. This equation happens to be an exact expression that holds for any fluid phase. As we shall see, such a precise relationship is not available for the other variables of interest. Also we note that equation (3.4) eliminates the need for independent information about only one of the N mass fractions in a phase. When we deal with a fluid composed of a single component, equation (3.4) confirms that the mass fraction of that species must be 1.

3.4.2 Mass Density and Pressure

A general postulate of the dependence of the mass density of fluid phase α composed of N chemical species would include dependence on the mass fractions, pressure, and temperature such that:

$$\rho^\alpha = \rho^\alpha(p^\alpha, T^\alpha, \omega^{i\alpha}) \quad i = 1, \ldots, N-1 \tag{3.5}$$

Note that dependence is indicated on only $N - 1$ of the species mass fractions rather than on all N because, as shown in equation (3.4), only $N - 1$ mass fractions are independent. Additionally, we will not be considering changes in or distributions of temperature within the system under study. Thus, with the temperature treated as a constant, we can differentiate equation (3.5) to obtain:

$$d\rho^\alpha = \frac{\partial \rho^\alpha}{\partial p^\alpha} dp^\alpha + \sum_{i=1}^{N-1} \frac{\partial \rho^\alpha}{\partial \omega^{i\alpha}} d\omega^{i\alpha} \tag{3.6}$$

Define the α phase *compressibility* at fixed composition as:

$$\beta^\alpha = \frac{1}{\rho^\alpha} \frac{\partial \rho^\alpha}{\partial p^\alpha} \tag{3.7}$$

The compressibility, β^α, is positive since the density will increase as the pressure increases. The inverse of the compressibility is called the *bulk modulus* and is commonly designated as κ^α for an α phase. The *concentration compressibility* for species i is defined as:

$$\beta^{i\alpha} = \frac{1}{\rho^\alpha} \frac{\partial \rho^\alpha}{\partial \omega^{i\alpha}} \tag{3.8}$$

This change in density with mass fraction of component i is at fixed pressure and other mass fractions except the Nth. For this reason, the Nth species is usually chosen as the one with the largest mass fraction. Because changes in the composition of the fluid phase can either increase or decrease the density, the sign of $\beta^{i\alpha}$ for any species could be positive or negative, depending on the fluid solution. Next divide equation (3.6) by the density and substitute in the definitions for the compressibilities to obtain:

$$\frac{d\rho^\alpha}{\rho^\alpha} = \beta^\alpha dp^\alpha + \sum_{i=1}^{N-1} \beta^{i\alpha} d\omega^{i\alpha} \tag{3.9}$$

When the compressibility coefficients are approximately constant over a range of change in density, pressure, and concentrations, equation (3.9) may be integrated to:

$$\ln\left(\frac{\rho^\alpha}{\rho_0^\alpha}\right) = \beta^\alpha(p^\alpha - p_0^\alpha) + \sum_{i=1}^{N-1} \beta^{i\alpha}(\omega^{i\alpha} - \omega_0^{i\alpha}) \tag{3.10}$$

where the subscript "0" is used to indicate some reference value. Equation (3.10) is a special case of the more general form given by equation (3.5).

In general, the *compressibility of a fluid*, β^α, will depend on the composition of the fluid. When the changes in concentrations of the various species in a phase are small or when the compressibility effects due to concentration are unimportant, the concentration compressibilities may be neglected. Even if they are important for a particular fluid phase composed of many species, it is likely that only one or two of the concentration changes will be large enough that the concentration compressibility will have to be considered. If one is modeling flow of water containing small amounts of contaminant, it may be appropriate to use the *compressibility of pure water*, $\beta^w = 4.65 \times 10^{-10}\,\mathrm{m^2/N} = 4.65 \times 10^{-8}\,\mathrm{mbar^{-1}}$. Although this compressibility is small, approximately four orders of magnitude smaller than the compressibility of a gas, the physical behavior of large natural aquifers can be influenced by the compressibility of water.

In some instances, it may be useful to have expressions for pressure in terms of density, temperature, and mass fraction, or:

$$p^\alpha = p^\alpha(\rho^\alpha, T^\alpha, \omega^{i\alpha}) \quad i = 1, \ldots, N-1 \tag{3.11}$$

This equation is not independent of equation (3.5) but is a rearrangement, or inversion of the variables. Equation (3.1) may be rearranged to obtain a particular constitutive form of equation (3.11) that applies at constant temperature:

$$p^\alpha - p_0^\alpha = \kappa^\alpha \ln\left(\frac{\rho^\alpha}{\rho_0^\alpha}\right) - \kappa^\alpha \sum_{i=1}^{N-1} \beta^{i\alpha}(\omega^{i\alpha} - \omega_0^{i\alpha}) \tag{3.12}$$

This inversion is possible because the density of any real fluid always increases with pressure such that $\beta^\alpha = 1/\kappa^\alpha$ is always positive. If an idealized case of an incompressible fluid is considered such that density does not change with pressure, a constitutive equation for pressure as a function of density, temperature, and composition does not exist because $\beta^\alpha = 0$.

3.4.3 Fluid Viscosity

The *dynamic viscosity* of a fluid α phase, $\mu^\alpha[M/(TL)]$, is a measure of a fluid's resistance to deformation when subjected to shearing. It relates to the interaction of the molecules within a fluid. For high viscosity fluids, the molecules do not easily slide by each other when a shear stress is applied (e.g., molasses), but for a low viscosity fluid, the molecules have limited interaction (e.g., acetone).

The viscosity of a fluid is generally taken to be a function of pressure, temperature, and composition (see [34] and [40] for specific examples), such that:

$$\mu^\alpha = \mu^\alpha(p^\alpha, T^\alpha, \omega^{i\alpha}) \quad i = 1, \ldots, N-1 \tag{3.13}$$

Experimental studies, backed by some theoretical studies in simplified cases, have shown that viscosity is rather insensitive to pressure, except at very high pressures. Furthermore, the viscosity of a liquid tends to decrease with temperature while the viscosity of a gas increases with temperature. Thus the temperature of the system is important to selecting the viscosity value while the changes in pressure are relatively unimportant. The viscosity of pure water at $293\,°\text{K}$ is $\mu^\alpha = 1.0 \times 10^{-3}\,\text{N sec/m}^2 = 1.0$ centipoise. This is about 3 times the viscosity of acetone, but roughly from 1% to 0.01% of the viscosity of machine and motor oils. The viscosity of water is 72% less at $372\,°\text{K}$ than at $293\,°\text{K}$.

The composition of the fluid is of importance in determining the viscosity, and the change in viscosity with composition must be considered. If we neglect the dependence of viscosity on pressure and consider a fixed temperature:

$$\mathrm{d}\mu^\alpha = \sum_{i=1}^{N-1} \frac{\partial \mu^\alpha}{\partial \omega^{i\alpha}} \mathrm{d}\omega^{i\alpha} \tag{3.14}$$

We may define the *viscosity-composition coefficient* with respect to species i as:

$$\beta_\mu^{i\alpha} = \frac{1}{\mu^\alpha} \frac{\partial \mu^\alpha}{\partial \omega^{i\alpha}} \tag{3.15}$$

where all mass fractions are constant except those of species i and N. Then equation (3.14) is expressed as:

$$\frac{\mathrm{d}\mu^\alpha}{\mu^\alpha} = \sum_{i=1}^{N-1} \beta_\mu^{i\alpha} \mathrm{d}\omega^{i\alpha} \tag{3.16}$$

If the coefficients $\beta_\mu^{i\alpha}$ are treated as constants over the range of mass fractions being considered, this equation may be integrated to obtain:

$$\ln\left(\frac{\mu^\alpha}{\mu_0^\alpha}\right) = \sum_{i=1}^{N-1} \beta_\mu^{i\alpha}(\omega^{i\alpha} - \omega_0^{i\alpha}) \tag{3.17}$$

where the subscript "0" refers to some reference situation. Equations of this form for each phase, tabulated data, or some other expression based on experimental data can be used to provide values of the fluid viscosities to be used in a modeling exercise.

3.5 HYDRAULIC POTENTIAL

In addition to the quantities defined in the last section, we need an understanding of the mechanisms that cause flow in porous media to occur. A general approach to obtaining this quantity would be to formulate a momentum balance on a small element of fluid. The momentum of the fluid would be altered whenever the sum of the forces acting on the fluid is not zero. This approach shares some of the characteristics of the analysis that was performed in the last chapter to obtain an equation for mass balance. However, the momentum equation is conceptually and mathematically more difficult. Therefore, the approach that has been traditionally applied for flow modeling is to infer the equation of flow from consideration of Darcy's experiments as described in Section 3.2.

The most useful observation from Darcy's experiment is that flow occurs when the water levels, or head, in the manometers are not equal. The head is thus a potential for flow that causes fluid to move from regions of higher potential to regions of lower potential. This concept is similar to the observation that temperature differences in a wire cause heat to be conducted from high temperature to low temperature.

The Darcy experiment is very simple in that the properties of the water are essentially constant. If we are going to define a hydraulic potential for flow, it will generally be necessary to consider the influence of variations in density, composition, temperature, and pressure. Here, we will develop an expression for the hydraulic head based on an examination of the balance of forces at equilibrium. Then these considerations will be extended in light of Darcy's experiment to obtain an approximate equation that describes the flow velocity when the forces acting on the fluid do not balance. This extension is not the development of a conservation equation but, rather, involves hypothesizing an expression that seems to have possibilities for being useful in describing the flow. If subsequent experimental studies indicate that, indeed, the hypothesis has merit in some instances, it can be used in those instances. Fortuitously, the simplest sort of expressions that can be hypothesized in the context of porous media turn out to have a wide range of applicability. But this is getting ahead of the story. The first step is to derive an expression for hydraulic potential or hydraulic head.

In standard terminology, the hydraulic potential has dimensions of energy per mass. Division of the hydraulic potential by gravity yields a quantity with units of length called the hydraulic head. Since the head and the potential differ by a constant factor, these terms can be used interchangeably without confusion.

3.5.1 Hydrostatic Force and Hydraulic Head

Although we are concerned with flow in a porous medium, we can develop the concept of the hydraulic head by first considering fluid w in a column without the solid material. For our discussion, consider the cylinder with cross-sectional area A containing a fluid at rest as in Figure 3.2. Assume that the fluid properties do not

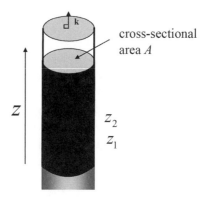

Figure 3.2: Cylinder containing fluid with levels z_1 and z_2 indicated.

vary in a horizontal cross-section. We are interested in the force exerted on the fluid at two different levels in the column designated as z_1 and z_2, where the z coordinate is chosen as positive upward and the unit vector in the z direction is $\mathbf{k}$. Note that we will be analyzing physical phenomena, and such phenomena are not affected by the coordinate system used in the analysis. Thus, the z coordinate could be chosen as positive downward if one wished.

The magnitude of the force exerted at any level in the cylinder is the pressure at the level multiplied by the cross-sectional area. The pressure at z_2 will be less than that at z_1 because the fluid between the two positions does not contribute to p_w at z_2. We can therefore write the hydrostatic relation between the two pressures as:

$$p_w|_1 A = p_w|_2 A - \rho_w|_{\text{avg}} \mathbf{g} \cdot \mathbf{k} A (z_2 - z_1) \tag{3.18}$$

where $\rho_w|_{\text{avg}}$ is the average fluid density in the volume between z_1 and z_2 and $\mathbf{g}$ is the gravity vector. Thus the difference in magnitude of the forces acting at z_1 and z_2 is the weight of the fluid contained in the region between z_1 and z_2. The cross-sectional area may be eliminated from the equation which can be rearranged to:

$$\frac{p_w|_2 - p_w|_1}{z_2 - z_1} = \rho_w|_{\text{avg}} \mathbf{g} \cdot \mathbf{k} \tag{3.19}$$

In the limit as the distance between z_1 and z_2 becomes small, the left side of this equation is a derivative, and the average value of density simply becomes the density at the location being studied such that:

$$\frac{dp_w}{dz} = \rho_w \mathbf{g} \cdot \mathbf{k} \tag{3.20}$$

This equation applies at any location in the column and can be rearranged and integrated from a vertical location z_0 where the pressure is equal to p_0 to an arbitrary vertical location z to obtain:

$$\int\limits_{p_0}^{p_w} \frac{dp'_w}{\rho_w g} - \int\limits_{z_0}^{z} \frac{\mathbf{g} \cdot \mathbf{k}}{g} dz' = 0 \tag{3.21}$$

where the primes have been employed to indicate the variables of integration whereas the limits of the integration are not primed. The second integral may be evaluated easily since the integrand is a constant.

To evaluate the first integral, we need an equation of state for density in terms of pressure. For illustrative purposes, we will use a general form as provided by equation (3.5) written in terms of microscale variables in phase w. Substitution of this form with implicit understanding that the subscript i implies dependence on $N - 1$ of the mass fractions yields:

$$\int\limits_{p_0}^{p_w} \frac{dp'_w}{\rho_w(p'_w, T_w, \omega_{iw}) g} - \frac{\mathbf{g} \cdot \mathbf{k}}{g} (z - z_0) = 0 \tag{3.22}$$

In considering field problems, it is convenient to introduce a *reference pressure*, p_{ref}, that might be encountered rather than p_0, which is based on some thermodynamic reference condition. We can introduce a reference pressure by changing the pressure integral into a two-part integral, one from p_0 to p_{ref} and a second from p_{ref} to p_w:

$$\int\limits_{p_0}^{p_{\text{ref}}} \frac{dp'_w}{\rho_w(p'_w, T_w, \omega_{iw}) g} + \int\limits_{p_{\text{ref}}}^{p_w} \frac{dp'_w}{\rho_w(p'_w, T_w, \omega_{iw}) g} - \frac{\mathbf{g} \cdot \mathbf{k}}{g} (z - z_0) = 0 \tag{3.23}$$

Now introduce a fixed reference elevation datum z_{ref} and rearrange the components of this expression to obtain:

$$\int\limits_{p_{\text{ref}}}^{p_0} \frac{dp'_w}{\rho_w(p'_w, T_w, \omega_{iw}) g} - \frac{\mathbf{g} \cdot \mathbf{k}}{g} (z_0 - z_{\text{ref}}) = \int\limits_{p_{\text{ref}}}^{p_w} \frac{dp'_w}{\rho_w(p'_w, T_w, \omega_{iw}) g} - \frac{\mathbf{g} \cdot \mathbf{k}}{g} (z - z_{\text{ref}}) \tag{3.24}$$

We previously indicated that z_0 is an arbitrarily selected location in the column and p_0 is the pressure at that elevation. Equation (3.24) indicates that, regardless of the location z_0 with its corresponding pressure, the value of the expression on the left will always equal the right side of the equation. Subject to the constraint that, at equilibrium, the head will have the same value at every point in the system, this observation suggests definition of the hydraulic head as:

$$h_w = \int\limits_{p_{\text{ref}}}^{p_w} \frac{dp'_w}{\rho_w(p'_w, T_w, \omega_{iw}) g} - \frac{\mathbf{g} \cdot \mathbf{k}}{g} (z - z_{\text{ref}}) \tag{3.25}$$

The integral term is commonly referred to as the *pressure head* while the second term is called the *elevation head*.

Although equation (3.25) provides a concise mathematical definition of the hydraulic head, it still presents some problems. First, the derivation applies for microscale properties. The relation will apply for a connected fluid within a porous medium as well as to a pool of fluid with no solid phase present. Modeling of porous media systems involves formulation of the equations at a macroscale, however. To this end we will assume that definition (3.25) applies when written in terms of macroscale quantities. Thus, the macroscale head relation employed is:

$$h^w = \int_{p_{ref}}^{p^w} \frac{dp^{w\prime}}{\rho^w\left(p^{w\prime}, T^w, \omega^{iw}\right)g} - \frac{\mathbf{g} \cdot \mathbf{k}}{g}\left(z^w - z_{ref}\right) \tag{3.26}$$

In this equation, the error involved in replacing microscale quantities with macroscale counterparts is small. An instance where this approximation might be important is if the fluid distribution in the averaging volume is not uniform such that z^w, the vertical coordinate of the centroid of phase w, is significantly difference from the vertical coordinate of the centroid of the total averaging region. This difference is typically small and will be neglected here such that we can write equation (3.26) as:

$$h^w = \int_{p_{ref}}^{p^w} \frac{dp^{w\prime}}{\rho^w\left(p^{w\prime}, T^w, \omega^{iw}\right)g} - \frac{\mathbf{g} \cdot \mathbf{k}}{g}\left(z - z_{ref}\right) \tag{3.27}$$

A second problem lies in the fact that expressions for density in equation (3.25) or in equation (3.27) are equations of state. Equations of state do not involve spatial locations but simply express equilibrium relations among variables, in this case, density, pressure, temperature, and the mass fractions. If the density depends only on pressure, the pressure head integral may be evaluated. However, in a more general equation of state for use when the temperature or the mass fractions are not constant, the density will be impacted as will the value of the integral. In other words, a value of the integral, and therefore for the head, may be obtained only for the cases where the density is constant or is a function only of the pressure. When the density in a system of interest depends, for example, on a nonconstant temperature as well as pressure, the integral would be impacted.

It is important, though perhaps somewhat confusing, to note that the integral in equation (3.27) involves integration of an equation of state and not of the actual distribution of density, temperature, and concentration in a physical system of interest. However, for the special cases where the temperature and mass fractions are constant or their variation is small enough that they do not impact the fluid density in a study system, the dependence of the equation of state on these variables may be neglected such that the expression for the head may be written:

$$h^w = \frac{\Phi^w}{g} = \int_{p_{ref}}^{p^w} \frac{dp^{w\prime}}{\rho^w\left(p^{w\prime}\right)g} - \frac{\mathbf{g} \cdot \mathbf{k}}{g}\left(z - z_{ref}\right) \tag{3.28}$$

The head given in this form can be calculated uniquely when the state equation for density as a function of pressure, the reference pressure, and reference elevation are selected. For air-water systems in soil, the reference pressure in equation (3.28) is typically selected to be the atmospheric pressure. In this case, Φ^w is called the *Hubbert potnetial*. For instances when the density may be assumed to be constant, the integral in the definition for head may be evaluated directly to obtain:

$$h^w = \frac{p^w - p_{\text{ref}}}{\rho^w g} - \frac{\mathbf{g} \cdot \mathbf{k}}{g}(z - z_{\text{ref}}) \tag{3.29}$$

This last expression is worth considering in light of Darcy's experiment. In his study, the changes in pressure were small enough that the density of the fluid was essentially constant. Now consider the conditions at the top of the manometer. The pressure is atmospheric so that at this location $p^w = p_{\text{atm}}$. Selection of the reference pressure to be atmospheric then eliminates the pressure term. Also, with the z-coordinate pointing upward, $\mathbf{g} \cdot \mathbf{k} = -g$. Substitution of these relations into equation (3.29) provides the expression for the head at the top of the manometer:

$$h^w = z|_{\text{top}} - z_{\text{ref}} \tag{3.30}$$

This expression confirms that in the simplest case, the head is the elevation of the fluid in a manometer above a reference datum. When the system is at equilibrium, this value of head applies at any point in the system as evidenced by the fact that the fluid will rise to the same level in the manometer regardless of where it is inserted into the system. The important observation from Darcy' experiments is that when the fluid is not at equilibrium, the fluid level in manometers in different parts of the system are not necessarily equal. Changes in head with time and space are important for describing systems not at equilibrium. Expressions for these derivatives will be obtained next.

3.5.2 Derivatives of Hydraulic Head

The fact that the hydraulic head defined in equation (3.26) cannot be evaluated generally when temperature and concentration effects are important means that for purposes of flow modeling, hydraulic head is best used directly only when these effects are unimportant.

At equilibrium, h^w is constant throughout phase w. Away from equilibrium, h^w will vary with time and space. It will be useful to take the time derivative and the gradient of head. To do this, we will make use of the Leibnitz rule for differentiation of an integral:

$$\frac{\partial}{\partial \xi} \int_{a(\xi)}^{b(\xi)} f(\xi, \zeta) \, d\zeta = \int_{a(\xi)}^{b(\xi)} \frac{\partial f(\xi, \zeta)}{\partial \xi} \, d\zeta + f(\xi, b)\frac{\partial b}{\partial \xi} - f(\xi, a)\frac{\partial a}{\partial \xi} \tag{3.31}$$

where a and b are limits of integration that may depend on independent variables, f is an integrand, ζ is the variable of integration, and the derivative is being taken with respect to another independent variable, ξ. If, for example, ξ is a spatial coordinate, equation (3.31) states that the derivative of an integral with respect to that coordinate is equal to the integral of the derivative of the integrand plus terms involving the integrand evaluated at the limits of integration and the derivatives of those limits. A second form of the Leibnitz rule that follows directly involves the gradient operator such that differentiation is with respect to the three spatial coordinates. In this instance, the Leibnitz rule extends to:

$$\nabla \int_{a(\mathbf{x})}^{b(\mathbf{x})} f(\mathbf{x}, \zeta) \, \mathrm{d}\zeta = \int_{a(\mathbf{x})}^{b(\mathbf{x})} \nabla f(\mathbf{x}, \zeta) \, \mathrm{d}\zeta + f(\mathbf{x}, b) \nabla b - f(\mathbf{x}, a) \nabla a \qquad (3.32)$$

The upper limit of the integral in equation (3.27), p^w, may depend on time and spatial coordinates while the lower limit, p_{ref}, is a constant. The integrand is independent of space and time since it is a constitutive equation. The quantity z is the vertical spatial coordinate. Thus the expressions for the partial time derivative and for the gradient of h^w may be obtained, respectively, from equations (3.31) and (3.32) as:

$$\frac{\partial h^w}{\partial t} = \frac{1}{\rho^w g}\bigg|_{p^w} \frac{\partial p^w}{\partial t} \qquad (3.33)$$

and:

$$\nabla h^w = \frac{1}{\rho^w g}\bigg|_{p^w} \nabla p^w - \frac{\mathbf{g}}{g} \qquad (3.34)$$

In these equations, ρ^w is evaluated at the same location as p^w. Since a location is involved, the temperature and mass fractions at that location can be measured so that the density is specified unambiguously. For notational convenience, the vertical bar indicating density is evaluated at the location where p^w is assessed will not be employed so that the expressions for the gradient and time derivative of hydraulic head are:

$$\frac{\partial h^w}{\partial t} = \frac{1}{\rho^w g} \frac{\partial p^w}{\partial t} \qquad (3.35)$$

and:

$$\nabla h^w = \frac{1}{\rho^w g} \nabla p^w - \frac{\mathbf{g}}{g} \qquad (3.36)$$

These expressions will be useful subsequently as we hypothesize equations for flow.

3.6   SINGLE-PHASE FLUID FLOW

The simplest case of flow in porous media is flow of a single homogenous fluid phase through a porous solid. At the beginning of this chapter, we stated the point conservation of mass equation for this situation as equation (3.1). In fact, since this is a mass conservation equation for a phase, it can be particularized to both the fluid phase, indicated as w, and the solid phase, designated as s. The mass conservation equations for the fluid and solid phases are, respectively:

$$\frac{\partial(\varepsilon\rho^w)}{\partial t} + \nabla\cdot(\varepsilon\rho^w\mathbf{v}^w) = e_{ws}^w + \sum_{W=1}^{N_W}\rho_W^w Q_W^w \delta(\mathbf{x}-\mathbf{x}^W) \qquad (3.37)$$

and:

$$\frac{\partial[(1-\varepsilon)\rho^s]}{\partial t} + \nabla\cdot[(1-\varepsilon)\rho^s\mathbf{v}^s] = e_{ws}^s \qquad (3.38)$$

where ε is the porosity, the single fluid phase w occupies the entire pore space such that $\varepsilon^w = \varepsilon$, the solid phase volume fraction, ε^s, is equal to $1 - \varepsilon$, and the assumption is made that no solid phase is removed from or added to the system though the wells. The mass exchange terms e_{ws}^w and e_{ws}^s account for dissolution of the solid phase or adsorption of fluid onto the solid. Because these exchanges do not produce mass, they must satisfy the constraint $e_{ws}^w + e_{ws}^s = 0$. The specification of these terms typically requires that one consider chemical species interaction or melting/freezing phenomena. If the phases are treated without study of the chemical interactions and in the absence of heat effects, these terms would have to be specified as known quantities. We will not consider thermal effects here and will postpone chemical interactions until the next chapter. We will thus assume that the exchange terms are specified. When conditions at the wells are also specified, equations (3.37) and (3.38) contain nine unknowns (ρ^w, ρ^s, ε, and the three components of each of the two velocity vectors $\mathbf{v}^w$ and $\mathbf{v}^s$) which must be accounted for in trying to solve the equations. Darcy's experiments and the definition of hydraulic head provide us with some insights that will be used to eliminate some of the variables. The objective here is to make use of the information developed in the preceding sections to close the mass conservation equations, i.e., to provide enough supplementary information to make the equations solvable. The information we will develop for single phase flow and the approach to obtaining that information will be useful later in the discussion of multiphase flow.

3.6.1 Darcy's Law

From Darcy's experiments, we have the observation that flow occurs when the head in a system is not constant. In fact, equation (3.2), a correlation obtained from Darcy's experiments, suggests that the flow is proportional to a difference in head divided by the distance between the locations where the head is measured. Note, however, that the movement of fluid we are discussing is movement relative to the solid. For example, one could have water within a rock core with the head being constant. Picking up this core and carrying it across a laboratory causes the fluid to have a velocity relative to the laboratory floor, but it has no velocity relative to the rock. Thus if we are going to be precise we should say that fluid velocity relative to the solid phase in a porous medium is related to head differences within the porous medium. In his experiments, Darcy did not have to consider the solid motion because the experimental column was small. However, in large subsurface systems, or in the study of soil deformation such as encountered in soil mechanics, the movement of the solid can be important to the description of the system.

The next order of business is to extend the results of Darcy by making some plausible mathematical hypotheses. In Darcy's experiments, Q is the volumetric flow rate of the fluid through the column. This flow occurs only in the pore space. Thus the effective area of flow is not the entire column cross section, A, but is this area multiplied by the porosity, εA. Note that although the porosity ε is a volume fraction, it is also useful in determining an average effective area of flow. Thus the average speed of the macroscopically one-dimensional flow in a cross section of Darcy's column relative to the solid grains is:

$$|v^w - v^s| = \frac{Q}{\varepsilon A} \tag{3.39}$$

We are now going to extend the algebraic equation proposed by Darcy to a differential form. We consider that as the column length is reduced, the results of the experiment are unchanged. We therefore propose that:

$$\lim_{L \to 0} \frac{h_2 - h_1}{L} = \frac{dh}{dL} \tag{3.40}$$

Substitution of equations (3.39) and (3.40) into equation (3.2) yields the form:

$$\varepsilon |v^w - v^s| = K^w \left| \frac{dh}{dL} \right| \tag{3.41}$$

where the hydraulic conductivity, K^w, has been adorned with a superscript w in comparison to its presence in equation (3.2) to emphasize that in a particular medium, its value will depend on the properties of the fluid phase. With respect to Figure 3.2, the differential element dL may be replaced by a differential element denoted dz in the z direction. Since flow occurs from the location of higher head to lower head, we see that when dh/dz is positive, the flow will be downward (negative velocity) while when dh/dz is negative, the flow will be upward (positive velocity). Therefore, the absolute value signs in equation (3.41) can be removed if dL is replaced by dz and a negative sign is judiciously included such that:

$$\varepsilon(v^w - v^s) = -K^w \frac{dh}{dz} \tag{3.42}$$

Now as final steps, we assume that the head at a location, h, may be replaced by the averaged head for the location, h^w, as given by equation (3.27). Further, we assume that the spatial gradients in head in a particular direction give rise to flow in that direction only. Thus equation (3.42) may be written in terms of a partial derivative in the z direction, in a system where head changes as a function of all three dimensions, and the velocity component in that direction:

$$\varepsilon(\mathbf{v}^w - \mathbf{v}^s) \cdot \mathbf{k} = -K^w \frac{\partial h^w}{\partial z} \tag{3.43}$$

Corresponding equations may also be written in the x and y directions, and the three components of the equation may be combined to yield:

$$\varepsilon(\mathbf{v}^w - \mathbf{v}^s) = -K^w \nabla h^w \tag{3.44}$$

Implicit in this form of the equation is the assumption that the medium is *isotropic* in the sense that the hydraulic conductivity, K^w, is the same in all directions. In other words, the response to the fluid in light of a head gradient in the porous medium is independent of the direction of the gradient. Furthermore, flow occurs in the direction of the gradient only, such that the flow velocity vector and head gradient vector are *collinear*. A more extensive discussion of the hydraulic conductivity is in the next subsection.

The quantity on the left side of equation (3.44) is referred to as the *specific discharge* or *Darcy velocity* and is designated as $\mathbf{q}^w$, where:

$$\mathbf{q}^w = \varepsilon(\mathbf{v}^w - \mathbf{v}^s) \tag{3.45}$$

such that for the isotropic medium, the vector form of Darcy's law can be expressed:

$$\mathbf{q}^w = -K^w \nabla h^w \tag{3.46}$$

In cases when temperature or species gradients make definition of the hydraulic head ambiguous, we may use equation (3.36) to obtain Darcy's law explicitly in terms of the gradient of pressure and elevation heads as:

$$\mathbf{q}^w = -\frac{K^w}{\rho^w g}(\nabla p^w - \rho^w \mathbf{g}) \tag{3.47}$$

In the interest of full disclosure, we emphasize that the alternative forms of Darcy's "law" given as equations (3.46) and (3.47) are not actually laws. They are approximations based on observation of system behavior. We know that the equations are accurate when the head gradient is zero or when the pressure gradient is balanced by gravitational forces. For these equilibrium states, the Darcy velocity is zero. The equations propose that for a "small" imbalance in the gradients that drive the flow, the Darcy expressions reasonably describe the velocity. Confirming experimental evidence indicates that Darcy's correlation is a useful expression in describing flow, despite the fact that its ancestry is not nearly as distinguished as that of a full conservation equation.

Equation (3.45) makes it clear that the Darcy velocity is not a true velocity of the fluid but represents an effective flow rate through the porous medium. Chemical species carried with the flow move only in the pores and thus are transported relative to the solid at an average pore velocity $\mathbf{v}^w - \mathbf{v}^s$. The difference in an average pore between the Darcy velocity and what is called the *pore velocity* will be important if one is interested in transport of heat or a chemical species due to flow within a porous medium. Species transport is the subject of the next chapter. For the present, we turn to a discussion of the value of K^w in terms of the physical properties of the porous media system.

3.6.2 Hydraulic Conductivity and Permeability

The hydraulic conductivity, K^w, that appears in the various manifestations of Darcy's law in the last subsection is a measure of the ease with which a fluid will travel through a pore space. Inherent in the use of a scalar to represent the hydraulic conductivity is the assumption that at any point within the medium, the flow properties of the medium are the same in all directions (i.e., the medium is isotropic), although the properties may vary from point to point (i.e., the medium may be heterogeneous). The hydraulic conductivity is a function of the properties of the solid medium as well as of the flowing fluid. It would be convenient if the functional dependence of K^w on the fluid and the solid phases were known explicitly. In fact, a significant effort has gone into this problem. Experimental evidence has provided the insight that the hydraulic conductivity is proportional to the fluid weight per unit volume, $\rho^w g$, and inversely proportional to the dynamic viscosity of the fluid, μ^w. Thus we can define the *intrinsic permeability* (Table 3.1), denoted as k^s, as a property of the solid phase according to the relation:

$$K^w = \frac{k^s \rho^w g}{\mu^w} \qquad (3.48)$$

Note that K^w has dimensions of L/t, $\rho^w g$ has dimensions of $M/(L^2 t^2)$, and μ^w has dimensions $M/(tL)$. Therefore, for dimensional consistency, the intrinsic permeability has dimensions of L^2. The intrinsic permeability is only a function of the properties of the porous medium. The important macroscopic properties of the soil that control the value of k^s would seem to be both the porosity, ε, and some measure of pore structure. For example, if two soils have the same porosity, but one has many small pores while the other has a lesser number of large pores, it seems reasonable that the second soil would have a higher intrinsic permeability. In fact, the more contact a fluid has with the solid surface of the porous medium, the more resistance to flow it will encounter.

From this perspective, the question arises as to what characteristics of a solid matrix or soil impact the amount of surface area that a fluid will encounter as it

Table 3.1: Approximate range of values of intrinsic permeabilities for various materials. For unit conversion, 1 Darcy = 10^{-8} cm^2. To obtain hydraulic conductivity ($K^w = \rho g k^s / \mu^w$) of water at 60 °F (15.6 °C) use $\rho^w g / \mu^w$

Material	k^s cm^2
Well-sorted gravel	$10^{-3} - 10^{-4}$
Fractured rock	$10^{-3} - 10^{-6}$
Well-sorted sand	$10^{-5} - 10^{-7}$
Peat	$10^{-7} - 10^{-8}$
Fine sand, loam	$10^{-8} - 10^{-11}$
Layered clay	$10^{-9} - 10^{-11}$
Limestone	$10^{-12} - 10^{-13}$
Roof tile	$10^{-12} - 10^{-13}$
Granite	$10^{-14} - 10^{-15}$
Concrete	$10^{-14} - 10^{-16}$

flows in the porous medium. We note that we are searching for relations that may help describe parameters and are not bound by a stringent requirement of rigor. Thus, we envision the porous media flow as occurring in tubes of diameter d_p. As the flow takes place through the tubes it may wind through a complex path. To travel a macroscale distance L^p, it may actually travel a distance at the microscale of L_p. In colloquial terms, flow from point to point in a porous medium does not take place "as the crow flies" but follows a tortuous path dictated by the pore distribution and the ease with which the fluid can flow through the various pores when subjected to a head gradient. Note that the flow responds to the head gradient imposed only along the path it follows. Thus in traveling between two points, the head gradient along the direct path would be higher than that along the actual flow path.[1] We can define a dimensionless parameter, τ, as *the tortuosity*, which provides a measure of the directness of the path the fluid may take:

$$\tau = \frac{L_p}{L^p} \tag{3.49}$$

The tortuosity defined in this manner will always have a value greater than 1, and can be greater than 2 or even larger. Larger values of tortuosity would decrease the intrinsic permeability. We also know, even if just from sucking a drink through a straw, that the resistance to flow in a tube decreases as the diameter of the tube increases. Therefore, we can hypothesize that an expression for the intrinsic permeability that is consistent in having dimensions of L^2 would be proportional to the square of the pore diameter, D_p^2. Indeed, if our speculation is correct, a promising empirical relation would have the form:

$$k^s = C \frac{D_p^2}{\tau} \tag{3.50}$$

Although this relation provides insight into intrinsic permeability, it does not provide a relationship that can be employed directly without error. A problem with this relation lies in the fact that the flow paths are not tubes with a single diameter. In fact, the cross-sectional area of a flow path changes with position along the microscale flow, and the cross-sectional geometry is not circular. The actual flow path is very difficult to estimate and any tortuosity eventually employed would have to represent some average value. Nevertheless, the functional form provided gives some insight into the continuing search for reliable relations for intrinsic permeability.

A family of correlations for intrinsic permeability that seems to underlie current thinking is based on the work of Kozeny [29], who proposed in 1927 the form:

$$k^s = C \frac{\varepsilon^3}{a^{s2}} \tag{3.51}$$

---

[1] The concept of reducing the gradient by increasing the path length is used in switchbacks, where a winding road is used to permit vehicles to ascend a steep grade such as found in mountainous terrain.

where a^s is the solid area per volume and C is assigned the value 0.5, 0.562, or 0.597 depending on whether the pore shape is circular, square, or an equilateral triangle, respectively. Carman [9] improved on Kozeny's equation by eliminating the tunable parameter C to obtain:

$$k^s = \frac{D_m^2}{180} \frac{\varepsilon^3}{(1-\varepsilon)^2} \tag{3.52}$$

where D_m is some characteristic particle diameter, and thus still may require some fitting. In fact, the difficulty in identifying a single characteristic length—whether that be volume per area, some representative grain size, or some pore diameter—makes general use of the correlation for accurately determining the intrinsic permeability difficult. Information on grain size distributions for various soil types was presented in Section 1.3.

One approach to incorporating the dependence of intrinsic soil permeability on the full range of grain diameters is the formula of Fair and Hatch [18], which can be applied to a soil with its grain sizes separated using a sieve set:

$$k^s = \frac{1}{\beta}\left[\left(a\sum_{i=1}^{N_{\text{int}}} \frac{w_i}{D_i}\right)^2\right]^{-1} \frac{\varepsilon^3}{(1-\varepsilon)^2} \tag{3.53}$$

where β is a *packing factor* (shown by experiment to be about 5), a is a *sand shape factor* (varying from 6 for spherical grains to 7.7 for angular grains), N_{int} is the number of intervals between sieves (i.e., the number of sieves minus one) used to sort the soil sample, w_i is the fractional weight of sand in the interval between diameters of sieve i and sieve $i + 1$, and D_i is the geometric mean of the grain size of sieve i and sieve $i + 1$.

Besides the challenges of determining a value for the scalar intrinsic permeability, the problem is further complicated when a medium has a preferred direction of flow. Pore structure can impart directional dependence on permeability. For example, sediments deposited in a fluvial (or river) environment are deposited in horizontal layers with the soil grains oriented with the longest axes in the horizontal plane. This implies a different effective pore structure in the horizontal and vertical directions. One can also imagine that if holes are drilled in one direction into a rock sample, the permeability of the sample in that direction will be increased. The directional-dependence is called *anisotropy*.

To account for anisotropy, the intrinsic permeability scalar is generalized to a second rank tensor, a 3×3 matrix, denoted by $\mathbf{k}^s$, the components of which are dependent on the physical system as well as the orientation of the pore structure relative to the coordinate system. For the anisotropic case, the Darcy equations in terms of hydraulic head and in terms of pressure and elevation may be written respectively as:

$$\mathbf{q}^w = -\frac{\rho^w g \mathbf{k}^s}{\mu^w} \cdot \nabla h^w \tag{3.54}$$

and:

$$\mathbf{q}^w = -\frac{\mathbf{k}^s}{\mu^w} \cdot (\nabla p^w - \rho^w \mathbf{g}) \tag{3.55}$$

where the intrinsic permeability is:

$$\mathbf{k}^s = \begin{bmatrix} k_{xx}^s & k_{xy}^s & k_{xz}^s \\ k_{yx}^s & k_{yy}^s & k_{yz}^s \\ k_{zx}^s & k_{zy}^s & k_{zz}^s \end{bmatrix} \tag{3.56}$$

Without going into the mathematical intricacies of how this tensor was developed, we can make the following statements:

1. The matrix entry $k_{\xi\zeta}^s$ is the intrinsic permeability that allows for flow to occur in coordinate direction ξ as a result of a gradient in head in coordinate direction ζ.
2. The matrix $\mathbf{k}^s$ is symmetric such that $k_{\xi\zeta}^s = k_{\zeta\xi}^s$ for all coordinate subscript pairs $\xi\zeta$.
3. A coordinate system used for a study of a porous medium may be rotated such that all the off-diagonal elements of the matrix $\mathbf{k}^s$ are zero. For this case, the coordinate axes are said to be aligned with the principal directions of flow.
4. When the off-diagonal elements of $\mathbf{k}^s$ are zero and $k_{xx}^s = k_{yy}^s = k_{zz}^s$, the system is isotropic and a scalar intrinsic permeability may be used.

Some of the implications of these statements, particularly as they affect equation (3.54), will be explored further in the exercises at the end of the chapter. Here we note that if the unit vectors in Cartesian coordinate directions x, y, and z are denoted respectively as $\mathbf{i}$, $\mathbf{j}$, and $\mathbf{k}$, expansion of equation (3.54) into its component parts yields:

$$q_x^w = -\frac{\rho^w g k_{xx}^s}{\mu^w}\left(\frac{\partial h^w}{\partial x}\right) - \frac{\rho^w g k_{xy}^s}{\mu^w}\left(\frac{\partial h^w}{\partial y}\right) - \frac{\rho^w g k_{xz}^s}{\mu^w}\left(\frac{\partial h^w}{\partial z}\right) \tag{3.57a}$$

$$q_y^w = -\frac{\rho^w g k_{yx}^s}{\mu^w}\left(\frac{\partial h^w}{\partial x}\right) - \frac{\rho^w g k_{yy}^s}{\mu^w}\left(\frac{\partial h^w}{\partial y}\right) - \frac{\rho^w g k_{yz}^s}{\mu^w}\left(\frac{\partial h^w}{\partial z}\right) \tag{3.57b}$$

$$q_z^w = -\frac{\rho^w g k_{zx}^s}{\mu^w}\left(\frac{\partial h^w}{\partial x}\right) - \frac{\rho^w g k_{zy}^s}{\mu^w}\left(\frac{\partial h^w}{\partial y}\right) - \frac{\rho^w g k_{zz}^s}{\mu^w}\left(\frac{\partial h^w}{\partial z}\right) \tag{3.57c}$$

When the coordinate directions are aligned with the principal directions of flow, the off-diagonal elements of the intrinsic permeability tensor are zero and the component equations simplify to:

$$q_x^w = -\frac{\rho^w g k_{xx}^s}{\mu^w}\left(\frac{\partial h^w}{\partial x}\right) \tag{3.58a}$$

$$q_y^w = -\frac{\rho^w g k_{yy}^s}{\mu^w}\left(\frac{\partial h^w}{\partial y}\right) \tag{3.58b}$$

$$q_z^w = -\frac{\rho^w g k_{zz}^s}{\mu^w}\left(\frac{\partial h^w}{\partial z}\right) \tag{3.58c}$$

Furthermore, as noted above, if $k_{xx}^s = k_{yy}^s = k_{zz}^s = k^s$, the porous medium is isotropic and these last three equations may be combined into vector form equivalent to equation (3.46)

$$\mathbf{q}^w = -\frac{\rho^w g k^s}{\mu^w}\nabla h^w \tag{3.59}$$

This isotropic expression is identical to equation (3.54) for the special case where $\mathbf{k}^s = k^s\mathbf{I}$ and $\mathbf{I}$ is the unit tensor (matrix) with 1's on the diagonal and 0's in all the off-diagonal locations.

At this point in our excursion toward closed single phase flow equations, we have the mass conservation expressions for the fluid phase, equation (3.37), and for the solid phase, equation (3.38); we have the equations of state for phase w presented in Section 3.4; and we have the Darcy correlations that relate fluid velocity to gradients in head. To complete the derivation of the flow equations, we need to combine these disparate elements and also introduce some state equations and approximations relating to the behavior of the solid phase. This will be accomplished in the next subsection.

3.6.3 Derivation of Groundwater Flow Equation

In this subsection, we will combine the mass conservation equations and Darcy's law to obtain the expression that describes subsurface flow in a slightly deforming medium. To accomplish this, we need to rearrange the mass conservation equation for the fluid phase so that it is expressed in terms of the Darcy velocity, $\mathbf{q}^w$ rather than the pore velocity, $\mathbf{v}^w$. Rearrangement of equation (3.45) yields:

$$\varepsilon\mathbf{v}^w = \mathbf{q}^w + \varepsilon\mathbf{v}^s \tag{3.60}$$

This may be inserted into equation (3.37) to eliminate $\varepsilon\mathbf{v}^w$:

$$\frac{\partial(\varepsilon\rho^w)}{\partial t} + \nabla\cdot(\rho^w\mathbf{q}^w) + \nabla\cdot(\varepsilon\rho^w\mathbf{v}^s) = e_{ws}^w + \sum_{W=1}^{N_W}\rho_W^w Q_W^w\delta(\mathbf{x} - \mathbf{x}^W) \tag{3.61}$$

Now apply the product rule to the first and third terms to expand this equation to the form:

$$\rho^w\frac{\partial\varepsilon}{\partial t} + \varepsilon\frac{\partial\rho^w}{\partial t} + \rho^w\mathbf{v}^s\cdot\nabla\varepsilon + \varepsilon\mathbf{v}^s\cdot\nabla\rho^w + \varepsilon\rho^w\nabla\cdot\mathbf{v}^s + \nabla\cdot(\rho^w\mathbf{q}^w)$$
$$= e_{ws}^w + \sum_{W=1}^{N_W}\rho_W^w Q_W^w\delta(\mathbf{x} - \mathbf{x}^W) \tag{3.62}$$

Define the macroscale *material derivative* of some property f moving with the solid phase as:

$$\frac{D^s f}{Dt} = \frac{\partial f}{\partial t} + \mathbf{v}^s \cdot \nabla f \tag{3.63}$$

This material derivative provides the time rate of change of a property f measured while moving at velocity $\mathbf{v}^s$. The property of interest may be any property of the system, fluid or solid; the definition simply provides an expression indicating the motion experienced as the time rate of change is evaluated. We can use this definition and collect terms in equation (3.62) to obtain:

$$\rho^w \frac{D^s \varepsilon}{Dt} + \varepsilon \frac{D^s \rho^w}{Dt} + \varepsilon \rho^w \nabla \cdot \mathbf{v}^s + \nabla \cdot (\rho^w \mathbf{q}^w) = e_{ws}^w + \sum_{W=1}^{N_W} \rho_W^w Q_W^w \delta(\mathbf{x} - \mathbf{x}^W) \tag{3.64}$$

This equation for the fluid phase requires information about the solid phase velocity, some of which can be obtained from the mass balance equation for the solid, equation (3.38).

We can apply the product rule to the divergence term in equation (3.38) to obtain:

$$\frac{\partial \left[(1-\varepsilon)\rho^s\right]}{\partial t} + \mathbf{v}^s \cdot \nabla\left[(1-\varepsilon)\rho^s\right] + (1-\varepsilon)\rho^s \nabla \cdot \mathbf{v}^s = e_{ws}^s \tag{3.65}$$

Terms in this expression may be collected by making use of the definition of the material derivative given in equation (3.63):

$$\frac{D^s \left[(1-\varepsilon)\rho^s\right]}{Dt} + (1-\varepsilon)\rho^s \nabla \cdot \mathbf{v}^s = e_{ws}^s \tag{3.66}$$

Common practice in single-phase subsurface flow studies is to model the movement of the solid phase only implicitly. That is, rather than trying to obtain detailed information about movement of the solid phase, only its compression is considered. This approach is consistent with conditions for many groundwater flow problems. Of course, if one is concerned primarily with geotechnical questions such as slope stability or subsidence, the solid phase movement may become the predominant phenomenon of interest. However, in the study of subsurface flow and transport, equations (3.64) and (3.66) may be combined such that $\nabla \cdot \mathbf{v}^s$ is eliminated between them. To achieve this goal we rearrage equation (3.66) to yield:

$$\nabla \cdot \mathbf{v}^s = \frac{1}{(1-\varepsilon)\rho^s} e_{ws}^s - \frac{1}{(1-\varepsilon)\rho^s} \frac{D^s \left[(1-\varepsilon)\rho^s\right]}{Dt} \tag{3.67}$$

which we now introduce into equation (3.64) to eliminate $\nabla \cdot \mathbf{v}^s$. The result is:

$$\rho^w \frac{D^s \varepsilon}{Dt} + \varepsilon \frac{D^s \rho^w}{Dt} - \frac{\varepsilon \rho^w}{(1-\varepsilon)\rho^s} \frac{D^s \left[(1-\varepsilon)\rho^s\right]}{Dt} + \nabla \cdot (\rho^w \mathbf{q}^w)$$
$$= e_{ws}^w - \frac{\varepsilon \rho^w}{(1-\varepsilon)\rho^s} e_{ws}^s + \sum_{W=1}^{N_W} \rho_W^w Q_W^w \delta(\mathbf{x} - \mathbf{x}^W). \tag{3.68}$$

Consider now the first term in equation (3.68) which can be rewritten as:

$$\rho^w \frac{D^s \varepsilon}{Dt} = -\rho^w \frac{D^s(1-\varepsilon)}{Dt} = -\frac{\rho^w}{\rho^s}\left[\rho^s \frac{D^s(1-\varepsilon)}{Dt}\right] \qquad (3.69)$$

Now use the product rule to expand the last term as:

$$\rho^s \frac{D^s(1-\varepsilon)}{Dt} = \frac{D^s(1-\varepsilon)\rho^s}{Dt} - (1-\varepsilon)\frac{D^s \rho^s}{Dt} \qquad (3.70)$$

and introduce equation (3.70) into equation (3.69) to obtain:

$$\rho^w \frac{D^s \varepsilon}{Dt} = -\frac{\rho^w}{\rho^s}\left[\frac{D^s[(1-\varepsilon)\rho^s]}{Dt}\right] + \frac{(1-\varepsilon)\rho^w}{\rho^s}\frac{D^s \rho^s}{Dt} \qquad (3.71)$$

Next combine equations (3.71) and (3.68) to yield:

$$-\left[\frac{\varepsilon\rho^w}{(1-\varepsilon)\rho^s} + \frac{\rho^w}{\rho^s}\right]\frac{D^s[(1-\varepsilon)\rho^s]}{Dt} + \varepsilon\frac{D^s \rho^w}{Dt} + \frac{(1-\varepsilon)\rho^w}{\rho^s}\frac{D^s \rho^s}{Dt} + \nabla\cdot(\rho^w \mathbf{q}^w)$$

$$= e_{ws}^w - \frac{\varepsilon\rho^w}{(1-\varepsilon)\rho^s}e_{ws}^s + \sum_{W=1}^{N_W}\rho_W^w Q_W^w \delta(\mathbf{x}-\mathbf{x}^W) \qquad (3.72)$$

Rearranging the first term in equation (3.72) one obtains:

$$-\frac{\rho^w}{(1-\varepsilon)\rho^s}\frac{D^s[(1-\varepsilon)\rho^s]}{Dt} + \varepsilon\frac{D^s \rho^w}{Dt} + \frac{(1-\varepsilon)\rho^w}{\rho^s}\frac{D^s \rho^s}{Dt} + \nabla\cdot(\rho^w \mathbf{q}^w)$$

$$= e_{ws}^w - \frac{\varepsilon\rho^w}{(1-\varepsilon)\rho^s}e_{ws}^s + \sum_{W=1}^{N_W}\rho_W^w Q_W^w \delta(\mathbf{x}-\mathbf{x}^W) \qquad (3.73)$$

Substitution of the definition of the fraction of total mass at a macroscale point that is s phase, that is:

$$x_{\text{mass}}^s = \frac{(1-\varepsilon)\rho^s}{\varepsilon\rho^w + (1-\varepsilon)\rho^s} \qquad (3.74)$$

into equation (3.73) one obtains the following extremely important general formula that may subsequently be simplified for special cases:

$$-\frac{\rho^w}{(1-\varepsilon)\rho^s}\frac{D^s[(1-\varepsilon)\rho^s]}{Dt} + \varepsilon\frac{D^s \rho^w}{Dt} + \frac{(1-\varepsilon)\rho^w}{\rho^s}\frac{D^s \rho^s}{Dt} + \nabla\cdot(\rho^w \mathbf{q}^w)$$

$$= \frac{e_{ws}^w}{x_{\text{mass}}^s} + \sum_{W=1}^{N_W}\rho_W^w Q_W^w \delta(\mathbf{x}-\mathbf{x}^W) \qquad (3.75)$$

Next we will make use of state equations to convert the material derivatives of $(1-\varepsilon)\rho^s$, ρ^w, and ρ^s into material derivatives of pressure. For this derivation, we will consider the case where effects of temperature and chemical composition on the density are negligible. Therefore, the mass density is a function only of pressure with $\rho^\alpha = \rho^\alpha(p^\alpha)$. For the fluid phase, we make use of equation (3.7) with superscript α replaced by w to obtain:

$$\varepsilon\frac{\mathrm{D}^s\rho^w}{\mathrm{D}t} = \varepsilon\frac{\partial\rho^w}{\partial p^w}\frac{\mathrm{D}^s p^w}{\mathrm{D}t} = \varepsilon\rho^w\beta^w\frac{\mathrm{D}^s p^w}{\mathrm{D}t} \qquad (3.76)$$

where β^w is the *fluid compressibility*. Consider the solid to respond elastically and isotropically to the pressure, and note that the pressure exerted on the solid material is the pore pressure, the pressure of fluid w in the pore space within the medium. Thus, the state equation for the solid is $\rho^s = \rho^s(p^w)$ and we will use the functional form similar to equation (3.7):

$$\beta^s = \frac{1}{\rho^s}\frac{\partial\rho^s}{\partial p^w} \qquad (3.77)$$

which may be integrated to:

$$\ln\left(\frac{\rho^s}{\rho_0^s}\right) = \beta^s(p^w - p_0^w) \qquad (3.78)$$

when the compressibility of the solid phase, β^s, may be considered constant over the range of pressures considered and the subscript "0" denotes a reference state. The material derivative of ρ^s in equation (3.75) may therefore be expanded using the chain rule:

$$\frac{(1-\varepsilon)\rho^w}{\rho^s}\frac{\mathrm{D}^s\rho^s}{\mathrm{D}t} = \frac{(1-\varepsilon)\rho^w}{\rho^s}\frac{\partial\rho^s}{\partial p^w}\frac{\mathrm{D}^s p^w}{\mathrm{D}t} = (1-\varepsilon)\rho^w\beta^s\frac{\mathrm{D}^s p^w}{\mathrm{D}t} \qquad (3.79)$$

Additionally, we need an equation of state that relates the change in volume of the bulk solid material (i.e., the volume occupied by the porous medium) when the matrix is subjected to a stress. This stress will change the mass of solid per volume of the medium, the quantity $(1 - \varepsilon)\rho^s$. If we consider the stress exerted on the bulk solid to be the pressure of phase w, that the matrix is elastic, and also note that the fractional volume occupied by the solid will decrease as the pressure of the fluid increases, we can define the *bulk compressibility*, α^b, as:

$$\alpha^b = -\frac{1}{(1-\varepsilon)\rho^s}\frac{\partial[(1-\varepsilon)\rho^s]}{\partial p^w} \qquad (3.80)$$

If α^b is approximately constant over the range of pressures examined, this expression integrates to:

$$\ln\left[\frac{(1-\varepsilon)\rho^s}{(1-\varepsilon_0)\rho_0^s}\right] = -\alpha^b(p^w - p_0^w) \qquad (3.81)$$

where the subscript 0 refers to some reference situation. With condition (3.80), application of the chain rule to the material derivative of the solid mass per volume of porous medium in equation (3.75) provides:

$$\frac{\rho^w}{(1-\varepsilon)\rho^s} \frac{D^s[(1-\varepsilon)\rho^s]}{Dt} = \frac{\rho^w}{(1-\varepsilon)\rho^s} \frac{\partial[(1-\varepsilon)\rho^s]}{\partial p^w} \frac{Dp^w}{Dt} = -\rho^w \alpha^b \frac{Dp^w}{Dt} \qquad (3.82)$$

The transformation of the material derivatives in equation (3.75) to material derivatives of pressure is completed by substitution of the equations (3.76), (3.79), and (3.82) and collection of terms:

$$\rho^w[\alpha^b + \varepsilon\beta^w + (1-\varepsilon)\beta^s]\frac{D^s p^w}{Dt} + \nabla\cdot(\rho^w \mathbf{q}^w) = \frac{e_{ws}^w}{x_{mass}^s} + \sum_{W=1}^{N_W} \rho_W^w Q_W^w \delta(\mathbf{x} - \mathbf{x}^W) \qquad (3.83)$$

The group of terms multiplying the time derivative account for changes in the amount of water that an aquifer stores due to compressibility effects. We can define the *specific storage*, S_s, with dimensions of $1/L$ according to:

$$S_s = \rho^w g\left[\alpha^b + \varepsilon\beta^w + (1-\varepsilon)\beta^s\right] \qquad (3.84)$$

The specific storage is conceptually the volume of water released from or taken into a unit volume at a location in the porous medium due to expansion or contraction of water, the matrix, and of the solid when the pressure head $p^w/(\rho^w g)$ changes by a unit amount. For sand, α^b ranges from 10^{-7} to $5 \times 10^{-8}\,\mathrm{m^2/N}$. This is 2 to 3 orders of magnitude larger than the compressibility of water, and more than 3 orders of magnitude larger than the solid compressibility. Thus, the greatest contribution to storage is reorganization of the matrix structure. Because it is relatively small, the solid phase compressibility is typically neglected in determining the value of the specific storage.

We use the definition of specific storage in equation (3.83) and also eliminate the Darcy velocity in favor of the pressure by making use of equation (3.55):

$$\frac{S_s}{g}\frac{D^s p^w}{Dt} - \nabla\cdot\left[\rho^w \frac{\mathbf{k}^s}{\mu^w}\cdot(\nabla p^w - \rho^w \mathbf{g})\right] = \frac{e_{ws}^w}{x_{mass}^s} + \sum_{W=1}^{N_W} \rho_W^w Q_W^w \delta(\mathbf{x} - \mathbf{x}^W) \qquad (3.85)$$

where the anisotropic form of the intrinsic permeability is retained for generality. Two terms in this equation are further simplified by making use of the fact that some of their elements are small. First, the solid velocity is taken to be small so that the material derivative of pressure may be approximated using:

$$\frac{D^s p^w}{Dt} \approx \frac{\partial p^w}{\partial t} \qquad (3.86)$$

Also, the density gradient is small enough that the second term in equation (3.85) may be approximated as:

$$\nabla \cdot \left[\rho^w \frac{\mathbf{k}^s}{\mu^w} \cdot (\nabla p^w - \rho^w \mathbf{g}) \right] \approx \rho^w \nabla \cdot \left[\frac{\mathbf{k}^s}{\mu^w} \cdot (\nabla p^w - \rho^w \mathbf{g}) \right] \qquad (3.87)$$

Application of these approximations and division by ρ^w finally yields the groundwater flow equation in terms of pressure:

$$\frac{S_s}{\rho^w g} \frac{\partial p^w}{\partial t} - \nabla \cdot \left[\frac{\mathbf{k}^s}{\mu^w} \cdot (\nabla p^w - \rho^w \mathbf{g}) \right] = \frac{e_{ws}^w}{\rho^w x_{mass}^s} + \sum_{W=1}^{N_W} \frac{\rho_W^w}{\rho^w} Q_W^w \delta(\mathbf{x} - \mathbf{x}^W) \qquad (3.88)$$

When the pumping is out of a well, such that Q_W^w is negative, the mass density of the pumped water will be the same as the mass density of the water in the aquifer so that $\rho_W^w / \rho^w = 1$. However, if water is being injected into the aquifer, its density may be different from that in the aquifer so that the density ratio is different from 1. In most cases, the total amount of mass being transferred between phases is small enough that e_{ws}^w is negligible in the groundwater flow equation. However, in the next chapter when we consider transport of the chemical species that comprise a phase, the exchange of mass between phases is important in the equations that describe the distribution of the various chemical constituents within each phase.

The groundwater flow equation may alternatively be expressed in terms of hydraulic head rather than pressure by employing equations (3.35) and (3.36) to eliminate the derivatives of pressure from equation (3.88):

$$S_s \frac{\partial h^w}{\partial t} - \nabla \cdot (\mathbf{K}^w \cdot \nabla h^w) = \frac{e_{ws}^w}{\rho^w x_{mass}^s} + \sum_{W=1}^{N_W} \frac{\rho_W^w}{\rho^w} Q_W^w \delta(\mathbf{x} - \mathbf{x}^W) \qquad (3.89)$$

where the hydraulic conductivity tensor $\mathbf{K}^w = \rho^w g \mathbf{k}^s / \mu^w$ has been introduced for convenience. This equation may be solved for the hydraulic head after appropriate specification of boundary and initial conditions.

The goal of this derivation has been to obtain an equation that can be solved for the pressure or head distribution. The flow velocity can then be calculated from a Darcy equation. The goal has been reached with the alternative equations (3.88) in terms of pressure or (3.89) in terms of the head. The derivation has been somewhat involved. Thus, rather than proceeding directly to a consideration of boundary and initial conditions, we will recapitulate the important elements of the derivation that have led to the governing equations. In the same way that "instant replay" from different camera angles is employed to enhance viewing of a sporting event, the following subsection will examine the derivation from a different perspective.

3.6.4 Recapitulation of the Derivation

The derivation of the single-phase flow equation began with the mass conservation equations for the fluid and the solid, equations (3.37) and (3.38), respectively. These equations describe the fundamental concept that mass can be neither created nor destroyed. The two equations are combined to form the single equation (3.75), a general conservation expression for subsurface flow written in terms of the Darcy velocity. This equation contains nine unknowns described by three scalars and two vectors (ε, ρ^w, ρ^s, $\mathbf{q}^w$, and $\mathbf{v}^s$) that must be accounted for in the derivation because a single equation can be solved for only a single unknown. The number of unknowns is reduced by a set of constitutive relations (Darcy's law) and equations of state that inter-relate the variables and their derivatives, and by recognizing that some of the terms in the equation are small and can be neglected.

The simplest assumption is that the elastic solid deforms very slowly relative to the fluid motion so that the three components of $\mathbf{v}^s$ can be set to zero in the material derivatives. This assumption reduces the number of unknowns in the single governing equation to six: ε, ρ^w, ρ^s, and $\mathbf{q}^w$. The "trick" that makes this situation manageable is to introduce the pressure, p^w, thus raising the number of unknowns to seven, but then finding expressions for each of the other six variables in terms of this single new variable. The actual expressions used are subject to experimental verification. We used some standard forms of these functional expressions. For $\varepsilon(p^w)$, the governing expression is equation (3.81), which indicates how the elastic matrix compresses. We used the equation of state for the fluid density as a function of pressure as given by equation (3.10) without the terms involving dependence on chemical constituents. The equation of state used for ρ^s is equation (3.78).[2] Finally, we relate the Darcy velocity $\mathbf{q}^w$ to the gradient in p^w through the Darcy equation (3.55). When we make use of all these relations in the flow equation, we arrive at equation (3.88).

At first glance, it may seem that our manipulations have been for nought. We may have eliminated some of the extra variables, but we now have specific storage, S_s, fluid viscosity, μ^w, and the intrinsic permeability, $\mathbf{k}^s$. If the system is isotropic, the intrinsic permeability involves a single parameter; however, if the medium is anisotropic, five parameters must be specified. Of course, the pumping rates and the density of any fluid being injected must be specified. Despite the new parameters, equation (3.88) is preferable to equation (3.75) because it can be solved if the parameters are specified and the parameters are characteristic of the system. They may vary with location in the system, primarily due to system structural heterogeneities. The specific storage is directly related to the equations of state for the densities and the solid matrix; the intrinsic permeability is a function of the geometry of the pore space through the solid phase. Thus equation (3.88) actually does represent a significant advance over its precursors. When the system is homogeneous, the parameters will be constants. Expressing the flow equation in terms of head, as in equation (3.89), is not simpler than the form in terms of pressure; however, the hydraulic head is intuitively attractive since it relates directly to the level of the water in a manometer inserted into the system.

Let us emphasize, yet again, that the mass conservation equations are fundamental principles while the additional relationships employed here are approximations subject to revision and improvement based on experimental observation or the need to adequately describe the study region. Although we have developed the single-phase flow equation, we also need to specify initial and boundary conditions if it is to be solved. This will be the topic of the next subsection.

---

[2] This is a somewhat simplified form as the solid density is more generally taken to be a function of the effective stress exerted on the particles which is only approximated as p^w. Nevertheless, for our purposes, the equation of state presented here is sufficient.

3.6.5 Initial and Boundary Conditions

Equations (3.88) and (3.89) are differential equations in pressure and head, respectively, that provide alternative descriptions of the physics that govern single-phase flow. Since they are written in terms of derivatives, they describe how the pressure or head changes from point to point in the system as a function of time or position. This is not enough information, however, to determine the value of pressure or head at a location. The equations indicate the changes, but they do not indicate from what values the changes are. For example, if we fill a jug of known volume by pouring water into it at a set rate, we will not know the time it will take for the jug to be filled unless we know how much water was in the jug when the filling process began. Similarly, knowing the rate of change of head in a porous medium, either in time or in space, does not mean we know the value of head at any location unless we have some information about a starting value. Furthermore, going back to the example of a jug, if there is a crack in the wall of the jug such that some water escapes at that boundary, we will need to take that into account in calculating the progress toward filling the jug. To solve a single-phase flow equation we need both a specification of the initial state of the system, for a time-dependent problem, and some information about what is happening at the boundary of the system. We will not delve into the theory of differential equations to determine the necessary information needed but will specify the conditions based on consideration of the physical system.

It should not come as a surprise that since the time and space rate of changes of head and pressure are related, respectively, by equations (3.35) and (3.36), the conditions needed to solve the flow equations in terms of head or pressure are also similar. Therefore, the following discussion will be applicable to both the pressure equation (3.88) and the head equation (3.89) although it will be framed in terms of pressure. The specific auxiliary conditions that are needed will be stated explicitly in terms of both head and pressure.

If the flow equation is to be solved, the coefficients and forcing functions appearing in the equation must be specified as functions of time and location. These coefficients include, depending on the form of the equation being considered, ρ^w, μ^w, S_s, $\mathbf{k}^s$, and $\mathbf{K}^w$. Some of these parameters, for example ρ^w, may be specified as functions of pressure so that the change in density as the solution to the flow equation evolves is accounted for. Additionally, it is necessary to specify the mass of fluid per time, $\rho_w^w Q_w^w$, being injected into or extracted from each well and the locations of the wells, $\mathbf{x}^W$.

Consider, next, the implications of describing a *transient* flow problem, one in which the pressure at a location in the system is a function of time. Solution of equation (3.88) for the pressure requires that an initial value of the pressure be specified at all locations in the domain. Then the differential equation describes changes from this initial distribution. Therefore, solution of equation (3.88) requires the initial condition for the pressure field $p^w(t, \mathbf{x})$:

$$p^w(t_0, \mathbf{x}) = P(\mathbf{x}) \tag{3.90}$$

where t_0 is the initial time and $P(\mathbf{x})$ is the specified pressure distribution at all points in the domain. For equation (3.89), the required initial condition is:

$$h^w(t_0, \mathbf{x}) = H(\mathbf{x}) \tag{3.91}$$

where $H(\mathbf{x})$ is the initial head distribution. However, recall the definition of hydraulic head given by equation (3.27) and the subsequent discussion. The head can only be determined unambiguously if ρ^w is either constant or a function only of pressure. If this stipulation is not satisfied, then $H(\mathbf{x})$ cannot be specified, and one is only able to solve the pressure form of the flow equation.[3] If the problem being solved is at *steady state* such that the pressure distribution in the system is independent of time, no initial condition is specified.

A boundary condition must be specified at each point on the spatial boundary of the system under consideration. We will designate the spatial coordinates of the boundary as $\mathbf{x}_b^w$. At the boundary several alternative boundary conditions may be specified, and different conditions may be specified on different parts of the boundary. Only one boundary condition is specified at a boundary location for the groundwater flow equation.

The simplest condition is specification of the dependent variable at a boundary location. This specified value may be a function of time. For the pressure equation, this condition is expressed:

$$p^w(t, \mathbf{x}_b^w) = P^b(t, \mathbf{x}_b^w) \tag{3.92}$$

where $P^b(t, \mathbf{x}_b^w)$ is some specified function of time and location on the boundary. Conditions of this form are known as *first type conditions* or *Dirichlet conditions*. For a steady state problem, either a Dirichlet condition or a third type condition (to be discused shortly) must be specified at at least one point on the boundary. In terms of head, the Dirichlet condition takes the form:

$$h^w(t, \mathbf{x}_b^w) = H^b(t, \mathbf{x}_b^w) \tag{3.93}$$

in cases where the head can be determined.

A different condition that may be specified is commonly referred to as a *flux condition*. With this condition, the fluid flux through the boundary is specified. This flux is the Darcy velocity, $\mathbf{q}^w$ in the direction normal to the boundary, $\mathbf{n}$, where $\mathbf{n}$ is a unit vector normal to the boundary and oriented pointing outward from the boundary. The *normal flux* is $\mathbf{n} \cdot \mathbf{q}^w$ so that this condition is:

$$(\mathbf{n} \cdot \mathbf{q}^w)\big|_{\mathbf{x}_b^w} = q^b(t, \mathbf{x}_b^w) \tag{3.94}$$

where $q^b(t, \mathbf{x}_b^w)$ is the specified flux at the location being considered. Since Darcy's law provides the expression for $\mathbf{q}^w$ in terms of the pressure, the condition is written in terms of the pressure as:

$$-\left[\mathbf{n} \cdot \frac{\mathbf{k}^s}{\mu^w} \cdot (\nabla p^w - \rho^w \mathbf{g}) \right]_{\mathbf{x}_b^w} = q^b(t, \mathbf{x}_b^w) \tag{3.95}$$

---

[3] When density dependence on other state variables, such as temperature or chemical species concentrations, is important, differential equations for these quantities must also be solved. Species equations are considered in the next chapter, but heat transfer is beyond the scope of this text.

When $q^b(t, \mathbf{x}_b^w) = 0$, the boundary is a *no flow boundary* because no fluid enters or leaves the system across the boundary. For the particular case where the medium is isotropic such that $\mathbf{k}^s = k^s\mathbf{I}$, this equation may be rearranged to:

$$-\left[\frac{k^s}{\mu^w}\mathbf{n}\cdot(\nabla p^w - \rho^w\mathbf{g})\right]\Bigg|_{\mathbf{x}_b^w} = q^b(t, \mathbf{x}_b^w) \tag{3.96}$$

In this instance, the boundary condition specifies the gradient of pressure in the direction normal to the boundary. The gradients in directions tangent to the boundary do not appear in the expression. This is called a *second type boundary condition* or *Neumann condition*. Written in terms of head, the conditions corresponding to equations (3.95) and (3.96) are, respectively:

$$-\left(\mathbf{n}\cdot\frac{\mathbf{k}^s}{\mu^w}\cdot\nabla h^w\right)\Bigg|_{\mathbf{x}_b^w} = q^b(t, \mathbf{x}_b^w) \tag{3.97}$$

and:

$$-\left(\frac{k^s}{\mu^w}\mathbf{n}\cdot\nabla h^w\right)\Bigg|_{\mathbf{x}_b^w} = q^b(t, \mathbf{x}_b^w) \tag{3.98}$$

Flux conditions may be specified at any point on the boundary $\mathbf{x}_b^w$, but for a steady state problem, the pressure (head) must appear in the boundary condition at at least one point on the boundary, for example in a Dirichlet boundary condition.

A boundary condition other than a Dirichlet condition that invokes the pressure at the boundary is employed where the flux at the boundary of the system is proportional to the difference between the pressure at the boundary and some other external pressure. For this case, the condition would take the form:

$$(\mathbf{n}\cdot\mathbf{q}^w)\big|_{\mathbf{x}_b^w} = \kappa^b\left[p^w(t, \mathbf{x}_b^w) - P_{ext}(t, \mathbf{x}_b^w)\right] \tag{3.99}$$

where κ^b is a measure of the permeability of the boundary to the flow. When $\kappa^b = 0$, the boundary is impermeable and the boundary condition is a Neumann condition for a no flow boundary. When κ^b is very large, this condition degenerates to a Dirichlet form with $p^w(t, \mathbf{x}_b^w) = P_{ext}(t, \mathbf{x}_b^w)$. For intermediate values of κ^b, however, this condition written in terms of pressure is:

$$-\left[\mathbf{n}\cdot\frac{\mathbf{k}^s}{\mu^w}\cdot(\nabla p^w - \rho^w\mathbf{g})\right]\Bigg|_{\mathbf{x}_b^w} = \kappa^b\left[p^w(t, \mathbf{x}_b^w) - P_{ext}(t, \mathbf{x}_b^w)\right] \tag{3.100}$$

When the medium is isotropic, this condition becomes:

$$-\left[\frac{k^s}{\mu^w}\mathbf{n}\cdot(\nabla p^w - \rho^w\mathbf{g})\right]\Bigg|_{\mathbf{x}_b^w} = \kappa^b\left[p^w(t, \mathbf{x}_b^w) - P_{ext}(t, \mathbf{x}_b^w)\right] \tag{3.101}$$

This is called a *third type boundary condition* or *Robin condition*. This sort of boundary condition is useful, in practice, for portions of the boundary where leakage between an external geologic formation and the system under study is important. The head difference determines the rate of exchange of fluid between the two formations. Written in terms of head, the leakage boundary conditions corresponding to equations (3.100) and (3.101) are, respectively:

$$-\left(\mathbf{n}\cdot\frac{\mathbf{k}^{s}}{\mu^{w}}\cdot\nabla h^{w}\right)\bigg|_{\mathbf{x}_{b}^{w}} = \kappa^{b}\left[h^{w}(t,\mathbf{x}_{b}^{w})-H_{\text{ext}}(t,\mathbf{x}_{b}^{w})\right] \tag{3.102}$$

and:

$$-\left(\frac{k^{s}}{\mu^{w}}\mathbf{n}\cdot\nabla h^{w}\right)\bigg|_{\mathbf{x}_{b}^{w}} = \kappa^{b}\left[h^{w}(t,\mathbf{x}_{b}^{w})-H_{\text{ext}}(t,\mathbf{x}_{b}^{w})\right] \tag{3.103}$$

where $H_{\text{ext}}(t,\mathbf{x}_{b}^{w})$ is a measure of the head external to the region of study at the boundary that influences the flow at time t at location $\mathbf{x}_{b}^{w}$.

Although the governing equation with the auxiliary conditions is sufficient to describe a single-phase flow problem, the equation cannot be solved analytically except for some special cases. Major factors complicating the solution process are heterogeneity of the hydraulic conductivity and the specific storage, modeling a region whose boundaries are not parallel with coordinate directions, the time dependence of the pumping at the wells, the mix of first, second, and third type boundary conditions, and the time and space dependence of the boundary conditions. Nevertheless, the ability to properly formulate the differential equation and auxiliary conditions that describe a problem of interest is essential. Computational methods can be employed to obtain approximate solutions. However, if the problem formulation does not describe the physics, neither an analytic nor a numerical solution can be used to provide anything more than an attractive looking set of graphics depicting the solution to the wrong problem.

3.6.6 Two-Dimensional Flow

One special case of the single fluid subsurface flow problem that is of practical importance is when the vertical flow within the study region contributes negligibly to the system dynamics. For this case, the groundwater flow equation may be integrated through the vertical to obtain a problem description that depends on time and the two lateral spatial coordinates. The derivation of the equation is presented here starting from equation (3.89) with equation (3.46) substituted in. Because it is typically negligible, the mass exchange term between the fluid and solid is neglected. Integration over the vertical is thus:

$$\int_{z_{B}}^{z_{T}} S_{s}\frac{\partial h^{w}}{\partial t}dz + \int_{z_{B}}^{z_{T}} \nabla\cdot\mathbf{q}^{w}dz = \int_{z_{B}}^{z_{T}}\sum_{W=1}^{N_{W}}\frac{\rho_{W}^{w}}{\rho^{w}}Q_{W}^{w}\delta(\mathbf{x}-\mathbf{x}^{W})dz \tag{3.104}$$

where z_{B} and z_{T} are the bottom and top of the flow region respectively.

The simplification to two dimensions is useful for analysis of essentially horizontal flow in an aquifer. This is often a useful approximation in light of the large areal extent of an aquifer in comparison to its vertical thickness. When the flow within the system is negligible in the vertical direction, h^w may be considered to be independent of the vertical coordinate, designated here as the z coordinate, but will vary with the horizontal coordinates, x and y. The assumption of essentially horizontal flow is known as the *Dupuit assumption*.

Although the flow is considered to be horizontal, the thickness of the aquifer may be a function of position. This fact is incorporated into the approximate two-dimensional equation by integrating the flow equation through the vertical rather than just eliminating derivatives with respect to z. Furthermore, the specific storage, S_s, hydraulic conductivity, $\mathbf{K}^w$, and the lateral Darcy velocity, $\mathbf{q}^w$, are all considered to be independent of vertical position to the degree that they can be well-represented by their vertical averages.

The theorems for vertical integration needed are [21]:

$$\int_{z_B}^{z_T} \frac{\partial h^w}{\partial t}\, dz = \frac{\partial}{\partial t} \int_{z_B}^{z_T} h^w \, dz - \left(\frac{\mathbf{n}\cdot\mathbf{w}h^w}{\mathbf{n}\cdot\mathbf{k}}\right)\bigg|_{z_T} + \left(\frac{\mathbf{n}\cdot\mathbf{w}h^w}{\mathbf{n}\cdot\mathbf{k}}\right)\bigg|_{z_B} \tag{3.105}$$

and:

$$\int_{z_B}^{z_T} \nabla\cdot\mathbf{q}^w\, dz = \nabla'\cdot\int_{z_B}^{z_T} \mathbf{q}^{w\prime}\, dz + \left(\frac{\mathbf{n}\cdot\mathbf{q}^w}{\mathbf{n}\cdot\mathbf{k}}\right)\bigg|_{z_T} - \left(\frac{\mathbf{n}\cdot\mathbf{q}^w}{\mathbf{n}\cdot\mathbf{k}}\right)\bigg|_{z_B} \tag{3.106}$$

where $b = z_T - z_B$; the superscript $'$ indicates that the operator or vector is two-dimensional in the lateral directions; $\mathbf{k}$ is the unit vector in the z direction, positive upward; $\mathbf{w}$ is the velocity of the upper or lower boundary of the flow region, as indicated; and $\mathbf{n}$ is the outward unit normal vector.

We can simplify this equation by noting that h^w and $\mathbf{q}^{w\prime}$ are approximately constant with changes in z and thus can be removed from the integrals on the right sides. Also, when the slope of the top (bottom) of the flow region is small, $\mathbf{n}\cdot\mathbf{k} \approx 1$ ($\mathbf{n}\cdot\mathbf{k} \approx -1$). Thus the theorems used simplify to:

$$\int_{z_B}^{z_T} \frac{\partial h^w}{\partial t}\, dz = \frac{\partial(bh^w)}{\partial t} - h^w\left[(\mathbf{n}\cdot\mathbf{w})\big|_{z_T} + (\mathbf{n}\cdot\mathbf{w})\big|_{z_B}\right] \tag{3.107}$$

and:

$$\int_{z_B}^{z_T} \nabla\cdot\mathbf{q}^w\, dz = \nabla'\cdot\left(b\mathbf{q}^{w\prime}\right) + (\mathbf{n}\cdot\mathbf{q}^w)\big|_{z_T} + (\mathbf{n}\cdot\mathbf{q}^w)\big|_{z_B} \tag{3.108}$$

The last term in equation (3.107) accounts for the change in the thickness of the region being studied and involves the dot products of the outward unit normal vectors with the velocity of the top and the bottom surfaces of the region. Therefore, we can make the approximation:

$$(\mathbf{n}\cdot\mathbf{w})\big|_{z_T} + (\mathbf{n}\cdot\mathbf{w})\big|_{z_B} \approx \frac{\partial b}{\partial t} \tag{3.109}$$

so that equation (3.107) reduces to:

$$\int\limits_{z_B}^{z_T} \frac{\partial h^w}{\partial t} \, dz = b \frac{\partial h^w}{\partial t} \tag{3.110}$$

The last two terms in equation (3.108) account for flow at the top and bottom of the study region. These terms will be discussed further after the final equation is obtained.

Application of the equations (3.108) and (3.110) to the flow equation (3.104) when S_s can be considered independent of z yields:

$$S_s b \frac{\partial h^w}{\partial t} + \nabla' \cdot \left(b \mathbf{q}^{w'} \right) + \left(\mathbf{n} \cdot \mathbf{q}^w \right)\big|_{z_T} + \left(\mathbf{n} \cdot \mathbf{q}^w \right)\big|_{z_B} = \int\limits_{z_B}^{z_T} \sum_{W=1}^{N_W} \frac{\rho_W^w}{\rho^w} Q_W^w \delta(\mathbf{x} - \mathbf{x}^W) \, dz \tag{3.111}$$

The horizontal Darcy velocity maybe eliminated by using Darcy's equation in the form $\mathbf{q}^{w'} = -\mathbf{K}^{w''} \cdot \nabla' h^w$ which recognizes that the head is independent of z and the lateral head gradients are also independent of z. Equation (3.111) may then be written:

$$S \frac{\partial h^w}{\partial t} - \nabla' \cdot \left(\mathbf{T}^{w''} \cdot \nabla' h^w \right) + \left(\mathbf{n} \cdot \mathbf{q}^w \right)\big|_{z_T} + \left(\mathbf{n} \cdot \mathbf{q}^w \right)\big|_{z_B} = \sum_{W=1}^{N_W} Q_W^{w'} \tag{3.112}$$

where $S = S_s b$ is the *storativity* or *storage coefficient*, $\mathbf{T}^{w''} = b\mathbf{K}^{w''}$ is the two-dimensional *transmissivity*:

$$\mathbf{T}^{w''} = \begin{bmatrix} bK_{xx}^w & bK_{xy}^w \\ bK_{yx}^w & bK_{yy}^w \end{bmatrix} = \begin{bmatrix} T_{xx}^w & T_{xy}^w \\ T_{yx}^w & T_{yy}^w \end{bmatrix} \tag{3.113}$$

and $Q_W^{w'}$ is the pumping volume per unit time per cross-sectional area at a location:

$$Q_W^{w'} = \int\limits_{z_B}^{z_T} \frac{\rho_W^w}{\rho^w} Q_W^w \delta(\mathbf{x} - \mathbf{x}^W) \, dz \tag{3.114}$$

The expansion of $\nabla' \cdot \left(\mathbf{T}^{w''} \cdot \nabla' h^w \right)$ in both Cartesian and cylindrical coordinates is given in Table 3.2. The cylindrical coordinate form is convenient when a pumping well is located at the origin.

The last term on the left side of equation (3.112) accounts for the exchange of fluid at the bottom surface of the aquifer and is positive when fluid leaves the aquifer but negative when fluid is added from the underlying formation. When the region being analyzed is a *confined aquifer* and its hydraulic head is above the base of the overlying formation, $(\mathbf{n} \cdot \mathbf{q}^w)\big|_{z_T}$ is positive when leakage occurs from the study region into the overlying formation and is negative when leak age is from the overlying formation into the study aquifer.

For an unconfined system, the upper surface location is equal to the head. This surface is called a *water table* and is also the location where the pressure is atmospheric. For an unconfined system, $(\mathbf{n} \cdot \mathbf{q}^w)\big|_{z_T}$ corresponds to the flow at the upper surface. Whether this flow is addition or subtraction of fluid to or from the aquifer

Table 3.2: Expansions of the vector quantity $\nabla' \cdot \left(\mathbf{T}^{w\prime\prime} \cdot \nabla' h^w \right)$ **in both Cartesian (x and y are areal coordinates) and cylindrical coordinates (r is the radial coordinate and θ is the angular coordinate)**

Expansions of $\nabla' \cdot \left(\mathbf{T}^{w\prime\prime} \cdot \nabla' h^w \right)$

Cartesian Coordinates

Anisotropic form	$\dfrac{\partial}{\partial x}\left(T_{xx}^{w}\dfrac{\partial h^w}{\partial x} + T_{xy}^{w}\dfrac{\partial h^w}{\partial y} \right) + \dfrac{\partial}{\partial y}\left(T_{yx}^{w}\dfrac{\partial h^w}{\partial x} + T_{yy}^{w}\dfrac{\partial h^w}{\partial y} \right)$
Constant $\mathbf{T}^{w\prime\prime}$	$T_{xx}^{w}\dfrac{\partial^2 h^w}{\partial x^2} + (T_{xy}^{w} + T_{yx}^{w})\dfrac{\partial^2 h^w}{\partial x \partial y} + T_{yy}^{w}\dfrac{\partial^2 h^w}{\partial y^2}$
Isotropic, $\mathbf{T}^{w\prime\prime} = T^{w}\mathbf{I}''$	$\dfrac{\partial}{\partial x}\left(T^{w}\dfrac{\partial h^w}{\partial x} \right) + \dfrac{\partial}{\partial y}\left(T^{w}\dfrac{\partial h^w}{\partial y} \right)$
Isotropic, constant T^{w}	$T^{w}\left(\dfrac{\partial^2 h^w}{\partial x^2} + \dfrac{\partial^2 h^w}{\partial y^2} \right)$

Cylindrical Coordinates

Anisotropic form	$\dfrac{1}{r}\dfrac{\partial}{\partial r}\left(r T_{rr}^{w}\dfrac{\partial h^w}{\partial r} + T_{r\theta}^{w}\dfrac{\partial h^w}{\partial \theta} \right) + \dfrac{1}{r}\dfrac{\partial}{\partial \theta}\left(r T_{\theta r}^{w}\dfrac{\partial h^w}{\partial r} + T_{\theta\theta}^{w}\dfrac{1}{r}\dfrac{\partial h^w}{\partial \theta} \right)$
Constant $\mathbf{T}^{w\prime\prime}$	$T_{rr}^{w}\dfrac{1}{r}\dfrac{\partial}{\partial r}\left(r\dfrac{\partial h^w}{\partial r} \right) + (T_{r\theta}^{w} + T_{\theta r}^{w})\dfrac{1}{r}\left(\dfrac{\partial^2 h^w}{\partial r \partial \theta} \right) + T_{\theta\theta}^{w}\dfrac{1}{r^2}\left(\dfrac{\partial^2 h^w}{\partial \theta^2} \right)$
Isotropic, $\mathbf{T}^{w\prime\prime} = T^{w}\mathbf{I}''$	$\dfrac{1}{r}\dfrac{\partial}{\partial r}\left(r T^{w}\dfrac{\partial h^w}{\partial r} \right) + \dfrac{1}{r^2}\dfrac{\partial}{\partial \theta}\left(T^{w}\dfrac{\partial h^w}{\partial \theta} \right)$
Isotropic, constant T^{w}	$T^{w}\left(\dfrac{\partial^2 h^w}{\partial r^2} + \dfrac{1}{r}\dfrac{\partial h^w}{\partial r} + \dfrac{1}{r^2}\dfrac{\partial^2 h^w}{\partial \theta^2} \right)$

depends on the magnitude of this flow velocity relative to the normal velocity of the water table. The fluid moves in the pore space. However, since some of the fluid may be held in the formation against gravity and some air may occupy the pore space, the upward or downward flow may not occur in the entire pore space. Therefore, we approximate the movement of the water table as:

$$\left[\varepsilon_{\text{eff}} \mathbf{n} \cdot (\mathbf{w} - \mathbf{v}^{s}) \right]\Big|_{z_{\text{T}}} \approx \varepsilon_{\text{eff}}\frac{\partial h^w}{\partial t} \qquad (3.115)$$

where ε_{eff} is the *effective porosity*, the porosity through which the water flows at the upper surface of the aquifer, with $\varepsilon_{\text{eff}} \leq \varepsilon$, and $\left[\mathbf{n} \cdot (\mathbf{w} - \mathbf{v}^{s}) \right]\big|_{z_{\text{T}}}$ is the normal velocity of the upper surface of the system, i.e., of the water table, relative to the grains. Thus gain or loss of fluid at the top surface may be expressed in terms of $\left(\mathbf{n} \cdot \mathbf{q}^{w*} \right)\big|_{z_{\text{T}}}$, where flow out of the unconfined aquifer occurs if $\left(\mathbf{n} \cdot \mathbf{q}^{w*} \right)\big|_{z_{\text{T}}} > 0$ and is into the aquifer if $\left(\mathbf{n} \cdot \mathbf{q}^{w*} \right)\big|_{z_{\text{T}}} < 0$, with:

$$\left(\mathbf{n} \cdot \mathbf{q}^{w*} \right)\Big|_{z_{\text{T}}} = \left[\mathbf{n} \cdot \mathbf{q}^{w} - \varepsilon_{\text{eff}} \mathbf{n} \cdot (\mathbf{w} - \mathbf{v}^{s}) \right]\Big|_{z_{\text{T}}} \qquad (3.116)$$

After substitution of equation (3.115) into this equation to express the velocity of the top surface in terms of change in head we obtain:

$$\left(\mathbf{n}\cdot\mathbf{q}^{w*}\right)\big|_{z_\mathrm{T}} = \left[\mathbf{n}\cdot\mathbf{q}^{w}\right]\big|_{z_\mathrm{T}} - \varepsilon_{\mathrm{eff}}\frac{\partial h^{w}}{\partial t} \tag{3.117}$$

This equation can be used to eliminate $\left[\mathbf{n}\cdot\mathbf{q}^{w}\right]\big|_{z_\mathrm{T}}$ from equation (3.112) to obtain the equation for the unconfined case:

$$S_y\frac{\partial h^{w}}{\partial t} - \nabla'\cdot\left(\mathbf{T}^{w\prime\prime}\cdot\nabla'h^{w}\right) + \left(\mathbf{n}\cdot\mathbf{q}^{w*}\right)\big|_{z_\mathrm{T}} + \left(\mathbf{n}\cdot\mathbf{q}^{w}\right)\big|_{z_\mathrm{B}} = \sum_{\mathrm{W}=1}^{N_\mathrm{W}} Q_\mathrm{W}^{w\prime} \tag{3.118}$$

where $S_y = S + \varepsilon_{\mathrm{eff}}$ is called the *specific yield*.

Although the vertically integrated equation (3.118) for an unconfined aquifer has a term-by-term correspondence with the form of equation (3.112) that can be applied directly for the confined case, there are two subtle differences. Both of these differences stem from the fact that the upper surface of the flow region may move substantially for the unconfined case while it is relatively fixed for the confined case. The implication of this observation is that a change in head causes upward movement of the top surface of the flow region for the unconfined case so that the specific yield, S_y, which is approximately equal to the porosity, replaces the storativity, S, in the equation. Both S_y and S are dimensionless and physically account for the volume of fluid released by a system per unit cross section of area per unit drop in head. In the unconfined case, this release is due primarily to flow in the pores; in the confined case, the release comes exclusively from the compressibility of the fluid, solid, and matrix. Thus S_y is several orders of magnitude larger than S. The other difference is the presence of the superscript asterisk * in the vertical flow term for the unsaturated case. This accounts for the fact that flow which leaves or enters the system at the top surface must be relative to the velocity of the top surface rather than just relative to the solid grains, as with the confined case. For use in applications, the differences in the equations basically reduce to specification of different values for parameters. We note, additionally, that the assumption has been made that vertical flow within the region is negligible such that the vertical head gradient can be neglected. This assumption may be inapplicable for the unconfined case when the slope of the water table can be significant, such as in the vicinity of a river, or as a consequence of pumping.

Solution of the vertically averaged flow equation requires the specification of boundary conditions at the lateral boundaries. The discussion of needed conditions corresponds essentially to that for the fully three-dimensional equation in Subsection 3.6.5. The effect of the integration procedure through the vertical is to replace the need for boundary conditions at the top and bottom of the region with incorporation of those conditions directly into the equation as leakage terms for the confined case and a bottom leakage term and specification of the specific yield for the unconfined case. The vertically integrated equation is often used for the study of essentially horizontal flow to a pumping well in an aquifer. Some aspects of the two- and three-dimensional formulations for single-phase flow will be explored in the problems at the end of the chapter.

3.7   TWO-PHASE IMMISCIBLE FLOW

Modeling the flow of two immiscible fluids in a porous medium requires that the concepts employed in the previous section be extended. We will have mass conservation equations for each phase as described by equation (3.1). However, to solve the equations, we will again need additional conditions that will lead to having the same number of unknowns as equations. Conceptually, there are two major differences between two-phase flow and single-phase flow that must be accounted for by the closure relations.

First, with single-phase flow, a single fluid, w, occupies all the pore space such that its volume fraction is equal to the porosity, $\varepsilon^w = \varepsilon$. Because the porosity, ε, changes only slightly with time, the volume fraction of the single fluid phase at a point in the porous medium will be approximately constant. With two-phase flow, the total fluid fraction will again change only slightly with time, but the relative amounts of each fluid can change. One of the phases can range from completely filling the pore space to being completely absent. This observation causes us to make use of the concept of saturation, as introduced in Section 1.4.

The fraction of pore space occupied by phase w is designated as the saturation, s^w. Therefore, $\varepsilon^w = s^w \varepsilon$. For single-phase flow, $s^w = 1$. For the two-phase case being considered here $0 \leq s^w \leq 1$. Also if the second fluid phase is designated as phase n, its saturation, $s^n = 1 - s^w$ and $\varepsilon^n = s^n \varepsilon$.[4] In the present discussion, we will assume that phase w preferentially wets the solid, while phase n is referred to as nonwetting in that it has a lesser attraction to the solid. Within an REV, the solid phase surface can have contact with both phases. In some real systems with a heterogeneous solid, a particular fluid phase may be wetting in one part of the region and nonwetting in the other. This situation will not be considered here. In modeling two-phase flow in a porous medium, the constitutive equations employed must account for the time and space variability of the saturation.

A second feature of two-phase flow is the fact that each fluid phase interacts with another fluid phase in addition to interacting with the solid phase. The dynamics of the fluid-fluid interface must be accounted for in describing the system. The issues of surface tension and capillary pressure at the interface between fluids has been presented at the microscale in Section 1.6. These phenomena must be accounted for in the macroscale closure conditions for two-phase flow.

In this section, we will first examine the extensions that arise in the mass conservation equations for the two fluid phases as opposed to a single phase. Then conditions will be developed for closure of the equations so that they may be solved.

3.7.1   Derivation of Flow Equations

The macroscale point conservation of mass equation for a phase in a porous medium has been derived as equation (3.1). Here we will work with the forms of this equation that apply to the solid phase, designated as the s phase, and the two fluid phases, designated as the w and n phases. For the fluid phases, we will also make use of the

---

[4] When a system contains more than two fluid phases, the constraint on the fluid saturations is that their sum over all fluid phases must equal 1. The two-fluid-phase case is a special case of this more general situation.

saturation times the porosity to account for the volume fraction of the phase. Therefore, for the two-fluid system, the mass conservation equations for the *w, n,* and *s* phases are, respectively:

$$\frac{\partial(s^w \varepsilon \rho^w)}{\partial t} + \nabla \cdot (s^w \varepsilon \rho^w \mathbf{v}^w) = e_{ws}^w + e_{wn}^w + \sum_{W=1}^{N_W} \rho_W^w Q_W^w \delta(\mathbf{x} - \mathbf{x}^W) \qquad (3.119)$$

$$\frac{\partial(s^n \varepsilon \rho^n)}{\partial t} + \nabla \cdot (s^n \varepsilon \rho^n \mathbf{v}^n) = e_{ns}^n + e_{wn}^n + \sum_{W=1}^{N_W} \rho_W^n Q_W^w \delta(\mathbf{x} - \mathbf{x}^W) \qquad (3.120)$$

and:

$$\frac{\partial[(1-\varepsilon)\rho^s]}{\partial t} + \nabla \cdot [(1-\varepsilon)\rho^s \mathbf{v}^s] = e_{ws}^s + e_{ns}^s \qquad (3.121)$$

The equation for the solid phase is identical to that used for single-phase flow, equation (3.38), including the assumption that no solid phase is pumped into or from the system. The mass exchange between the solid and each of the fluid phases is designated separately. The equations for both fluid phases are similar in form; they differ from the single fluid phase flow equation (3.37) only in the presence of the saturation. To streamline the subsequent derivation, we will write the two fluid equations using the superscript α, which can refer to either phase, according to the single formula:

$$\frac{\partial(s^\alpha \varepsilon \rho^\alpha)}{\partial t} + \nabla \cdot (s^\alpha \varepsilon \rho^\alpha \mathbf{v}^\alpha) = e_{\alpha s}^\alpha + e_{wn}^\alpha + \sum_{W=1}^{N_W} \rho_W^\alpha Q_W^\alpha \delta(\mathbf{x} - \mathbf{x}^W) \quad \alpha = w, n \quad (3.122)$$

When convenient, we can replace s^n with $1 - s^w$, and this emphasizes the fact that s^w and s^n are not independent of each other.

However, the three equations for the three phases contain 14 unknowns (ρ^w, ρ^n, ρ^s, ε, s^w, and the three components of each of the velocity vectors $\mathbf{v}^w$, $\mathbf{v}^n$, and $\mathbf{v}^s$). We are in need of 11 conditions that relate these variables, as well as initial and boundary conditions, if we are to be able to use these equations to describe a particular two-phase flow system. We will make use of state equations and constitutive assumptions to obtain the conditions needed. Parts of the derivation that are similar to that in Subsection 3.6.3 for single-phase flow will be abbreviated, hopefully without loss of clarity, in favor of providing details of the new issues that arise. Additionally, even though Darcy's experiments were employed for single-phase systems, considerations from those experiments will be applied to this more complex case.

We can define a Darcy velocity for each of the fluid phases relative to the solid that accounts for the volume fraction of each phase as:

$$\mathbf{q}^\alpha = s^\alpha \varepsilon (\mathbf{v}^\alpha - \mathbf{v}^s) \quad \alpha = w, n \qquad (3.123)$$

Substitution of this expression into equation (3.122) to eliminate $\mathbf{v}^\alpha$ yields:

$$\frac{\partial\left(s^{\alpha}\varepsilon\rho^{\alpha}\right)}{\partial t}+\nabla\cdot\left(\rho^{\alpha}\mathbf{q}^{\alpha}\right)+\nabla\cdot\left(s^{\alpha}\varepsilon\rho^{\alpha}\mathbf{v}^{s}\right)=e_{\alpha s}^{\alpha}+e_{wn}^{\alpha}+\sum_{W=1}^{N_{W}}\rho_{w}^{\alpha}Q_{W}^{\alpha}\delta\left(\mathbf{x}-\mathbf{x}^{W}\right)$$

$$\alpha=w,n \quad (3.124)$$

Next we apply the product rule to the first and third terms, collect terms, and introduce the material derivative with respect to the velocity of the solid phase to obtain:

$$\varepsilon\rho^{\alpha}\frac{D^{s}s^{\alpha}}{Dt}+s^{\alpha}\rho^{\alpha}\frac{D^{s}\varepsilon}{Dt}+s^{\alpha}\varepsilon\frac{D^{s}\rho^{\alpha}}{Dt}+s^{\alpha}\varepsilon\rho^{\alpha}\nabla\cdot\mathbf{v}^{s}+\nabla\cdot\left(\rho^{\alpha}\mathbf{q}^{\alpha}\right)$$

$$=e_{\alpha s}^{\alpha}+e_{wn}^{\alpha}+\sum_{W=1}^{N_{W}}\rho_{W}^{\alpha}Q_{W}^{\alpha}\delta\left(\mathbf{x}-\mathbf{x}^{W}\right)\quad\alpha=w,n \quad (3.125)$$

We can eliminate $\nabla\cdot\mathbf{v}^{s}$ between this equation and the mass balance equation for the solid phase, equation (3.121), to obtain:

$$\varepsilon\rho^{\alpha}\frac{D^{s}s^{\alpha}}{Dt}-\frac{s^{\alpha}\rho^{\alpha}}{(1-\varepsilon)\rho^{s}}\frac{D^{s}\left[(1-\varepsilon)\rho^{s}\right]}{Dt}+\frac{s^{\alpha}\rho^{\alpha}(1-\varepsilon)}{\rho^{s}}\frac{D^{s}\rho^{s}}{Dt}$$

$$+s^{\alpha}\varepsilon\frac{D^{s}\rho^{\alpha}}{Dt}+\nabla\cdot\left(\rho^{\alpha}\mathbf{q}^{\alpha}\right)=e_{\alpha s}^{\alpha}+e_{wn}^{\alpha}-\frac{x_{\text{mass}}^{\alpha}}{x_{\text{mass}}^{s}}\left(e_{ws}^{s}+e_{ns}^{s}\right)$$

$$+\sum_{W=1}^{N_{W}}\rho_{W}^{\alpha}Q_{W}^{\alpha}\delta\left(\mathbf{x}-\mathbf{x}^{W}\right)\quad\alpha=w,n \quad (3.126)$$

where:

$$\frac{x_{\text{mass}}^{\alpha}}{x_{\text{mass}}^{s}}=\frac{\rho^{\alpha}\varepsilon s^{\alpha}}{\left[\rho^{s}(1-\varepsilon)\right]}$$

is the ratio of the fractions of the total mass that are α and s phase.

When dealing with single-phase flow, we made the approximation that the force acting on the solid phase could be approximated as the pressure of the fluid acting on the solid surface. However, when two fluids are present in the pore space, the force will be due to some combination of the fluid pressures. A detailed derivation of the force acting on the solid may be found in [23], but here we adopt a more classical approach. Use the symbol p^{s} to represent the force per area exerted on the solid by the two fluids and stipulate that:

$$p^{s}=\chi p^{w}+(1-\chi)p^{n} \quad (3.127)$$

where χ is called the *Bishop parameter* and is considered to be a function of saturation that satisfies the condition $0\le\chi\le 1$ so that p^{s} has a value between p^{n} and p^{w}. From a physical perspective, the Bishop parameter is a measure of the fraction of the solid phase surface in contact with phase w while $1-\chi$ is a measure of the fraction of the phase s surface in contact with the n phase. Thus $\chi=s^{w}$ when s^{w} equals

0 or 1, but at intermediate saturations we need a relation between χ and s^w. Since w refers to the wetting phase, it seems reasonable that $s^w \leq \chi \leq 1$. If the nonwetting phase is not in contact with the solid, $\chi = 1$; in many cases, for lack of an easily implemented alternative, the assumption is made that $\chi = s^w$.

We now define the *solid and matrix compressibility* at fixed composition and temperature, respectively, as:

$$\beta^s = \frac{1}{\rho^s}\frac{\partial \rho^s}{\partial p^s} \tag{3.128}$$

and:

$$\alpha^b = \frac{1}{(1-\varepsilon)\rho^s}\frac{\partial\left[(1-\varepsilon)\rho^s\right]}{\partial p^s} \tag{3.129}$$

Substitution of these definitions back into equation (3.126), evocation of the condition that the matrix deforms slowly such that the material derivatives may be replaced by partial derivatives, and collection of terms provides:

$$\varepsilon\rho^\alpha\frac{\partial s^\alpha}{\partial t} + s^\alpha\rho^\alpha\left\{[\alpha^b + (1-\varepsilon)\beta^s]\frac{\partial p^s}{\partial t} + \varepsilon\beta^\alpha\frac{\partial p^\alpha}{\partial t}\right\} + \nabla\cdot(\rho^\alpha\mathbf{q}^\alpha)$$

$$= e^\alpha_{as} + e^\alpha_{wn} - \frac{x^\alpha_{\text{mass}}}{x^s_{\text{mass}}}(e^s_{ws} + e^s_{ns}) + \sum_{W=1}^{N_W}\rho^\alpha_W Q^\alpha_W \delta(\mathbf{x}-\mathbf{x}^W) \quad \alpha = w, n \tag{3.130}$$

It is reassuring to note that when $s^w = 1$ such that there is only a single phase present, this equation becomes identical to equation (3.83), the equation derived previously for the single-phase case.

We also recall that we began this derivation with three mass balance equations (one for each phase) and 14 variables. In eliminating $\nabla\cdot\mathbf{v}^s$ from the problem, we also reduced the number of equations we are studying to two, the mass balance equation for each fluid. The parameters α^b, β^w, β^n, and β^s must be specified. Also, the equations of state for $\rho^w(p^w)$ and $\rho^n(p^n)$ must be available, e.g., from a handbook or from integration of the compressibility equation. The Bishop parameter must be specified as a function of s^w. Thus the variables for which we need to solve are reduced to nine: s^w, p^n, p^w, and the three components of the Darcy velocity vectors $\mathbf{q}^w$ and $\mathbf{q}^n$. We are still in need of seven conditions that relate the nine unknown variables in order to have a solvable set of equations.

The search for the seven additional conditions first turns to Darcy's experiments. Although these involved only single-phase flow, we might imagine that each fluid phase in a multiple-phase flow would respond roughly the same way. We are in a search for useful conditions. Darcy's work may be a good place to start. After all, if what is tried does not work, we can try something else. Thus, let us propose that each fluid phase satisfies a Darcy equation similar to equation (3.55) according to:

$$\mathbf{q}^\alpha = -\frac{\mathbf{k}^{s\alpha}}{\mu^\alpha}\cdot(\nabla p^\alpha - \rho^\alpha\mathbf{g}) \quad \alpha = w, n \tag{3.131}$$

The difference between this equation and the single-phase form is the presence of $\mathbf{k}^{s\alpha}$ rather than the intrinsic permeability, $\mathbf{k}^s$. The extra superscript, α, is appended to designate the idea that the apparent intrinsic permeability for each phase in two-phase flow will depend on more than just the solid properties; it may also depend on the properties of the other fluid and certainly on the fact that the presence of the other fluid decreases the space available for flow of each phase. Of course, when $s^\alpha = 1$, only the α phase is present and $\mathbf{k}^{s\alpha} = \mathbf{k}^s$. When $s^\alpha < 1$, the pore space available to the α phase is less than for the single-phase case and $\mathbf{k}^{s\alpha} < \mathbf{k}^s$.

Sometimes, this last condition is expressed as $\mathbf{k}^{s\alpha} = k_{rel}^\alpha \mathbf{k}^s$ where k_{rel}^α is called the *relative permeability* and is a function of s^α that ranges from 0 to 1. However, this representation fails to account for the fact that the anisotropy of the flow region may be altered by a change in s^w (i.e., the directional dependence of $\mathbf{k}^{s\alpha}$ can change) in addition to its elements being scaled down. The relative permeability concept is appropriate when the flow region is isotropic and remains so after the fluids are introduced such that $\mathbf{k}^{s\alpha} = k^{s\alpha}\mathbf{I}$. For a two-fluid system, the relative permeability k_{rel}^α depends on s^w and is defined:

$$k^{s\alpha} = k^s k_{rel}^\alpha \quad \alpha = w, n \tag{3.132}$$

subject to:

$$0 \le k_{rel}^\alpha \left(s^w\right) \le 1 \quad \alpha = w, n \tag{3.133}$$

Also, $k_{rel}^\alpha = s^\alpha$ when $s^\alpha = 0$ or 1.

Substitution of equation (3.131) into equation (3.130) yields:

$$\varepsilon \rho^\alpha \frac{\partial s^\alpha}{\partial t} + s^\alpha \rho^\alpha \left\{ \left[\alpha^b + (1-\varepsilon)\beta^s \right] \frac{\partial p^s}{\partial t} + \varepsilon \beta^\alpha \frac{\partial p^\alpha}{\partial t} \right\}$$
$$- \nabla \cdot \left[\rho^\alpha \frac{\mathbf{k}^{s\alpha}}{\mu^\alpha} \cdot \left(\nabla \rho^\alpha - \rho^\alpha \mathbf{g} \right) \right] = e_{\alpha s}^\alpha + e_{wn}^\alpha - \frac{x_{mass}^\alpha}{x_{mass}^s} \left(e_{ws}^s + e_{ns}^s \right)$$
$$+ \sum_{W=1}^{N_w} \rho_W^\alpha Q_W^\alpha \delta \left(\mathbf{x} - \mathbf{x}^W \right) \quad \alpha = w, n \tag{3.134}$$

These two equations now contain only three unknowns, p^w, p^n, and s^w (with p^s given by equation (3.127)); one more condition is needed that relates these three variables. This additional condition is chosen as the capillary pressure, p^c, which is taken to be a function of s^w and is also equal to the pressure difference between the nonwetting and wetting phases:

$$p^c \left(s^w\right) = p^n - p^w \tag{3.135}$$

This equation should be compared with equation (1.37), which is its microscale counterpart. The microscale version involves pressure at the interface and the interface curvature while the macroscale version involves pressures averaged over the phase volumes and uses saturation as a surrogate for the curvature. Because of this, one might guess that this macroscale representation of capillary pressure may not be as robust as the microscale version. Additionally, the microscale result was obtained by examining an interface at equilibrium. Equation (3.135) will be used in modeling systems away from equilibrium, and the lack of equilibrium may also detract from the accuracy of this assumed relation.

In some instances it is convenient to make use of what is called the *capillary head* defined, based on equation (3.135) as:

$$h^c = \frac{p^n - p^w}{\rho^w g} \tag{3.136}$$

This quantity is particulary useful when studying air-water systems where the air phase pressure is essentially constant. It is extremely important to remember that although capillary pressure is defined in terms of phase pressure differences, $h^c \neq h^n - h^w$ where the heads in each phase are given, for example, by equation (3.29) unless $\rho^w = \rho^n$.

This completes the derivation of the general two-phase flow equations. Equation (3.134) is applied to each of the two fluid phases to solve for p^w and p^n. These two equations are supplemented by equation (3.135) that relates the difference in these pressures to the saturation. Additionally, we have equation (3.127), which relates p^s to the fluid pressures with the requirement that we make a reasonable specification of $\chi(s^w)$. Also, we must specify α^b, β^s, β^w, and β^n. Lastly, we specify state equations for $\rho^s(1 - \varepsilon)$ as a function of p^s, $\rho^w(p^w)$, $\rho^n(p^n)$, $\mathbf{k}^{sw}$, and $\mathbf{k}^{sn}$. Indeed, this is a significant amount of information that must be specified to determine the differential equations of multiphase flow. Boundary conditions for each phase, with similar considerations to those for single-phase flow, must be specified as well as the pumping information. The requirements of information plus the fact that the Darcy-type flow equation and the capillary pressure equation are approximate suggests that a solution to the equations that matches a physical system might be more serendipitous than scientific. However, before examining the merits of such a determination, as will be done in Chapter 5, it is useful to gain understanding of three of the important aspects of approximations that are employed: the $p^c(s^w)$ relation and the permeability functions $\mathbf{k}^{sw}$ and $\mathbf{k}^{sn}$.

In the subsections to follow, we will first examine the constitutive relationships for capillary pressure as a function of saturation (referred to herein as the p^c-s^w model but most commonly verbalized as a "p, c, s" model). The impact of saturation changes on a tensorial permeability is very difficult to quantify. For this reason, if the medium is anisotropic the permeability is typically modeled as $\mathbf{k}^{s\alpha} \approx k_{rel}^{\alpha}\mathbf{k}^s$ despite the limitations of this approach. When the solid is isotropic, the permeability is modeled using $k^{s\alpha} = k_{rel}^{\alpha}k^s$. Thus, for both isotropic and anistropic systems, the additional complication is the relative permeability. We will examine the constitutive relations that provide k_{rel}^{α} as a function of s^w and will refer to these as k_{rel}^{α}-s^w models (commonly verbalized as "k, s" models).

The relationships to be covered here are strictly applicable only to *granular soils*: those soils which contain few or no clay particles. Granular soils result from deposition in an active geologic environment, e.g. glacial, fluvial, coastal beach, and wind generated deposits. These soils typically have a high percentage of quartz grains which are rounded and are more or less chemically inert (they do not interact with the soil fluids). The rounded nature of the grains is responsible for the correlation between soil texture and soil structure.

On the other hand, a high clay content in a soil indicates deposition in a geologically quiet environment, e.g., a lake bottom. These soils tend to have a high organic content and the clay particles have a complex mineralogy and a large surface area. These attributes combine to create a porous medium which is chemically active. As a result, some of the assumptions inherent in what follows, such as uniform structure and degree of preferential wetting by one of the fluids, do not apply. In these instances, relations for p^c and $k_{\mathrm{rel}}^{\alpha}$ as functions only of s^w do not apply. Further information on the behavior of clay may be found, for example, in [26] or other soil physics references.

3.7.2 Observations on the p^c-s^w Relationship

Let us emphasize that the search for a macroscale p^c-s^w relationship has its roots in observations of the behavior of fluids in capillary tubes and in common observations of water interacting with soil. These observations typically involve water as a wetting phase and air as the nonwetting phase, but variations can occur by changing the material used for the capillary tube or by changing the fluid from water to mercury or oil. In any event, our objective is to convert the basic observations of behavior of an immiscible flow of two fluids in a porous medium into a quantitative statement that relates capillary pressure to the saturation. There is a large heuristic element to this conversion, and a need for improved understanding and relationships still exists. Thus, one must realize that this is an area for fruitful additional research; the relations to be presented are constitutive approximations that do not carry the same cachet of authority as do mass conservation equations such as equations (3.1) or (3.130). The reliability and robustness of the p^c-s^w relations in considering real systems with heterogeneous soils and undergoing cycles of wetting and drying is one of the weak links in the development of good models. With this disclaimer firmly in place, we proceed to considerations that have informed and support the current state of p^c-s^w relations.

In Chapter 1, we examined microscale capillary pressure. Equation (1.63) provided quantitative confirmation of the observation that if one inserts the tips of glass capillary tubes into a pool of water, the water will rise higher in tubes of smaller diameters. The capillary pressure at the interface between the water and air is greater in the smaller tubes and the curvature of the interface in these tubes is also greater. This is an example of wetting fluid moving into pores of smaller diameters. A wetting fluid will be drawn from one capillary tube into a tube of smaller diameter.

Now consider the case where a w phase and an n phase are both present in a porous medium made up of sand grains. In such a medium, there is a distribution of the diameter of the pores depending on how the grains pack together. One issue of interest is how the two fluids distribute in this medium. Based on the preceding observations, we would expect that phase w will be drawn into the smaller pores. When phase w is in these pores, the interfacial curvature will be greater than when it is in the larger pores, so that the capillary pressure will be higher. The smaller the relative amount of phase w in the medium, the more high-curvature interfaces there will be between the two fluid phases. Thus, from a macroscopic perspective, we expect that as s^w decreases, the macroscale measure of capillary pressure, p^c, will increase. The basis of this understanding is depicted in Figure 3.3.

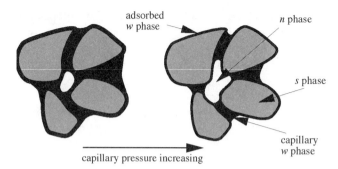

Figure 3.3: At increased capillary pressure, the wetting phase, w, is pulled into smaller portions of the pore space and the curvature of the interface between the fluids increases. The w phase is held in the pore space by capillary forces as well as adsorption and spreads over the solid phases, forming a thin film (after [26]).

From a slightly different point of view, the preceding suggests that the preference of the wetting phase to exist in smaller pores, tends to draw the solid grains together. This phenomenon is what allows one to build a sand castle. If one tries to build an elegant structure at the beach using dry sand, frustration will soon set in. Also, trying to build a structure in a quiescent pool of water such that the sand is fully saturated will also be futile. Having the right mix of air and water in the sand is crucial to developing just the right capillary and surface adsorptive forces to hold the sand grains together so that a beach construction project can be sizable and well sculpted.

Another aspect of multiphase flow in porous media can be observed by dipping the edge of a lump of sugar into a cup of tea. We are interested here in the flow that takes place and not in a longer observation that would include dissolution of the sugar and deterioration of the structure of the solid phase, although that is a fascinating problem. The surface of the cup of tea is large such that it is flat (except at the edges where the tea meets the cup wall). Thus the pressure at the surface is atmospheric, equal to the pressure of the air that is in the sugar cube. Nevertheless, as soon as the cube touches the surface, the tea will flow upward into the cube. This happens because when the small pores of the solid cube are put in contact with the wetting tea, a curved interface forms such that, to reach a new equilibrium state where the microscale interface equilibrium condition, $p_w + p_c = p_n$, is satisfied, phase w will move upward into the solid. The movement of a phase w into a dry porous medium occurs as soon as it is put in contact with the medium.

On the other hand, if the "sugar cube" in the previous example were actually made of some hydrophobic material, the tea would not move into the cube at first contact. The cube would have to be pushed down into the tea to a depth such that the pressure in the tea is high enough to overcome preference of the cube material for the air phase. The tea would enter first into the larger pores, where capillary forces are smaller, and would enter into smaller pores only if the cube were submerged deeply enough into the cup to overcome the capillary forces in those pores. These observations give rise to the concept of *entry* or *threshold pressure*. The entry pressure is the macroscale pressure difference, $p^n - p^w$, that is required to enable a nonwetting phase to begin to invade a porous medium saturated with the wetting phase, i.e., a porous medium with $s^w = 1$. The entry pressure is related to the largest pore diameter since invasion will take place first into the largest pores where capillary pressure effects will be smallest.

Based on the preceding, we list the following conditions that a macroscale p^c-s^w relation must satisfy:

- The onset of drainage of a porous medium saturated with a phase w occurs only when an entry pressure $p^n - p^w > 0$ is applied. The magnitude of the entry pressure is related to the largest diameter pore.
- Imbibition of a phase w into a medium with $s^w = 0$ occurs as soon as the medium is placed in contact with the wetting phase and does not require an entry pressure.
- At intermediate saturations at equilibrium, phase w tends to occupy the smallest pores with highest capillary pressures. Therefore, the functional form of p^c should decrease as s^w increases.

These expectations have been uncovered through very basic considerations. However, they also raise additional questions. We know that if we take a single glass capillary tube with a constant diameter and dip its tip into a reservoir of water, the water will rise into the tube to the same height each time we perform the experiment. Furthermore, no matter how far we lower the end of the tube into the reservoir, the height of the water in the tube above the reservoir surface will not change. If the diameter of the tube were to change along the tube axis, however, the height of water rise above the reservoir surface would depend on how far the tube is inserted. Furthermore, the level in the tube would depend on whether the tube is being inserted into the reservoir, with imbibition into the tube taking place, or is being withdrawn, with drainage of water taking place. We can discuss the influence of a spatially varying pore diameter in the context of the experiment depicted in Figure 3.4.

The figure depicts a capillary tube with a diameter that depends on axial position x. The capillary tube is oriented horizontally for convenience so that the effects of gravity do not have to be considered. Seven axial positions along the tube are indicated corresponding to important locations where the tube radius is changing. Additionally, the diameter of the tube is indicated at four locations of importance. The tube is connected on the left side to a reservoir containing wetting phase w and on the right to a reservoir containing nonwetting phase n. Below the tube is a rough sketch of the value of p^c as a function of s^w that is expected from this system. Here, $p^c = p^n - p^w$ and p^n and p^w are each constant at any equilibrium state of the system. The saturation, s^w is the fraction of the tube volume between x_1 and x_7 filled with fluid w.

Let the tube be initially filled with fluid phase w. We want to examine how the system responds during drainage caused by sequential incremental increases of the pressure of fluid n, i.e., how the system behaves as p^c at the interface between the fluids is changed. Let us stress that we will only be examining equilibrium states of the system, the states that are achieved after the system readjusts following a small incremental change in pressure. The system will initially be at equilibrium when p^c is sufficient to prevent fluid w from draining. When this value of p^c is

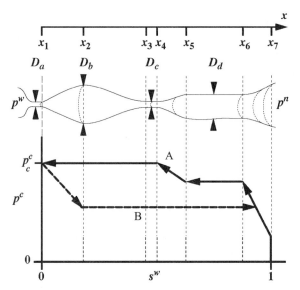

Figure 3.4: Schematic diagram of p^c-s^w relation based on equilibrium states achieved during drainage (solid line) and imbibition (dashed line) for a horizontal capillary tube with nonconstant diameter [25].

exceeded by a small increment, the meniscus will move into the tube as fluid n will displace fluid w. Since the tube diameter is decreasing, a new equilibrium location of the wn interface will be established between x_6 and x_7. A series of incremental increases in p^c will lead to the interface reaching position x_6. The double-segment arrow in the graph captures this behavior. In the first segment p^c increases from 0 to the entry value with no change in saturation from $s^w = 1$. Then in the second segment the saturation decreases as the pressure increases to achieve equilibrium with the interface at x_6.

Between x_5 and x_6 the capillary tube diameter is constant. Therefore, once the interface reaches position x_6 any infinitesimal increase in p^c will shift the interface to location x_5, where the diameter of the tube again decreases. In this situation, x_6 is an unstable equilibrium location, as are all the locations between x_5 and x_6. No stable equilibrium exists between these two locations. The phenomenon whereby the interface jumps from x_6 to x_5 is called a *Haines jump* corresponding to a locally abrupt change in saturation. Further incremental increases in p^c will drive the wn interface to location x_4, where the tube diameter reaches value D_c and remains at this value until position x_3. Thus, a Haines jump occurs to move the interface x_4 to x_3. Since the tube diameter is greater than D_c between x_3 and x_1, the interface will continue through this bulging section of the tube to x_1 if no adjustment is made to p_c.[5] The schematic curve indicated by the solid line in the figure is a *drainage curve*. The saturations associated with the horizontal parts of the drainage curve for the capillary tube considered here are not equilibrium states. For a porous medium, rather than just a capillary tube, all locations on the drainage curve are equilibrium

---

[5] If p^c is decreased after the interface has cleared location x_2, it is possible to establish an equilibrium state prior to the interface reaching x_1. This is not considered here as we develop the drainage scenario by allowing only sequential increases in p^c.

points. Haines jumps occurring at an individual pore generally do not significantly impact the effort to obtain equilibrium saturations involving a large number of pores. However, if the porous medium is composed such that it has a narrow distribution in pore sizes, the drainage curve will have a relatively flat section with large changes in s^w resulting from small changes in p^c.

With the capillary tube drained so that $s^w = 0$, let us reverse the drainage process by decreasing p^c starting with the *wn* interface at x_1. Small incremental decreases in p^c will allow the interface to reach equilibrium locations between x_1 and x_2. At x_2, the capillary tube diameter reaches its largest value. Thus the equilibrium capillary pressure at this location is the smallest in the tube. Therefore an infinitesimal decrease in p^c will allow the *wn* interface to jump to the right, all the way to the exit. The fact that we are running this experiment by only allowing incremental decreases in p^c constrains the system so that no equilibrium locations of the interface are found between x_2 and x_7.

The schematic curve for this experiment is an *imbibition curve* and is indicated as the dashed line in Figure 3.4. For this simple experiment, the positions on the sloped part of the imbibition curve are equilibrium locations. The locations on the flat part of the curve are not equilibrium positions. For a complex porous medium with a distribution of pore sizes, an equilibrium imbibition plot of p^c vs. s^w may be developed over the range of saturations.

The fact that the plots of p^c vs. s^w are different for imbibition and drainage is called *hysteresis*, where this term refers to different behavior of the process depending on the direction in which the process is occurring. The difference occurs because the small diameter pores control drainage events while the larger diameter regions control imbibition. The result is that, for a particular value of s^w, the capillary pressure is greater for drainage than for imbibition. The paths of drainage and imbibition from two end points are different, and the path of the full cycle is called a *hysteresis loop* (the path defined by letters A and B in Figure 3.4). We note, also, that it is possible to stop a drainage process at intermediate values of s^w by decreasing the capillary pressure. In such an experiment, the full loop depicted in Figure 3.4 would not be traversed. Rather, some path cutting across the loop would be followed depending on where on the upper curve the drainage process is halted. Similar behavior is observed when reversing an imbibition process at an intermediate value of saturation by increasing the capillary pressure, although the route across the middle of the loop would be different. These "shortcuts" across the hysteresis loop are referred to as *scanning curves*.[6] A set of scanning curves is illustrated in Figure 3.4. Hysteretic effects are important, in a practical sense, for the case of water infiltration and removal from soil across intermediate ranges of saturation.

The expectation that study of a single capillary tube with varying diameter would reveal all the subtleties of two-fluid-phase flow in a porous medium would have to be deemed somewhere between naive and optimistic. Nevertheless, it gives us a good start. An effect observed in porous media composed of sand or a solid matrix

[6] An everyday example of hysteresis occurs if one drives between home and the store using a network of one way streets seeking the shortest route. The path taken between the two destinations would be different depending on direction, and the two paths together would form a hysteresis loop. Also, if one were to get halfway to the store and then discover that he had left his money at home, the fastest path to return home from that point might involve a different route from the one taken if starting from the store. This alternative path is a scanning curve.

is the fact that some fluid becomes trapped in the system. It is not possible to drain all fluid w from a porous system simply by flushing with fluid n. This fact can be observed, for example, when trying to dry a porous solid simply by passing air through it. A sizable fraction of the water may be removed, but some remains trapped in small pores and is only released very slowly over large time scales due to evaporation of the water into the flowing air. Also, it is not possible to drain all of the n phase from a porous medium by flushing with a wetting fluid. This statement is confirmed by studies of organic liquids that have been spilled into the soil. Use of water to remove the organics, either by direct action or in letting infiltration of rain flush the system, is not effective. The organics persist and slowly dissolve into the water over many years thereby existing as continuing sources of contamination. Thus, it seems reasonable that a p^c-s^w hysteresis loop would not cover the entire range of $0 < s^w < 1$ but would operate over some more limited range. Some qualitative aspects of this situation will now be discussed.

An attempt to drain completely phase w from a porous medium by introduction of a phase n leaves some residual phase w in the pore network at some low saturation called the *irreducible saturation*, denoted s_i^w. At the irreducible saturation, phase w molecules are strongly adsorbed onto the solid surface and the *capillary wetting phase* is tightly held in the corners and crevices of the pores such that phase w will no longer flow in respone to a gradient in hydraulic head. Technically, phase w is considered to remain hydraulically connected as it coats the soil grains with a thin film such that a tortuous path for flow in the Darcy sense still remains.[7] However, the effective conductivity, k^{sw}, of the wetting phase is virtually zero at this low saturation given the time scale of natural forcing (for example, the next rainfall event or evaporation mass transfer). Therefore, for practical purposes, this volume of phase w at residual saturation can be thought of as becoming hydraulically disconnected.

To facilitate the ensuing discussion of fluid entrapment in a sample of porous medium, let us define the *trapped-phase volume* of a fluid as that volume which will no longer respond to a gradient in hydraulic potential. We also recognize that in a porous medium some fluid may be hydraulically connected and able to respond to a gradient in potential while pockets of the same fluid are trapped and unable to respond. In terms of α-phase saturation this distinction is written:

$$s^\alpha = s_f^\alpha + s_t^\alpha \tag{3.137}$$

where the subscripts f and t refer to the free and trapped portions, respectively. At the irreducible saturation, $s_f^\alpha = 0$. Because the extension of Darcy's law to two-phase flow is made without making a distinction between free and trapped elements of a phase, in essence the trapped portion is effectively part of the solid phase in that it helps outline paths available for flow.

The mechanisms that cause the nonwetting phase to become entrapped are different from those that cause wetting-phase entrapment. Two primary mechanisms have been identified: *snap-off* and *bypassing* [43]. When a wetting phase is imbibed into a porous medium, it tends to flow spontaneously along the pore walls as it displaces the nonwetting fluid. The ratio of the large diameter to small diameter in a

---

[7] Application of Darcy's law requires that a phase be spatially continuous.

pore (also referred to as ratio of the pore body diameter to pore throat diameter) is called the aspect ratio. When this ratio is small (on the order of 1), phase w essentially displaces phase n by pushing it out as depicted in Figure 3.5a. However when the aspect ratio is high (e.g., greater than 2)), the advancing wetting phase is able to move along the wall, overtaking some of phase n. Thus if one were to look at a cross section of the pore, there would be a ring of wetting phase surrounding the nonwetting phase. Then, where the pore narrows, phase n is "snapped-off" and unable to move forward through the constriction since that area is taken up by the wetting phase. Thus phase n becomes segmented with parts of the phase becoming isolated blobs. This phenomenon, called *snap-off*, is illustrated in Figure 3.5b. Besides being influenced by aspect ratios, snap-off is impacted by the contact angle of the wetting phase with the solid. Snap-off is enhanced when the contact angle is small, i.e., when the attraction of phase w to the solid relative to phase n is greatest.

Entrapment of phase n by the bypassing mechanism is described using the conceptual model referred to as a *pore doublet*. A pore doublet is a flow channel that splits into two pores of different geometry and then rejoins. Figure 3.6 illustrates three scenarios for displacement of a nonwetting phase by a wetting phase for different geometry pore doublets. A pressure differential is imposed such that flow occurs into the region fully occupied by phase n in each case. For case (a), the upper branch has a smaller diameter than the lower branch; but both branches have a

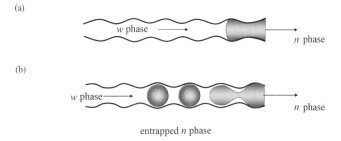

Figure 3.5: Illustration of (a) low and (b) high aspect ratio pores and the effect of aspect ratio in causing snap-off of phase n during imbibition (after [43]).

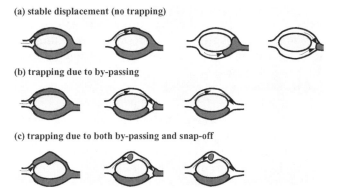

Figure 3.6: Displacement scenarios of a nonwetting fluid (dark grey) by a wetting fluid (white) in different geometry pore doublets (after [43]).

small aspect ratio. Having a smaller diameter, the top pore drains first to the junction where its radius gets slightly larger than the bottom pore. The bottom pore then drains. Since the exit diameters of the flow regions are similar at the downstream node, the interface is stable and the nonwetting phase drains completely from the doublet.

The geometry for case (b) is similar to that of case (a) except that the downstream pore diameter of the top branch constricts to a very small pore. Therefore, when the top pore drainage interface approaches the downstream junction, it continues to flow through the node leaving phase n fluid trapped in the bottom pore and hydraulically disconnected from any flowing fluid n. Flow in the system bypasses the lower pore. Case (c) is similar to case (b) except now the upper pore, also has a high aspect ratio. Therefore, in addition to bypassing in the lower pore, snap-off occurs in the upper pore, further isolating some of phase n.

While the amount of wetting fluid remaining in a porous medium subjected to drainage is called the *irreducible saturation*, the total amount of nonwetting phase that becomes entrapped as a result of imbibition is called the *residual saturation* and is denoted s_r^n. The magnitude of the residual saturation is a function of the following factors [43]:

- the geometry of the pore network;
- the properties of the fluids, primarily interfacial tension, density ratio, and viscosity ratio;
- the applied potential gradient causing the displacement process.

The effect that these factors have on the displacement process can be incorporated into two dimensionless quantities, the *capillary number*, N_c, and the *Bond number*, N_b defined, respectively, as:

$$N_c = \frac{|\mathbf{q}^w|\mu^w}{\gamma^{wn}} \tag{3.138}$$

$$N_b = \frac{k^s(\rho^w - \rho^n)g}{\gamma^{wn}} \tag{3.139}$$

where N_c is the ratio of viscous to capillary forces, and N_b is the ratio of gravitational to capillary forces. For $N_c \lesssim 10^{-4}$ capillary forces dominate over viscous forces and the trapping phenomenon is enhanced. A residual nonwetting phase saturation can be reduced by increasing N_c, for example by increasing $|\mathbf{q}^w|$ or decreasing γ^{wn} through the use of a surfactant.

Another factor that affects the magnitude of entrapped phase n is related to what is called *incomplete displacement*. In general, during drainage, the entering phase n tends to fill the larger pores first. When s^n becomes large, a wider range of pore types will become filled thus providing opportunity for entrapment during subsequent imbibition of phase w. An empirical model to quantify this process has been developed to estimate s_r^n as a function of flow history [31]. This model is based on the assumption that the maximum possible residual saturation, $s_{r\max}^n$, will result if the initial condition of the imbibition process is $s^n = 1$. If the maximum value of s^n achieved in the porous medium, $s_{\max}^n$, is less than 1, the value of s_r^n resulting from imbibition will also be less than $s_{r\max}^n$. Correlation of experimental data for sand led to the relation:

$$\frac{1}{s_r^n} = \frac{1}{s_{\max}^n} + \frac{1}{s_{r\max}^n} - 1 \tag{3.140}$$

The preceding observations provide part of the motivation for the definition of *effective saturation* of the wetting phase, s_e^w, as:

$$s_e^w = \frac{s^w - s_i^w}{1 - s_i^w} \tag{3.141}$$

In this expression, the numerator is saturation reduced to the part contributing to flow, and the denominator is the maximum value of the saturation that contributes to flow. With this definition, a drainage experiment will start with a fully saturated porous medium such that $s^w = s_e^w = 1$. The drainage process, if run to completion, will lead to $s^w = s_i^w$ such that $s_e^w = 0$. Subsequent imbibition will not return the system to a state where $s^w = 1$ because of trapping of phase n in the imbibition process.

The observations and definitions presented in this section provide bases for proposing forms of the p^c-s^w relationship. In the next subsection, we present some of the most commonly employed formulas.

3.7.3 Formulas for the p^c-s^w Relationship

The most common example of two-phase immiscible flow in a natural system is the case of water and air flow in the part of the soil column above the water table. This region is called the *vadose zone* (also called the *zone of aeration* or *unsaturated zone*). In the vadose zone, $s^w < 1$. Because this problem has been widely studied due to its importance for agriculture, much of the work for developing p^c-s^w relations is based in studies of the unsaturated zone. The discussion presented here will also deal explicitly with this case. The discussion, in general, is readily adaptable to any two-phase immiscible fluid system. The correlations developed that express p^c-s^w are particular to an air-water system. Some suggestions as to how these correlations might be scaled to systems containing other fluids are provided at the end of the subsection.

Figure 3.7 provides schematic figures of the expected p^c-s^w relationship for two different soil samples. The curves presented are envisioned as being obtained by incremental draining of a saturated soil which is allowed to equilibrate between each incremental change. The sketched curves correspond to p^c vs. s^w values that would be encountered in working with the soil.

The figure illustrates an entry pressure, an increase in the capillary pressure that must be imposed to allow phase n to begin to enter the soil sample. The difference in values of entry pressure corresponds to larger pores for the well-graded soil. For the well-sorted soil, a soil that has a narrow range of grain sizes and an implied narrow range of pore sizes, the p^c-s^w curve is relatively flat signifying that drainage occurs rather sharply over a narrow range of p^c. The broad distribution of grain sizes

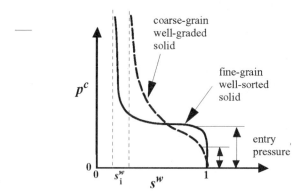

Figure 3.7: Sketches of expected forms of the p^c-s^w relationship that would be encountered for a drainage experiment in an initially saturated soil.

for the graded soil yields a mild transition from saturated to unsaturated conditions since drainage occurs over a greater range of capillary pressure. Furthermore, the graded soil contains more small pores and locations in which the wetting fluid can reside such that the irreducible saturation in that soil is greater.

Besides providing typical p^c-s^w curve shapes, Figure 3.7 also provides the vertical distribution of saturation in a soil column following drainage. This is due to the fact that the air pressure can be considered constant (atmospheric) throughout the soil column. The water phase pressure will be hydrostatic and a linear function of elevation. For a system at equilibrium, the wetting phase hydraulic head is constant and each change in elevation head will be balanced by a change in pressure head. Since the air pressure head is constant, and p^w is a linear function of elevation, $p^c = p^n - p^w$ must also be a linear function of elevation. Thus a plot of s^w vs. p^c, in this instance, is equivalent to a scaled plot of s^w vs. elevation and is called a *moisture retention curve*. As the elevation increases, p^c also increases corresponding to an increasing curvature of the *wn* interface from the microscale perspective, and a decreasing s^w from the macroscale perspective. The key factor in the p^c-s^w plot providing the saturation profile is the assumption that the air phase pressure is constant. For a more general case involving two fluids of comparable density, such an assumption would not apply. However, at equilibrium, if both of these fluid phases were connected, hydrostatic conditions would apply in each phase such that a s^w profile could be derived from the p^c-s^w relation.

For an air-water system, some particular definitions apply. With respect to Figure 3.7, the location where $p^c = 0$ such that phase w pressure is atmospheric is called the *phreatic surface* or *water table*. Water below this region is not bound in capillaries and is free to flow. In dealing with air-water systems, it is common to designate the air phase pressure as being 0, its gauge value, such that $p^c = -p^w$.[8] Immediately above the water table is the transition zone between the saturated and unsaturated soil region where $s^w = 1$, but $p^c > 0$. This region is called the *capillary fringe*. The thickness of the capillary fringe is equal to the entry pressure divided by $p^w g$.

---

[8] The capillary pressure is a positive quantity because $p^w < p^n = 0$ in the unsaturated zone. Note that p^w is subatmospheric in the unsaturated region and therefore is negative relative to the gauge atmospheric pressure.

Each of the moisture retention curves depicted in Figure 3.7 is referred to as a *primary drainage curve* (PDC) if the porous medium under consideration is homogeneous. The PDC represents the p^c-s^w relation obtained from draining a porous solid in a number of small, sequential, equilibrium steps beginning with a fully saturated system, $s^w = 1$ and proceeding to the irreducible saturation s_i^w. Since the moisture retention curve is a function of soil texture and structure, one needs to develop a primary drainage curve for each texturally and structurally similar soil type. For example, if the soil column in the vadose zone were made up of five different soil horizons, then a primary drainage curve would have to be determined for each. Even with this additional information, the primary drainage curve tells an incomplete story because the impact of residual saturation, s_r^w, is not included; nothing has been said about the curve that results during imbibition; and the hysteretic effects discussed previously have not been taken into account. The full representation of p^c-s^w must be developed experimentally.

A PDC for an air-water system is measured by placing a fully saturated soil sample in a cell such that it is confined on the sides and is in contact with air on the top and with water, through a porous plate, on the bottom. The height of the sample is small enough that gravitational effects may be ignored. The water pressure is decreased by sequential increments while the air phase pressure remains constant (i.e., the capillary pressure is increased incrementally). After each incremental change, the system is allowed to equilibrate and s^w is measured. The result is a set of p^c-s^w data points that can be plotted and interpolated to form the PDC as in Figure 3.8. The interpolation relation is then a quantitative model of primary drainage.

Experiments may also be performed starting with an initial condition other than $s^w = 1$ to obtain additional drainage and scanning curves that illustrate the hysteretic behavior expected based on the capillary tube example in Figure 3.4 and the subsequent discussion. Figure 3.8 provides a representative illustration of the hysteretic saturation-pressure curve-type relationship for the case when both capillary and entrapment effects are included. The nomenclature for the curve-type name describes whether the flow path is draining (D) or imbibing (I) with respect to the wetting phase, and the process that produced the curve. The curves illustrated may be summarized:

- *Primary*: a curve which begins with only one phase present in the pore space (i.e., the initial $s^w = 1$ for PDC while the initial $s^w = 0$ for PIC);
- *Main*: a curve which begins at a saturation for which only one phase is mobile (i.e., the initial values of saturation that apply are: $1 - s_r^n < s^w < 1$ for MDC*, $s^w = 1 - s_r^n$ for MDC, $0 < s^w < s_i^w$ for MIC*, and $s^w = s_i^w$ for MIC);
- *Scanning*: a curve that begins at a value of saturation for which both phases are mobile (i.e., $s_i^w < s^w < 1 - s_r^n$ for both SDC and SIC).

These curve types may be examined with respect to both capillary effects and entrapment.[9]

---

[9] It is important to remember, but easy to overlook, the fact that all curves are created by interpolating equilibrium data obtained from a sequence of incremental changes in p^c followed by allowing the system to reach a new equilibrium state.

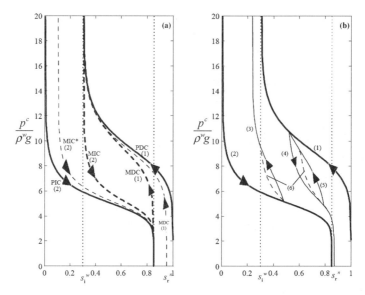

Figure 3.8: Definition plot of the hysteretic relationship between saturation and capillary pressure. Curve position and shape are governed by the mobility of the phases, the initial phase saturations when the process begins, and whether the wetting phase is draining (D) or imbibing (I). The curve-type numbering is such that odd numbers relate to drainage and even numbers to imbibition. Primary (P) and Main (M) curves are designated as "1" for drainage processes begun with $1 - s_r^n \leq s^w \leq 1$ and "2" for imbibition processes begun with $0 \leq s^w \leq s_i^w$. Scanning (S) curves indicate behavior after a drainage process is reversed (curves "4" and "6") or an imbibition process is reversed (curves "3" and "5") such that the initial saturation of the new process is $s_i^w < s^w < 1 - s_r^n$ [25].

The closed loop made by the combined MDC and MIC curves is called the *bounding loop* and it can be retraced over and over again by cyclically imbibing until $s^w = 1 - s_r^n$ and draining until $s^w = s_i^w$. Cyclic changes in operating strategies often lead to closed and reproducible hysteretic loops such as the scanning loop formed by SDC 5 in conjunction with 4 in part (b) of the figure. Finally, note that scanning curve 4 indicates that phase n being displaced will reduce to a residual saturation less than the residual indicated on the bounding loop. This attribute can be checked against the Land trapping model equation (3.140) for phase w with $s_{\max}^w$ occurring at the intersection of curves 1 and 4, $s_{r\max}^w$ being the value of s^w when the MIC intersects with $p^c = 0$, and s_r^w being the value where curve 4 intersects with $p^c = 0$.

Based on this discussion of the six different curves, it is easy to see that the laboratory work required to fully characterize the p^c-s^w relation for a soil could be daunting. Besides having to fill in the drainage and scanning curves within the bounding loop, one must wait for the system to equilibrate before taking any measurements. The time commitment alone to characterize a single soil could be weeks or months, depending on the level of detail required. Therefore, based on the observation that some correlation between soil texture and structure with the hydraulic properties of soils exists, one might ask the question, "Is there an empirical relationship that can represent the primary drainage curve or other aspects of the p^c-s^w relation using routinely measured soil data such as a grain-size distribution, bulk density, and grain density?"

The answer is a qualified, "Yes," provided the pore-size distribution of the soil can be determined and one focuses on the PDC. Consider the following statistical model of a porous medium that relies on the assumption that the soil can be characterized by one variable, the *pore-size distribution function*. From this perspective, a porous medium is assumed to be composed of a set of randomly distributed interconnected pores analogous to a bundle of capillary tubes of various radii, that are cut into many thin slices and randomly reassembled. The pores are characterized by a length scale, typically taken to be the *pore radius, r*. The pores are described in statistical terms by the pore-size distribution function, $f(r)$, where $f(r)dr$ is the ratio of the volume of pores with radii between r and $r + dr$ to the total pore volume. This distribution function satisfies the constraint:

$$\int_0^\infty f(r)\,dr = 1 \tag{3.142}$$

which states that the sum of pore volume fractions over pores of radii ranging from $0 \leq r \leq \infty$ is 1. Equations (1.36) and (1.62) can be combined to obtain the pore radius that supports a particular value of microscale capillary pressure as:

$$r_c = \frac{2}{p_c}\gamma_{wn}\cos\theta \tag{3.143}$$

where θ is the contact angle. We then make the assumption that all pores larger than r_c will be filled with phase n and that the ideas supporting equation (3.143) can be extended by the assumption that microscale capillary pressure is approximately equal to macroscale capillary pressure, $p_c \approx p^c$, such that $r_c = r_c(p^c)$.

These definitions allow us to relate the effective saturation, s_e^w, defined in equation (3.141) to r_c by:

$$s_e^w(r_c) = \frac{\int_{r_{\min}}^{r_c} f(r)\,dr}{\int_{r_{\min}}^{r_{\max}} f(r)\,dr} \tag{3.144}$$

where $r_{\max}$ and $r_{\min}$ are the maximum and minimum pore radii contributing to flow, respectively. The combination of equations (3.144) and (3.143) with the assumption that macroscale capillary pressure can be used without loss of accuracy establishes the relationship between pore-size distribution, $f(r)$, and $s_e^w(p^c)$ since p^c has been inversely related to r_c.

In natural soils a similarity between the cumulative grain-size distribution curve and the moisture retention curve is often observed [3, 2]. This situation can be used as a basis for developing an empirical model that translates the grain-size distribution curve into an equivalent pore-size distribution model. This method produces qualitatively good results for relatively coarse-grained soils.

Despite efforts to put the derivation of p^c-s^w relations on firm theoretical footing, complex pore-geometry and adsorption effects make prediction of the moisture retention properties from basic soil properties very difficult. An alternative approach is to simply correlate measured p^c-s^w data with a parametric model that seems to have potential for fitting the data well. These *parametric models* have the key attribute that they are continuous over the span of saturation and can therefore be

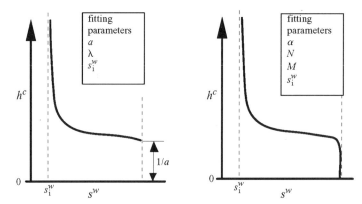

Figure 3.9: Schematic diagram of soil moisture retention curves as predicted by the Brooks-Corey model (left) and the van Genuchten model (right).

differentiated or used in other equations to develop other properties of interest.[10] While many such parametric models have been proposed, two stand out as being the most popular for application to air-water systems: the models of Brooks and Corey [7] and of van Genuchten [42]. These relations are typically written in terms of the capillary pressure head, $h^c = p^c/\rho^w g$.

The Brooks-Corey relationship [7] may be written:

$$s_e^w = \begin{cases} (ah^c)^{-\lambda} & ah^c > 1 \\ 1 & ah^c \leq 1 \end{cases} \tag{3.145}$$

The parameter a [$1/L$] is a fitting parameter equivalent to the inverse of the entry pressure head, and λ is a dimensionless fitting parameter related to the soil pore-size distribution that affects the slope of the retention curve. Buried in the definition of the effective saturation, equation (3.141), is specification of the irreducible saturation, s_i^w. In fact, s_i^w is difficult to measure since water continues to drain, even if extremely slowly, at very low saturations. As a result, when using equation (3.145), s_i^w is actually treated as a fitting parameter without much physical meaning. An example of the function defined by equation (3.145) is shown in Figure 3.9. Equation (3.145), with its three fitting parameters, a, λ, and s_i^w, has been shown to provide accurate results for relatively coarse-textured soils characterized by narrow grain- (or pore-) size distributions (large values of λ and a well-defined entry pressure). The model is less accurate for fine textured and undisturbed field soils, as these soils do not have a well-defined entry pressure.

The van Genuchten model [42] has one parameter in addition to the Brooks-Corey model and takes the form:

$$s_e^w = \left[1 + (\alpha h^c)^N \right]^{-M} \tag{3.146}$$

---

[10]Subsequently, we will see that the link between a parametric p^c-s^w model and the pore-size distribution model allows one to generate a k_{rel}^α-s^w model.

where α [$1/L$] is related to the inverse entry pressure head, N is a dimensionless parameter related to the pore-size distribution, and M is usually expressed in terms of N, e.g., $M = 1 - 1/N$. As with the Brooks-Corey model, s_i^w is treated as a fourth fitting parameter. An example of the function defined by equation (3.146) is shown in Figure 3.9. In general, equation (3.146) has a broader range of applicability than equation (3.145) because it has one additional fitting parameter, and because it is continuously differentiable over the span of saturations.

Thus far we have considered the relationship for saturation-pressure resulting from a monotonic drainage event, specifically primary drainage. For purposes of modeling, we need an extended p^c-s^w relation that is capable of modeling curves 2 through 6, as well as curve 1 in Figure 3.8. Such a model can be proposed by making empirical changes to a parametric form that has proven useful for primary drainage. Unfortunately there has been no consensus on the best procedure for obtaining these additional curves. Thus, rather than presenting a single form, we note that the search for effective rules for modeling hysteresis is an active and important research area.

Once a p^c-s^w relation has been determined, it would be useful to be able to apply that information to a system composed of different fluid and different solid material. Ideally, this can be done through some scaling of the data; however a consistent and workable approach to scaling the data must be uncovered. Here we present a scaling procedure recognizing that we seek a compromise between rigorous incorporation of all effects and simplicity in an equation that can be employed with reasonable accuracy.

We begin our hunt for a scaling relation with Darcy equation (3.54) for single fluid flow for an isotropic system:

$$\varepsilon\left(\mathbf{v}^w - \mathbf{v}^s\right) = -\frac{\rho^w g k^s}{\mu^w} \nabla h^w \tag{3.147}$$

where the Darcy velocity has been expressed explicitly in terms of the fluid and solid velocities. Take the magnitude of this vector equation and rearrange to:

$$\frac{k^s}{\varepsilon} = \frac{\mu^w \left|\mathbf{v}^w - \mathbf{v}^s\right|}{\rho^w g \left|\nabla h^w\right|} \tag{3.148}$$

This equation has the interesting feature that the quantities on the left depend on the solid matrix while the quantities on the right involve fluid properties and fluid flow in response to a gradient in fluid head. This observation suggests that k^s/ε accounts for all the solid phase properties and is a good candidate parameter to include in a scaling operation.

As a next step, recall that equation (3.50) gives a relation for k^s as a function of pore diameter. We invoke this equation to obtain:

$$\frac{k^s}{\varepsilon} = \frac{C}{\varepsilon \tau} D_p^2 \tag{3.149}$$

Continuing, we turn to equation (3.143), the *Laplace equation for capillary pressure*, and rewrite it in terms of pore diameter rather than radius and with the assumption that macroscale capillary pressure and surface tension can be used in place of their microscale counterparts:

$$D_{\mathrm{p}} = \frac{4}{p^c}\gamma^{wn}\cos\theta \tag{3.150}$$

Eliminate D_{p} between the last two equations to obtain:

$$4\left(\frac{C}{\varepsilon\tau}\right)^{1/2}\cos\theta = \frac{p^c}{\gamma^{wn}}\left(\frac{k^s}{\varepsilon}\right)^{1/2} \tag{3.151}$$

The next assumption to be made is that the quantity on the left depends primarily on saturation and is called the *Leverett J-function* such that the scaling equation is [33]:

$$J\left(s^w\right) = \frac{p^c}{\gamma^{wn}}\left(\frac{k^s}{\varepsilon}\right)^{1/2} \tag{3.152}$$

The capillary pressure on the right is a function of s^w and the other quantities are functions of the solid and fluid. Therefore, the magnitude of J is proportional to p^c.

Leverett plotted the retention data for different sands in terms of J vs. s^w and found that the data points fell, essentially, on a single curve. Thus the utility of this scaling method lies in the fact that it works. Certainly many other efforts to obtain good scaling relations have been made and are still being made. Some of these serve no useful purpose and are discarded; others make their way into the literature. The J-function is perhaps the most widely used scaling function.

In theory, use of the J-function to scale capillary pressure allows one to use p^c-s^w data measured for one particular soil type with parameters k_1^s and ε_1 and the fluids w_1 and n_1 with interfacial tension γ_1^{wn}, to predict the p^c-s^w retention curve for other similar soil types and fluid pairs. For example, if the second system is differentiated from the first by use of "2" in the subscripts, equality of the J-function between the two systems provides capillary pressure as a function of saturation for the second system as:

$$p_2^c\left(s^w\right) = p_1^c\left(s^w\right)\frac{\gamma_2^{wn}}{\gamma_1^{wn}}\left(\frac{k_1^s}{k_2^s}\right)^{1/2}\left(\frac{\varepsilon_2}{\varepsilon_1}\right)^{1/2} \tag{3.153}$$

Capillary pressure scaling is very useful when data is lacking, for example when modeling flow in a heterogeneous soil where only a few of the soil types have been tested. It is also useful when one wants to model capillary pressure dependence on fluid-phase composition by including the appropriate constitutive model for γ^{wn}.

The discussion of retention curves indicates some of the challenges that must be faced in obtaining p^c-s^w relations needed to facilitate use of equation (3.134) for

modeling multiphase systems. The methods employed to fit available data are somewhat heuristic. Dealing with the complexities of hysteresis remains a challenge, both
in obtaining data sets and in using the data in numerical simulations. Unfortunately,
we are also in need of information concerning the relative permeabiliity, k_{rel}^{α}, if we
are to model a system. The subsequent section explores this problem.

3.7.4 Observations of the k_{rel}^{α}-s^w Relationship

The relative permeability defined in equation (3.132) is a scaling factor, $0 \leq k_{rel}^{\alpha} \leq 1$,
which accounts for the fact that the pore space is not entirely filled with the α phase
in multiphase flow. Thus the permeability of the medium to the α phase is reduced
from k^s when $s^{\alpha} < 1$. This definition assumes that k_{rel}^{α} is a function of saturation
only.

Since k_{rel}^{α} is a scaling factor for Darcy's law, it is not surprising that measurements
of this quantity are based on an extension of Darcy's experiment to multiple phases.
The simplest method, though not the fastest, for determining relative permeabilities
is a steady state procedure (e.g., described in [1]). A homogeneous, isotropic soil
sample is placed in an apparatus of length L and cross-sectional area A so that flow
through the sample can be facilitated. Initially, the soil is saturated with phase w.
Then, flow through the sample is induced by imposing a head difference across the
sample. After the system reaches steady state, the intrinsic permeability can be calculated from Darcy's law based on the measured volumetric flow rate Q^w and head
difference Δh^w as:

$$k^s = \frac{Q^w \mu^w L}{\rho^w g A |\Delta h^w|} \tag{3.154}$$

The inflow is then changed so that it contains phase w plus a small amount of phase
n. Again the system is allowed to reach a steady state such that the fraction of
wetting phase in both the inflow and outflow streams is the same. The constant value
of s^w in the column is measured. With constant saturation, the capillary pressure is
constant and the pressure drop across the flow channel for each phase will be the
same (i.e., $\rho^w g \Delta h^w = \rho^n g \Delta h^n = \Delta p^w = \Delta p^n$). Writing Darcy's law for each phase, we
obtain:

$$k_{rel}^{\alpha} = \frac{Q^{\alpha} \mu^{\alpha} L}{k^s \rho^{\alpha} g A |\Delta h^{\alpha}|} \quad \alpha = w, n \tag{3.155}$$

which provides the relative permeability for the specified saturation. Sequential
incremental adjustments can be made to the proportional mixture of phases n and
w to obtain values of relative permeability at equilibrium along a drainage cycle
and then for an imbibition event to see if hysteretic effects have to be accounted
for.

Figure 3.10 shows typical relative-permeability curves that would be obtained for
a pair of immiscible fluids. From this figure several qualitative attributes regarding
the behavior of the relative permeability can be identified [5, 4]:

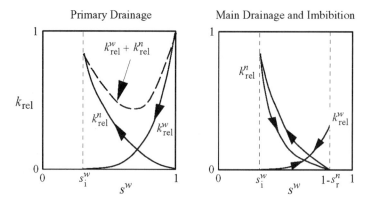

Primary Drainage   Main Drainage and Imbibition

Figure 3.10: Typical relative permeability curves for drainage and imbibition.

1. When both phases are present the sum of the relative permeabilities is usually less than 1. This phenomenon can be attributed to several factors: the presence of static *wn* interfaces that block flow channels which may otherwise contribute to flow; the pathways that each fluid takes are more tortuous than for the single fluid case; the fluids interfere with each other as they move due to differences in viscosity and fluid-solid interfacial friction.

2. The relative permeabilities of the two phases are not symmetric such that at a given saturation s^w:

$$k_{rel}^n\big|_{1-s^w} > k_{rel}^w\big|_{s^w} \qquad (3.156)$$

Asymmetry occurs because the wetting phase preferentially occupies the small pores. This behavior also explains why, as s^w decreases from 1 during primary drainage, k_{rel}^w is characterized by a rapid decline while k_{rel}^n is characterized by a rapid increase, and:

$$k_{rel}^n\big|_{s_i^w} > k_{rel}^w\big|_{1-s_r^n} \qquad (3.157)$$

3. The relative permeability of a phase goes to zero for values of saturation greater than zero. This coincides with the definitions of irreducible saturation for phase *w* and residual saturation for phase *n*. As a phase loses hydraulic connection, its permeability goes to zero.

4. The relative permeability for the nonwetting phase (*n*) exhibits hysteresis while hysteresis in the wetting phase (*w*) relative permeability is negligible. The lack of hysteresis in k_{rel}^w implies that phase *w* films left behind during drainage allow former flow channels to be efficiently re-established when s^w is subsequently increased. The fact that k_{rel}^n is higher for phase *w* drainage than for phase *w* imbibition implies that much of phase *n* trapping and re-connection processes occur soon after a flow reversal.

To the preceding comments relating to the features of the k_{rel}^α-s^w curves, we add a series of items relating to the fact that although k_{rel}^α may be predominately a function of saturation, other factors can be significant as well. These factors include fluid properties, soil properties, and the system forcing that causes flow [5, 4]. We note, in particular, the following properties of importance:

1. *Viscosity Ratio, μ^w/μ^n:* There are indications that a transfer of viscous forces across the *wn* interface in the pore may occur where the magnitude of the force depends on the viscosity ratio. When phase *w* coats the solid grains and phase *n* is thus in contact only with fluid *w*, phase *n* velocity at the *wn* interface may not be zero (a lubricant effect).

2. *Interfacial Tension, γ^{wn}, and the Contact Angle between the w and s Phases, θ:* These parameters together impact the magnitude of the capillary pressure, and the contact angle determines phase wettability (see Section 1.6). The contact angle θ varies from $0°$ (strongly wetting) to $90°$ (neutral wetting) depending on the interaction forces among the phases. Phase distribution with respect to pore class will correspondingly vary from all the wetting fluid being in the small pores and the nonwetting fluid in the large pores at $\theta = 0°$, to no preferred distribution whatsoever when $\theta = 90°$.

3. *Correlation between Pore-Size Distribution and Relative Permeability:* This implies that the model describing the k_{rel}^{α}-s^w relation should be influenced by the p^c-s^w relation since, in theory, the p^c-s^w relation reflects pore-size distribution. In fact, as will be seen, when k_{rel}^{α}-s^w data is unavailable, the k_{rel}^{α}-s^w model is predicted from the p^c-s^w model.

4. *Pressure Gradients:* Experimental evidence suggests that values of k_{rel}^n will be higher at higher flow rates. This may occur because an increased flow rate increases the ratio of viscous to capillary forces (i.e., increases the capillary number) such that flow-path segregation due to capillarity is minimized resulting in less tortuous flow paths.

This subsection is intended both to give an overview of the features of relative permeability for porous media flow of two immiscible fluids and to identify some of the complications that make use of relative permeabilities in field situations challenging. Anisotropic systems have not been considered. For modeling an inhomogeneous system, relative permeability curves have to be generated for each type of solid. When three fluid phases are present (for example with gas, water, and oil), three relative permeabilities are required, data sets relating capillary pressures for each of the three fluid pairs as a function of phase saturations are needed, and the ancillary variables that are neglected may also add errors. In addition, if there is some miscibility of the phases (e.g., small amounts of an organic contaminant that might dissolve in water), the compositional change of the phases will alter curves employed assuming complete immiscibility. Furthermore, we should not forget that the p^c-s^w curves are generated for equilibrium systems with no flow. System dynamics can alter these curves. Many outstanding issues concerning multiphase flow require additional research, and it is important to be cognizant of model approximations as possible sources of simulation errors. However, when applied to systems that satisfy the approximations on which they are based, relative permeability relationships are useful and important modeling tools. In the next subsection, we provide a brief overview of some of the formulas used to correlate k_{rel}^{α} data for flow of immiscible wetting and nonwetting phases in a porous medium.

3.7.5 Formulas for the $k_{\mathrm{rel}}^{\alpha}$-$s^w$ Relation

Visual inspection of Figure 3.10 suggests that a useful model of relative permeability might be based on a power law form. Note that the relative permeabilities are greater than zero only for $s^w \geq s_i^w$, and recall that the relative permeability satisfies the constraint $0 \leq k_{\mathrm{rel}}^{\alpha} \leq 1$. Thus we propose for the wetting phase:

$$k_{\mathrm{rel}}^{w} = \left(\frac{s^w - s_i^w}{1 - s_i^w} \right)^{m} \qquad s_i^w \leq s^w \leq 1 \tag{3.158}$$

The quantity in parentheses was defined in equation (3.141) as the effective saturation, s_e^w, so it is often convenient to re-express this equation in the equivalent form:

$$k_{\mathrm{rel}}^{w} = \left(s_e^w \right)^{m} \qquad 0 \leq s_e^w \leq 1 \tag{3.159}$$

For the nonwetting phase, even if the hysteresis loop is neglected, the behavior of $k_{\mathrm{rel}}^{\alpha}$ for $s^w \geq 1 - s_r^n$ depends on whether primary drainage or main drainage/imbibition is occurring. Therefore we propose two parts to the power law model for k_{rel}^{w}. First, for the primary drainage curve:

$$k_{\mathrm{rel}}^{n} \big|_{\mathrm{primary}} = k_{\mathrm{maxp}} \left(\frac{1 - s^w}{1 - s_i^w} \right)^{m_p} \qquad s_i^w \leq s^w \leq 1 \tag{3.160}$$

For the secondary drainage curve:

$$k_{\mathrm{rel}}^{n} \big|_{\mathrm{main}} = k_{\mathrm{maxm}} \left(\frac{1 - s_r^n - s^w}{1 - s_r^n - s_i^w} \right)^{m_m} \qquad s_i^w \leq s^w \leq 1 - s_r^n \tag{3.161}$$

In these equations, the exponents m, m_p, and m_m are obtained from fitting the data while the coefficients k_{maxp} and k_{maxm} are the maximum values of relative permeability obtained when $s^w = s_i^w$ on the respective curves. Experimental results [36] indicate that for a system composed of water and air, m is generally positive in granular porous media but is negative for unstructured soils of fine texture.

In addition to simply fitting relative permeability data with particular functions, some effort has been expended to find ways to predict relative permeabilities. The methods range from (1) characterizing a porous medium as a bundle of capillary tubes of different radii and examining the flow through these tubes algebraically based on Darcy's law and considerations of capillary pressure to (2) statistical models that make use of the pore-size distribution function developed in equation (3.142) for use in deriving p^c-s^w models. The statistical approach has been used widely for air-water flow in porous media. Here, we will outline one of the most common statistical approaches to this problem [35].

The model first conceptualizes the porous medium as a collection of interconnected cylindrical pores with radii ranging in size from r_{min} to r_{max} and lengths proportional to their radii. Laminar Poiseuille flow is considered to be occurring in each pore such that it can be described by the parabolic flow profile in a tube. An assumption is made about the correlation of pore sizes as a function of the distance of separation of the pores, and a factor is introduced to account for the change in correlation due to the effects of saturation and for tortuosity. With these assumptions, the relative permeability for the water phase is obtained as:

$$k_{rel}^w\left(s_e^w\right) = \left(s_e^w\right)^\kappa \left[\frac{\int_{r_{min}}^r r' f(r') \, dr'}{\int_{r_{min}}^{r_{max}} r' f(r') \, dr'}\right]^2 \tag{3.162}$$

where κ is a pore connectivity parameter related to tortuosity that may be positive or negative. Based on the definition of f, $f(r)dr = ds^w$. Also, since $p^c = -\rho^w g h^c$ is inversely proportional to a tube radius, the equation may be rewritten in terms of saturation and capillary head as:

$$k_{rel}^w\left(s_e^w\right) = \left(s_e^w\right)^\kappa \left[\frac{\int_0^{s_e^w} \dfrac{1}{h^c(s^w)} \, ds^w}{\int_0^1 \dfrac{1}{h^c(s^w)} \, ds^w}\right]^2 \tag{3.163}$$

By an analogous development as used to derive equation (3.163), the relative permeability of the nonwetting phase can be obtained as [7]:

$$k_{rel}^n\left(s_e^w\right) = \left(1 - s_e^w\right)^\varsigma \left[\frac{\int_{s_e^w}^1 \dfrac{1}{h^c(s^w)} \, ds^w}{\int_0^1 \dfrac{1}{h^c(s^w)} \, ds^w}\right]^2 \tag{3.164}$$

where ς is a pore connectivity parameter for the nonwetting phase, and the integration in the numerator is over all the pores whose radii are large enough that they are filled with the nonwetting phase.

At this point, if $h^c(s^w)$ is given in appropriate analytical form, then a closed-form expression for the relative permeabilities may be obtained from equations (3.163) and (3.164). For example, introduction of the Brooks-Corey p^c-s^w model of equation (3.145) gives:

$$k_{rel}^w\left(s_e^w\right) = \left(s_e^w\right)^{\kappa+2+2/\lambda} \tag{3.165}$$

$$k_{rel}^w\left(s_e^w\right) = \left(1 - s_e^w\right)^\varsigma \left[1 - \left(s_e^w\right)^{2+2/\lambda}\right] \tag{3.166}$$

Note that equation (3.165) is a power law expression for the wetting phase relative permeability and is identical to equation (3.159) when $m = \kappa + 2 + 2/\lambda$.

Introduction of the van Genuchten p^c-s^w model as given by equation (3.146) into equations (3.163) and (3.164) with the restriction that $N = 1/(1 - M)$ gives:

$$k_{rel}^w\left(s_e^w\right) = \left(s_e^w\right)^\kappa \left[1 - \left(1 - s_e^{w1/M}\right)^M\right]^2 \tag{3.167}$$

$$k_{rel}^n\left(s_e^w\right) = \left(1 - s_e^w\right)^\varsigma \left[1 - s_e^{w1/M}\right]^{2M} \tag{3.168}$$

Two important issues that limit the utility of obtaining relative permeability functions from pore-size distributions and p^c-s^w relations [11] are:

1. Derived expressions are valid only for s^w less than about 0.85, because the water-retention data does not reflect the pore-size distribution at high saturations since phase n is blocked from moving into all parts of the medium during drainage at high saturations.

2. The relative permeability coefficients were derived for cases of an isotropic medium. The scalar relative permeability does not account for anisotropy since the moisture-retention data does not reflect the directional properties of the medium. This issue was discussed at the end of Subsection 3.7.1.

The two pore connectivity fitting parameters, κ and ς, must be selected when fitting experimental data to the models of relative permeability presented above. These parameters reflect the impact of both the correlation between pore sizes and the tortuosity of the flow path. Based on the results obtained from water displacement of experiments for over fifty soils, Mualem [35] concluded that $\kappa = 0.5$ provides the best fit of the data, and this value should be used as a first guess when fitting wetting-phase data. In addition, it is reasonable to assume that the tortuous path followed by a nonwetting phase will be somewhat less than that followed by the wetting phase since it does not have an affinity for following the solid boundary or moving into the smallest pores. Therefore, ς is expected to be smaller than κ. For the air-water system, $\varsigma = 0.33$ is proposed as a first guess [39]. Marked deviation of the fitted parameters from these values is an indication that the p^c-s^w model is not an adequate predictor of pore-size distribution. As a result, the remaining parameters in the model can be used as fitting parameters rather than requiring they be obtained from the p^c-s^w model.

We now turn to the fact that hysteresis has not been included in any of the models of relative permeability presented. Recall from previous discussion that hysteresis in the p^c-s^w relationship is caused in part by fluid entrapment effects. Also, inherent in Darcy's law is the requirement that any fraction of a phase isolated from the coherent bulk phase must be considered, hydraulically, as part of the solid matrix because this fraction does not flow but participates in defining the flow channels. This attribute was the motivation for differentiating between free and trapped phase components in equation (3.137) and for defining the effective saturation in equation (3.141).

Hysteresis in the $k_{\mathrm{rel}}^{\alpha}$-$s^w$ relationship is generally accepted as being due, in large part, to fluid entrapment effects [31, 38, 32]. Therefore, if s_{t}^{α} could be quantified and subsequently assigned to the correct pore sizes, hysteresis in the $k_{\mathrm{rel}}^{\alpha}$-$s^w$ relationship should be quantifiable using known theory (for example by using the statistical approach introduced earlier). This process would involve taking out of play the pores where fluid is trapped and thus working with a revised flow channel structure.

A straightforward approximate way to try to include the effect of fluid entrapment on the $k_{\mathrm{rel}}^{\alpha}$-$s^w$ relationship is to redefine the effective saturations as follows:

$$s_{e_r}^{w} = \frac{s^w - s_t^w}{1 - s_t^w - s_t^n} \qquad (3.169)$$

and:

$$s_{er}^n = \frac{s^n - s_t^n}{1 - s_t^w - s_t^n} \tag{3.170}$$

such that $s_{er}^n = 1 - s_{er}^w$. The parameters s_t^w and s_t^w are functions of saturation and flow-path history. They are determined from the hysteretic p^c-s^w model.

For example, consider an air-water displacement experiment whereby a soil sample is initially saturated with water. Incremental increases in h_c followed by relaxation to equilibrium will produce data points on a primary drainage path, PDC in Figure 3.8. As the system drains, air forms a continuous phase, that is, $s^n - s_t^n$ (and $s_t^n = 0$) along the PDC. If part-way along the full PDC the flow were reversed by beginning to lower h_c, a scanning curve would be traced, such as curve 4 in Figure 3.8. At the saturation where the reversal began, let's call it $s_{left(f)}^n$, we know that $s_t^n = 0$. Then from the empirical Land equation (3.140) we know that phase n will be displaced to a residual saturation of s_{r0}^n, and this location at $p^c = 0$ will be the end of the scanning curve. To fill in the estimate of the trapped phase n along the scanning path from the start point to this end point, an empirical linear relation is generally used:

$$s_t^n = s_{r0}^n \frac{s^n - s_{left(f)}^n}{s_{r0}^n - s_{left(f)}^n} \tag{3.171}$$

This procedure, and a corresponding one for trapped wetting phase, is used to predict the trapped saturations in equations (3.169) and (3.170).

The new definitions of effective saturation, s_{er}^α, may now be used directly in the predictive model. For example, with the van Genuchten model of equations (3.167) and (3.168) and with the revised effective saturations, the resulting $k_{rel}^w(s^w)$ and $k_{rel}^n(s^w)$ curves for the data presented in Figure 3.8 are shown graphically in Figures 3.11 and 3.12, respectively.

3.7.6 Special Cases of Multiphase Flow

We return, now, to general multiphase flow equation (3.134) to examine two special cases. As was stated previously, solution of the flow equation requires support in the form of a p^c-s^w relation and the expression of the relative permeability as a function of s^w. Since these have been obtained, we can now formulate some specific problems. For these problems, the transfer of mass between phases is considered negligible such that the exchange terms e_{ws}^w, e_{wn}^w, e_{ns}^n, e_{wn}^n, e_{ws}^s, and e_{ns}^s are all ignored.

The first case involves a two-fluid phase system composed of a wetting and non-wetting phase. For this case, we will assume that the solid grains are coated by the wetting phase. This allows us to state that $p^s = p^w$ (i.e., $\chi = 1$ in the formula $p^s = \chi p^w + (1 - \chi)p^n$) since there is no direct contact between the n and s phases. With this assumption, equation (3.134) for the wetting phase becomes:

$$\varepsilon \rho^w \frac{\partial s^w}{\partial t} + s^w \rho^w [\alpha^b + (1-\varepsilon)\beta^s + \varepsilon \beta^w] \frac{\partial p^w}{\partial t}$$
$$- \nabla \cdot \left[\rho^w \frac{\mathbf{k}^{sw}}{\mu^w} \cdot (\nabla p^w - \rho^w \mathbf{g}) \right] = \sum_{W=1}^{N_W} \rho_W^w Q_W^w \delta(\mathbf{x} - \mathbf{x}^W) \tag{3.172}$$

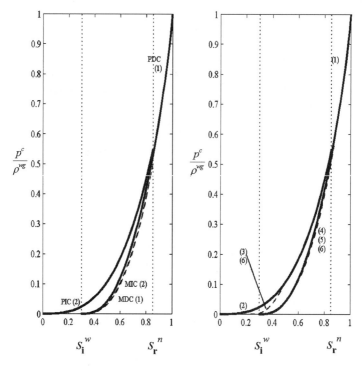

Figure 3.11: Hysteresis in the k_{rel}^w-s^w functionals for the wetting phase as a result of fluid entrapment effects only. The curves were generated using the data defining the p^c-s^w relationship in Figure 3.8. Note that the curves for MDC and MIC, shown as dashed lines, are practically coincident, and that the scanning curves, shown as dashed lines, are group-labeled because for the predictive model chosen they are coincident [25].

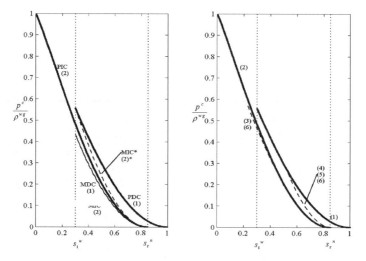

Figure 3.12: Hysteresis in the k_{rel}^n-s^w functionals for the nonwetting phase as a result of fluid entrapment effects only. The curves were generated using the data defining the p^c-s^w relationship in Figure 3.8. Note that the curve labeled MIC* is obtained upon reversal from a PDC where no nonwetting phase was previously trapped. Subsequent reversals follow the MDC and MIC curves which are practically coincident. Also note that the scanning curves are group-labeled because by model definition they are coincident [25].

We also make the assumption that if the product rule is applied to the third term, the gradient in ρ may be neglected (see equation (3.87) for the corresponding assumption with single-phase flow). Assume, also, that the medium is isotropic so that $\mathbf{k}^{sw} = k^s k^w_{rel} \mathbf{I}$. After division by ρ^w we obtain:

$$\varepsilon \frac{\partial s^w}{\partial t} + s^w [\alpha^b + (1-\varepsilon)\beta^s + \varepsilon\beta^w] \frac{\partial p^w}{\partial t} - \nabla \cdot \left[\frac{k^s k^w_{rel}}{\mu^w} (\nabla p^w - \rho^w \mathbf{g}) \right]$$

$$= \sum_{W=1}^{N_W} \frac{\rho^w_W}{\rho^w} Q^w_W \delta(\mathbf{x} - \mathbf{x}^W) \tag{3.173}$$

Application of the same assumptions to equation (3.134) for phase n yields:

$$\varepsilon \frac{\partial s^n}{\partial t} + s^n \left\{ [\alpha^b + (1-\varepsilon)\beta^s] \frac{\partial p^w}{\partial t} + \varepsilon\beta^n \frac{\partial p^n}{\partial t} \right\}$$

$$- \nabla \cdot \left[\frac{k^s k^n_{rel}}{\mu^n} (\nabla p^n - \rho^w \mathbf{g}) \right] = \sum_{W=1}^{N_W} \frac{\rho^n_W}{\rho^n} Q^n_W \delta(\mathbf{x} - \mathbf{x}^W) \tag{3.174}$$

We can use the relation $p^n = p^c + p^w$ to eliminate p^n from this equation and also make use of the fact that $s^n = 1 - s^w$ to eliminate s^n:

$$-\varepsilon \frac{\partial s^w}{\partial t} + (1-s^w) \left\{ [\alpha^b + (1-\varepsilon)\beta^s + \varepsilon\beta^n] \frac{\partial p^w}{\partial t} + \varepsilon\beta^n \frac{\partial p^c}{\partial t} \right\}$$

$$- \nabla \cdot \left[\frac{k^s k^n_{rel}}{\mu^n} (\nabla p^w + \nabla p^c - \rho^w \mathbf{g}) \right] = \sum_{W=1}^{N_W} \frac{\rho^n_W}{\rho^n} Q^n_W \delta(\mathbf{x} - \mathbf{x}^W) \tag{3.175}$$

Equations (3.173) and (3.175) together is called a *mixed formulation* because they are written in terms of derivatives of both p^c and s^w. One can differentiate p^c with respect to s^w to eliminate derivatives of p^c since:

$$\frac{\partial p^c}{\partial t} = \frac{dp^c}{ds^w} \frac{\partial s^w}{\partial t} \tag{3.176}$$

and:

$$\nabla p^c = \frac{dp^c}{ds^w} \nabla s^w \tag{3.177}$$

Substitution of these relations into equation (3.175) would provide the *saturation formulation*. Alternatively, one can eliminate derivatives of s^w by using the following relations:

$$\frac{\partial s^w}{\partial t} = \frac{ds^w}{dp^c} \frac{\partial p^c}{\partial t} \tag{3.178}$$

in equation (3.173) and (3.175), and:

$$\nabla s^w = \frac{ds^w}{dp^c} \nabla p^c \tag{3.179}$$

in equation (3.175). This would result in *capillary pressure formulation*. Having an analytic expression for the p^c-s^w relation, such as the Brooks Corey equation (3.145) or van Genuchten equation (3.146), makes evaluation of dp^c/ds^w easier than if one had to numverically differentiate and interpolate from data points. Regardless of the formulation selected, the equations are too complex and nonlinear to solve analytically. Mathematically, the mixed, saturation, and capillary forms of the equations are equivalent and should provide the same solutions. However, numerical solution introduces errors, and the propagation of those errors is different with the three equation forms. Additionally, the boundary conditions needed to solve this problem are of the same types as discussed for the single phase equation in Subsection 3.6.5. Depending on the problem, it may be easier to express the boundary conditions in terms of p^c, s^w, or some mixture of the two. Initial (for transient problems) and boundary conditions also must be specified for p^w. In any event, it should be obvious that, regardless of the formulation, solving two differential equations in conjunction with the p^c-s^w, k_{rel}^w-s^w, and k_{rel}^n-s^w constitutive equations is a formidable task. Inclusion of features such as hysteresis, inhomogeneity, and anisotropy serve to add to the challenge.

As a second special case, we will consider a natural air-water system used to study water infiltration into a soil in the unsaturated zone. This is classic problem, and we will obtain the classic form of the equations. This case is simplification of the previous system because of one key approximation: the pressure in the air phase is considered to be constant and atmospheric. This approximation ignores the pressure in any trapped air bubbles but indicates that the free air phase in the soil has pressure negligibly different from atmospheric so that the water phase may be modeled without need to explicity model the air phase. With this assumption, equation (3.134) does not need to be solved for the air phase pressure, which is already known. We will make the additional assumptions that the medium is isotropic and that the term involving $\nabla \rho^w$ that arises from applying the product rule is negligible. Also, for convenience, we will assume that there is no pumping. Therefore, equation (3.134) for $\alpha = w$ simplifies to:

$$\varepsilon \frac{\partial s^w}{\partial t} + s^w \left\{ [\alpha^b + (1-\varepsilon)\beta^s] \frac{\partial p^s}{\partial t} + \varepsilon \beta^w \frac{\partial p^w}{\partial t} \right\} - \nabla \cdot \left[\frac{k^s k_{\text{rel}}^w}{\mu^w} (\nabla p^w - \rho^w \mathbf{g}) \right] = 0 \tag{3.180}$$

Now the assumption is made that the fluid and solid compressibilities are negligible in comparison to the matrix compressibility. Quantitatively, this condition is imposed by setting $\beta^s \approx 0$ and $\beta^w \approx 0$. Additionally, the matrix compression is considered to be very small so that:

$$\varepsilon \left| \frac{\partial s^w}{\partial t} \right| \gg s^w \alpha^b \left| \frac{\partial p^s}{\partial t} \right| \tag{3.181}$$

These conditions allow the term in braces to be dropped from equation (3.180). With the change in porosity with time being very slow, ε may be moved inside the time derivative. Also, $\rho^w g$ can be factored out of the last term on the left of equation (3.180) so that we obtain:

$$\frac{\partial(\varepsilon s^w)}{\partial t} - \nabla \cdot \left\{ \frac{k^s \rho^w g k^w_{rel}}{\mu^w} \left[\frac{1}{\rho^w g} \nabla p^w - \frac{\mathbf{g}}{g} \right] \right\} = 0 \qquad (3.182)$$

It is standard notation in soil physics to designate εs^w by θ, the *water content* which is the volume of water per volume of porous medium. The hydraulic conductivity was defined in equation (3.48) as $K^w = k^s \rho^w g / \mu^w$. Equation (3.36) also indicates that the term in brackets is equal to the gradient in the water phase head. With these two notational modifications introduced, we obtain:

$$\frac{\partial \theta}{\partial t} - \nabla \cdot (K^w k^w_{rel} \nabla h^w) = 0 \qquad (3.183)$$

This formula is known as *Richards' equation*.[11] More specifically, it is called the *mixed form of Richards' equation* because there are two primary unknowns, θ and h^w, in this single equation.

An alternative form of the mixed Richards' equation may be obtained as follows. In equation (3.136) we defined the capillary head as $h^c = (p^n - p^w)/(\rho^w g)$. For the current problem, p^n and ρ^w are essentially constant. Therefore:

$$\nabla h^c = -\frac{1}{\rho^w g} \nabla p^w \qquad (3.184)$$

It is common to define the *suction head*, ψ, as the negative of the capillary head (i.e., $\psi = -h^c$). Thus equation (3.182) can be written in an alternative mixed form involving water content and suction:

$$\frac{\partial \theta}{\partial t} - \nabla \cdot \left[K^w k^w_{rel} \left(\nabla \psi - \frac{\mathbf{g}}{g} \right) \right] = 0 \qquad (3.185)$$

By the definitions employed, $p^c = -\rho^w g \psi$ and $s^w = \theta/\varepsilon$. Therefore the functions $p^c(s^w)$ and $k^w_{rel}(s^w)$ can be readily converted to $\psi(\theta)$ and $k^w_{rel}(\theta)$ or $k^w_{rel}(\psi)$, respectively. Making use of these expressions, we can convert the mixed form of Richards' equation into forms with dependence only on θ or dependence only on ψ.

First, we will derive the *water content form of Richards' equation*. This is most easily accomplished by first expanding out the divergence term in the mixed form of equation (3.185):

$$\frac{\partial \theta}{\partial t} - \nabla \cdot (K^w k^w_{rel} \nabla \psi) + \frac{\mathbf{g}}{g} \cdot \nabla (K^w k^w_{rel}) = 0 \qquad (3.186)$$

---

[11] Sometimes, the form of this equation with the spatial derivatives in the horizontal direction ignored is called Richards' equation. This is because the equation is often applied to the case of vertical infiltration over a large surface area such that the lateral movement of the infiltrating water can be neglected.

Since ψ may be expressed as a function only of θ, we can write:

$$K^w k_{\text{rel}}^w \nabla \psi = K^w k_{\text{rel}}^w \frac{\partial \psi}{\partial \theta} \nabla \theta \tag{3.187}$$

Introduce the *soil water diffusivity*, $D^w(\theta)$, defined as:

$$D^w(\theta) = K^w k_{\text{rel}}^w \frac{d\psi}{d\theta} \tag{3.188}$$

so that equation (3.186) is written in the water content form:

$$\frac{\partial \theta}{\partial t} - \nabla \cdot (D^w \nabla \theta) + \frac{\mathbf{g}}{g} \cdot \nabla (K^w k_{\text{rel}}^w) = 0 \tag{3.189}$$

For solution of this equation both D^w and $K^w k_{\text{rel}}^w$ are expressed as functions of θ. Thus, this equation contains a single unknown θ and can be solved subject to boundary conditions and being able to track the nonlinearities in the soil water diffusivity and relative permeability.

Next, we derive the *suction head form of Richards' equation* by eliminating the soil moisture content. We have the function $\psi(\theta)$ or, conversely $\theta(\psi)$. Therefore, we define the *water capacity function, C^w*, as:

$$C^w(\psi) = \frac{d\theta}{d\psi} \tag{3.190}$$

This function is useful for converting the time derivative of θ to a time derivative of ψ according to the equality:

$$\frac{\partial \theta}{\partial t} = C^w \frac{\partial \psi}{\partial t} \tag{3.191}$$

This is substituted into equation (3.186) to obtain the suction head Richards' equation:

$$C^w \frac{\partial \psi}{\partial t} - \nabla \cdot (K^w k_{\text{rel}}^w \nabla \psi) + \frac{\mathbf{g}}{g} \cdot \nabla (K^w k_{\text{rel}}^w) = 0 \tag{3.192}$$

where both C^w and $K^w k_{\text{rel}}^w$ are expressed as functions of ψ. This equation may be solved subject to specification of the appropriate boundary conditions for ψ.

As examples of the functions C^w and D^w we adapt the notation in van Genuchten equation (3.146) to what is used in the Richards equation and obtain:

$$\frac{\theta - \theta_r}{\theta_s - \theta_r} = \left[1 + (\alpha |\psi|)^N \right]^{-M} \tag{3.193}$$

where θ_r is the *residual water content* and θ_s is the *saturated water content*. The left side of this equation is obtained by multiplying both the numerator and the denominator of the expression for effective saturation by the porosity. The derivative $d\theta/d\psi$ can be calculated from equation (3.193) to provide the water capacity:

$$C^w(\psi) = (\theta_s - \theta_r)\alpha^N MN\left[1 + (\alpha|\psi|)^N\right]^{-M-1}|\psi|^{N-1} \tag{3.194}$$

Also, equation (3.193) can be used to obtain ψ as a function of θ:

$$\psi = -\frac{1}{\alpha}\left[\left(\frac{\theta - \theta_r}{\theta_s - \theta_r}\right)^{-1/M} - 1\right]^{1/N} \tag{3.195}$$

Differentiation to obtain $d\psi/d\theta$ allows us to evaluate the soil water diffusivity:

$$D^w(\theta) = \frac{K^w k_{\text{rel}}^w}{\alpha MN(\theta_s - \theta_r)}\left[\left(\frac{\theta - \theta_r}{\theta_s - \theta_r}\right)^{-1/M} - 1\right]^{-1+1/N}\left(\frac{\theta - \theta_r}{\theta_s - \theta_r}\right)^{-1-1/M} \tag{3.196}$$

The complex form of these equations, the fact that they do not account for hysteresis, and the nonlinearities they bring to the governing differential equation point to the fact that solution of Richards' equation, one of the simplest equations describing a case of two fluid phase flow in porous media, is difficult. Significant effort has been made, and continues to be made, to develop numerical algorithms that provide efficient and accurate solutions. Although the alternative forms of the equation are mathematically equivalent, the numerical challenges of the suction form revolve around problems in assuring that mass is conserved; the water content form is difficult to apply at low saturations because as θ approaches θ_r, D^w tends toward infinity. These issues revolve around the general modeling challenges of parameterizing data, obtaining robust constitutive equations, and accurately representing the physics of a problem of interest at an appropriate scale.

3.8   THE BUCKLEY-LEVERETT ANALYSIS

One approach to modeling two fluid phase flow makes use of some simplifying assumptions that lead to what is known as the *Buckley-Leverett equation*. Arguably, the Buckley-Leverett approach is the best known analytical approach to investigation of this topic. The key attribute of this approach is that the problem is formulated in terms of the flow of only one phase, the wetting phase, while the dynamics of the other phase are not totally neglected. The derivation of the Buckley-Leverett form first requires that we consider the fractional contributions to a Darcy velocity by each phase. Subsequently, this concept of fractional flow will be used in the derivation of the Buckley-Leverett equation.

3.8.1   Fractional Flow

Underpinning the Buckley-Leverett equation is a representation of the flow equation for the wetting phase in terms of fractions of each phase velocity contributing to the flow. We will consider the case of the flow of two immiscible fluids, a wetting phase (w) and a nonwetting phase (n). Compressibility effects and solid phase deformation are neglected. For flow in a horizontal column, Darcy's law for the w and n phases is expressed, respectively, as:

$$q^w = -\frac{k^s k_{rel}^w}{\mu^w} \frac{\partial p^w}{\partial x} \tag{3.197}$$

and:

$$q^n = -\frac{k^s k_{rel}^n}{\mu^n} \frac{\partial p^n}{\partial x} \tag{3.198}$$

The wetting phase pressure in equation (3.197) can be eliminated in favor of p^n and the capillary pressure p^c since $p^w = p^n - p^c$. Then we obtain:

$$q^w = -\frac{k^s k_{rel}^w}{\mu^w} \left(\frac{\partial p^n}{\partial x} - \frac{\partial p^c}{\partial x} \right) \tag{3.199}$$

Next we eliminate $\partial p^n / \partial x$ between equations (3.198) and (3.199) and rearrange the result to the form:

$$\frac{1}{k^s} \left(\frac{q^w \mu^w}{k_{rel}^w} - \frac{q^n \mu^n}{k_{rel}^n} \right) = \frac{\partial p^c}{\partial x} \tag{3.200}$$

A combined velocity in the column is given as a simple sum of the Darcy velocities for each phase. Note that this is not a total flux but is merely a mathematical construct. This quantity is designated as q_B and is defined as:

$$q_B = q^w + q^n \tag{3.201}$$

Using this definition, we now introduce the idea of fractional flow of the water phase, f^w, which is defined as:

$$q^w = f^w q_B \tag{3.202}$$

Thus for the nonwetting phase we obtain:

$$q^n = (1 - f^w) q_B \tag{3.203}$$

The quantity f^w is the fraction of the linear sum of the two phase velocities that is due to the water phase. Note that since q_B is not weighted in any way by the volume fractions of the two phases, it has limited physical significance while being mathematically convenient.

We now substitute equations (3.202) and (3.203) into equation (3.200) to obtain:

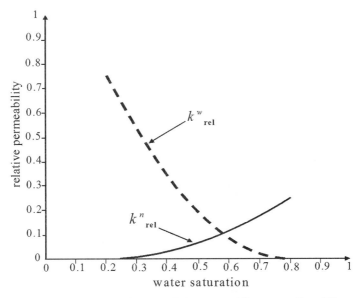

Figure 3.13: Typical oil-water relative permeability curves (from [2]).

$$\frac{q_B}{k^s}\left[\frac{f^w \mu^w}{k^w_{rel}} - \frac{(1-f^w)\mu^n}{k^n_{rel}}\right] = \frac{\partial p^c}{\partial x} \tag{3.204}$$

or, after multiplication by $k^s k^n_{rel}/(q_B \mu^n)$ and some rearrangement:

$$f^w = \left(1 + \frac{k^s k^n_{rel}}{q_B \mu^n}\frac{\partial p^c}{\partial x}\right)\bigg/\left(\frac{\mu^w}{\mu^n}\frac{k^n_{rel}}{k^w_{rel}} + 1\right) \tag{3.205}$$

If the magnitude of the term in the numerator involving the gradient of p^c is much smaller than 1, equation (3.205) reduces to:

$$f^w = 1\bigg/\left(\frac{\mu^w}{\mu^n}\frac{k^n_{rel}}{k^w_{rel}} + 1\right) \tag{3.206}$$

Figure 3.13 shows two typical relative permeability curves for an oil nonwetting phase and water. Figure 3.14 shows the corresponding fractional flow curve for the water phase, a plot of f^w vs. s^w as described by equation (3.206).

3.8.2 Derivation of the Buckley-Leverett Equation

The development of the Buckley-Leverett equation begins with a statement of the conservation of mass for the wetting phase. For the w phase in the horizontal column the macroscale mass balance at any location along the column axis is:

$$\frac{\partial(\varepsilon s^w \rho^w)}{\partial t} + \frac{\partial(\rho^w q^w)}{\partial x} = 0 \tag{3.207}$$

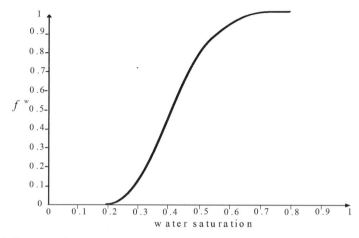

Figure 3.14: Fractional flow curve for the water phase based upon the relative permeability curves shown in Figure 3.13.

If ε, ρ^w, and q_B are considered to be constant, the substitution of equation (3.202) into equation (3.207) yields:

$$\frac{\varepsilon}{q_B}\frac{\partial s^w}{\partial t} + \frac{\partial f^w}{\partial x} = 0 \qquad (3.208)$$

However, since f^w is a function only of s^w, we can use the chain rule in differentiating equation (3.208) to obtain:

$$\frac{\varepsilon}{q_B}\frac{\partial s^w}{\partial t} + \frac{df^w}{ds^w}\frac{\partial s^w}{\partial x} = 0 \qquad (3.209)$$

which is the Buckley-Leverett equation [8].

3.8.3 Solution of the Buckley-Leverett Equation

Before solving the Buckley-Leverett equation, we note that we could follow the movement of the saturation front into the column in any number of ways. If we decide to monitor the position of a particular unchanging value of saturation, we are specifying that we are interested in:

$$\frac{Ds^w}{Dt} = \frac{\partial s^w}{\partial t} + \frac{dx}{dt}\frac{\partial s^w}{\partial x} = 0 \qquad (3.210)$$

where dx/dt is the velocity of the movement of the value of saturation of interest. A comparison of this equation with Buckley-Leverett equation (3.209) indicates that we can achieve this situation if we follow a value of saturation with:

$$\frac{dx}{dt} = \frac{q_B}{\varepsilon}\frac{df^w}{ds^w} \qquad (3.211)$$

For a fixed value of s^w, the right side of equation (3.211) is a constant. An example curve df^v/ds^w is given along with function $f^v(s^w)$ in Figure 3.15. With this information, we can integrate equation (3.211) to obtain:

$$x_f(s^w) - x_0(s^w) = \frac{q_B}{\varepsilon} \frac{df^w}{ds^w}(t_f - t_0) \tag{3.212}$$

Thus the value of s^w that is at location x_0 at time t_0 will translate to position x_f at time t_f. Using this functional form, Kleppe [30] evaluated equation (3.212) and obtained the saturation distribution shown in Figure 3.16. This plot shows that, for all locations, two different values of s^w are calculated. Of course, this is not physically possible but is the result of the simplifying assumptions used to derive Buckley-Leverett equation (3.209). This anomaly arises because of the dependence of dx/dt

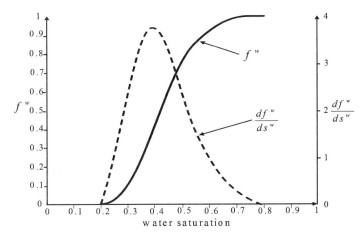

Figure 3.15: Fractional flow curve and its derivative (from [30]).

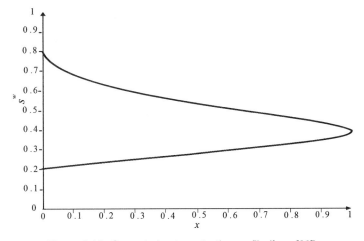

Figure 3.16: Computed water satuation profile (from [30]).

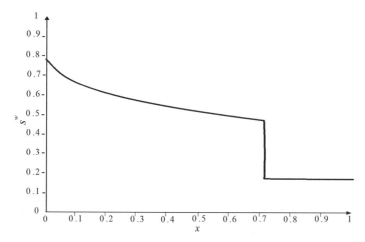

Figure 3.17: Final water saturation profile (from [30]).

on saturation that allows a fast-moving saturation to overtake a slow-moving saturation.

To eliminate this contradiction we impose the requirement that there must be a shock front at which there is a discontinuity in the s^w function. We locate the shock such that the velocity of the front is increasing with increasing saturation. Designate the value of s^w where df^w/ds^w begins to decrease as the saturation increases as $\widetilde{s^w}$. The shock front is then located at the location in the column where this saturation exists. We note that $\widetilde{s^w}$ is simply the value of s^w for which $d^2 f^w/ds^{w2} = 0$. The velocity of the shock front is:

$$\left.\frac{dx}{dt}\right|_{shock} = \frac{q_B}{\varepsilon} \left.\frac{df^w}{ds^w}\right|_{\tilde{s}^w} \tag{3.213}$$

The resulting function employed as the Buckley-Leverett solution is shown in Figure 3.17. Upstream of the shock, the saturation profile is evolving, while downstream of the shock, the initial saturation profile is preserved.

Finally, we note that the appearance of a shock in the Buckley-Leverett solution is a product of the assumptions and simplifications that have gone into the formulation of the Buckley-Leverett equation. In fact, if one observes the physical problem or solves the mass conservation equations for each fluid phase subject to Darcy's law, the sharp front does not appear. Nevertheless, the simplicity of the Buckley-Leverett solution in comparison to the full problem makes it a useful tool for approximating system behavior.

3.9 SUMMARY

The focus of this chapter was the development of the equations that describe the flow of multiphase fluids through porous media. We introduced the classic experiment carried out by Henry Darcy in 1856 which first quantified the relationship between fluid flux and fluid potential gradient. Next we discussed fluid properties, followed by a general discussion of their equations of state. A discussion of the concept of hydraulic potential, especially as it pertains to hydraulic head, was then presented. The subsequent two sections were devoted to a discussion of single-phase fluid flow and two-phase fluid flow respectively. In each section the constitutive equations (experimental relationships) and the balance equations are discussed and their integration presented. Particular attention is focused on the multiphase flow relationships. The last section is devoted to a presentation of the classic Buckley-Leverett problem, arguably the most often quoted analysis of multiphase flow in porous media available in the literature.

3.10 EXERCISES

1. Show that if Darcy had performed his experiments in columns sloped at an angle rather than in a vertical column, he still would have obtained the result given as equation (3.2).

2. An ideal gas, g, composed of a single chemical species with molecular weight M obeys the ideal gas law:

$$p^g V = nRT$$

where p^g is the pressure of the gas, V is the volume of the container containing the gas, n is the number of moles of gas in the container, R is the ideal gas constant, and T is the temperature. Obtain the expression for the gas phase compressibility, β^g.

3. Calculate the expression for the Hubbert potential of an ideal gas. Comment on the relative importance of the pressure head vs. the elevation head.

4. Consider single-phase flow of water in the subsurface. Using representative parameter values for the soil and fluid properties, determine the head gradient needed to develop the following Darcy velocities. Also, indicate field situations when such a velocity could arise. (a) $q^w = 30$ m/day in a sandy aquifer; (b) $q^w = 0.5$ m/day in a sandy aquifer; (c) $q^w = 30$ m/day in clay; (d) $q^w = 0.5$ m/day in clay.

5. Consider two-dimensional flow in an aquifer in the horizontal direction where x and y correspond to the principal horizontal coordinate axes of the intrinsic permeability tensor (such that for these coordinates, $k_{xy} = k_{yx} = 0$). Assume that $k_{xx}/k_{yy} = R$, where R is a constant. If a uniform head gradient is applied in the direction $30°$ from the x-axis, what will be the ratio of the flow in the x direction to that in the y direction.

6. Suppose a single well that fully penetrates a large, homogeneous, isotropic, confined aquifer pumps water at a rate Q^w. Also, assume that the background natural flow need not be considered. What form of the single-phase flow equation would you use to describe this problem. Indicate how you would simplify a general form.

7. Suppose that one is interested in modeling single-phase flow in an aquifer. The fluid phase compressibility, solid phase compressibility, and matrix compressibility are small, such that one is tempted to neglect them. Describe the difference in predicted behavior of a large confined aquifer subjected to pumping if these terms are neglected.

8. When we discussed microscale capillary pressure, p_c, there was no reference to hysteresis. However, for macroscale capillary pressure, p^c, hysteresis is important. Indicate the features of macroscale analysis that account for this additional complication.

9. Consider the capillary tube depicted in Figure 3.4. Flip it horizontally, keeping the fluid reservoirs at the same places, and generate the hysteresis loop starting with primary drainage. Distinguish between parts of the curve associated with Haines jumps and parts associated with equilibrium states.

10. Consider the special case of multiphase flow in an isotropic medium when the fluid and solid densities can be considered constant, but the matrix compaction is important. Obtain the particular form of equation (3.134) that applies to the phases w and n and find a mixed form in terms of h^w, h^c, and s^w.

11. Determine how the p^c-s^w relationship relates to the soil moisture profile above a water table obtained by draining from full saturation to equilibrium.

12. Data obtained from an experiment with a silty loam soil has been analyzed to obtain the van Genuchten soil moisture profile. For this soil, $\theta_r = 0.19$, $\theta_s = 0.41$, $\alpha = 0.47\,\mathrm{cm}^{-1}$, $M = 0.33$, and $N = 1/(1 - M)$. Plot the the soil moisture profile (ψ–θ or p^c–s^w), relative permeability (k_{rel}^w-θ or k_{rel}^w-s^w), soil water capacity (C^w–ψ), and soil water diffusivity(D^w–θ). Identify portions of these curves where a small change in one parameter can significantly alter the value of the other parameter. These portions of the curves are where complex numerical algorithms are needed to ensure accurate solution to the equations.

BIBLIOGRAPHY

[1] Abaci, S., J.S. Edwards, and B.N. Whitaker, Relative permeability measurements for two phase flow in unconsolidated sands, *Mine Water and the Environment*, **11**(2), 11–26, 1992.

[2] Arya, L.M., and T.S. Dierolf, Predicting soil moisture characteristics from particle-size distributions: an improved method to calculate pore radii from particle radii, in *Indirect Methods for Estimating the Hydraulic Properties of Unsaturated Soils* (M.Th. van Genuchten et al., eds.), University of California, Riverside, 115–124, 1992.

[3] Arya, L.M., and J.F. Paris, A physicoempirical model to predict the soil moisture characteristic from particle-size distribution and bulk density data, *Soil Sci. Soc. Am. J.*, **45**, 1023–1030, 1981.

[4] Bear, J., *Dynamics of Fluids in Porous Media*, Elsevier, New York, 1972.

[5] Bear, J., *Hydraulics of Groundwater*, McGraw-Hill, New York, 1979.

[6] Bear, J., C. Braster, and P.C. Menier, Effective and relative permeabilities in anisotropic porous media, *Transport in Porous Media*, **2**, 301–316, 1987.

[7] Brooks, R.H., and A.T. Corey, Properties of porous media affecting fluid flow, *J. of the Irrigation and Drainage Division Proc. ASCE*, No. IR2, **92**, 61–88, 1966.

[8] Buckley, S.E., and M.C. Leverett, Mechanism of fluid displacement in sands, *Trans. AIME*, **146**, 107–116, 1942.

[9] Carman, P.C., Fluid flow through granular beds, *Trans. Institute of Chem. Engineers, London*, **15**, 150–157, 1937.

[10] Freeze, R.A., and J.A. Cherry, *Groundwater*, Prentice-Hall, Englewood Cliffs, NJ, 1979.

[11] Corey, A., Pore-size distribution, in *Indirect Methods for Estimating the Hydraulic Properties of Unsaturated Soils* (M.Th. van Genuchten et al., eds.), University of California, Riverside, 37–43, 1992.

[12] Darcy, H., *Les Fontaines Publiques de la Ville de Dijon*, Dalmont, Paris, 1856.

[13] Darcy, H., Determination of the laws of flow of water through sand, in *Physical Hydrology* (R.A. Freeze and W. Back, eds.), Hutchinson Ross, Stroudsburg, PA, 1983.

[14] de Marsily, G., *Quantitative Hydrogeology: Groundwater Hydrology for Engineers*, Academic Press, San Diego, CA, 1986.

[15] Demond, A.H. and P.V. Roberts, An examination of relative permeability relations for two-phase flow in porous media, Water Resources Bulletin, *Am. Water Res. Assn.*, **23**(4), 617–628, 1987.

[16] Domenico, P.A., and F.W. Schwartz, *Physical and Chemical Hydrogeology*, John Wiley And Sons, New York, 1998.

[17] Dullien, F.A.L., *Porous Media: Fluid Transport and Pore Structure*, Academic Press, New York, 396 pages, 1979.

[18] Fair, G.M., and L.P. Hatch, Fundamental factors governing the streamline flow of water through sand, *J. Am. Water Works Assoc.*, **25**, 1551–1565, 1933.

[19] Fu, J., B. Li, and Z. Wang, Estimation of fluid-fluid interfacial tensions of multicomponent mixtures, *Chem. Eng. Sci.*, **41**(13), 2031–2046, 1986.

[20] Gray, W.G., and Hassanizadeh, S.M., Paradoxes and realities in unsaturated flow theory, *Water Resour. Res.*, **27**(8), 1847–1854, 1991.

[21] Gray, W.G., A. Leinjse, R.A. Kolar, and C.A. Blain, *Mathematical Tools for Changing Spatial Scales in the Analysis of Physical Systems*, CRC Press, Boca Raton, FL, 232 pages, 1993.

[22] Gray, W.G., and C.T. Miller, Examination of Darcy's law for flow in porous media with variable porosity, *Environ. Sci. Technol.*, **38**, 5895–5901, 2004.

[23] Gray, W.G., and B.A. Schrefler, Analysis of the solid phase stress tensor in multiphase porous media, *Int. J. Numer. Anal. Meth. Geomech.*, **31**(4), 541–581, 2007.

[24] Greenkorn, R.A., *Flow Phenomenon in Porous Media*, Marcel Dekker, New York, 550 pages, 1983.

[25] Guarnaccia, J., G. Pinder, and M. Fishman, NAPL: Simulator Documentation, EPA/600/SR-97/102, 1997.

[26] Hillel, D., *Environmental Soil Physics*, Academic Press, San Diego, 771 pages, 1998.

[27] Hubbert, M.K., The theory of groundwater motion, *J. Geol.*, **48**, 785–944, 1940.

[28] Kool, J.B., and J.C. Parker, Development and evaluation of closed-form expressions for hysteretic soil hydraulic properties, *Water Resources Res.*, **23**(1), 105–114, 1987.

[29] Kozeny, J., Über die kapillare Leitung des Wassers im Boden, *Sitzungsberichte der Akademie der Wissenschaften in Wien*, **136a**, 271–306, 1927.

[30] Kleppe, J., TPG4150 Reservoir Recovery Techniques 2006 Hand-out Note 4: Buckley-Leverett Analysis, www.ipt.ntnu.no/˜kleppe/TPG4150/BL.pdf, 2006.

[31] Land, C.S., Calculation of imbibition relative permeability for two- and three-phase flow from rock properties, *Trans. AIME*, **243**, 149–156, 1968.

[32] Lenhard, R., and J.C. Parker, A model for hysteretic constitutive relations governing multiphase flow: 2. Permeability-saturation relations, *Water Resources Res.*, **23**(12), 2197–2206, 1987.

[33] Leverett, M.C., Capillary behavior in porous media, *Trans. AIME*, **142**, 341–358, 1941.

[34] Lohrenz, J., B.G. Bray, and C.R. Clark, Calculating viscosities of reservoir fluids from their compositions, *J. Pet. Tech.*, October, 1171–1176, 1964.

[35] Mualem, Y., Hydraulic conductivity of unsaturated porous media: generalized macroscopic approach, *Water Resources Res.*, **12**(3), 513–521, 1976.

[36] Mualem, Y., A new model for predicting the hydraulic conductivity of unsaturated porous media, *Water Resources Res.*, **14**(2), 325–334, 1978.

[37] Mualem, Y., Modeling the hydraulic conductivity of unsaturated porous media, in *Indirect Methods for Estimating the Hydraulic Properties of Unsaturated Soils* (M.Th. van Genuchten et al., eds.), University of California, Riverside, 15–36, 1992.

[38] Narr, J., and R.J. Wygal, Three-phase imbibition relative permeability, *SPEJ*, **1**, 254–258, 1968.

[39] Nielsen, D.R., and L. Luckner, Theoretical aspects to estimate reasonable initial parameters and range limits in identification procedures for soil hydraulic properties, in *Indirect Methods for Estimating the Hydraulic Properties of Unsaturated Soils* (M.Th. van Genuchten et al., eds.), University of California, Riverside, 147–160, 1992.

[40] Ried, R.C., J.M. Prausnitz, and B.E. Poling, *The Properties of Gases and Liquids*, Fourth Edition, 741 pages, McGraw Hill, New York, 1987.

[41] Rose, W., and W.A. Bruce, Evaluation of capillary character in petroleum reservoir rock, *Trans. AIME*, **128**, 127–142, 1949.

[42] van Genuchten, M.Th., A closed-form equation for predicting the hydraulic conductivity of unsaturated soils, *Soil Sci. Soc. Am. J.*, **44**(5), 892–898, 1980.

[43] Wilson, J.L., S. Conrad, W. Mason, W. Peplinski, and E. Hagan, Laboratory Investigation of Residual Organics from Spills, Leaks, and the Disposal of Hazardous Wastes in Groundwater, EPA/600/6-90/004, April 1990.

4

MASS TRANSPORT EQUATIONS

4.1   INTRODUCTION

The last chapter was concerned with the equations that describe the flow of a phase as a whole. The mass conservation equation of the phase in conjunction with supplementary conditions among variables, such as Darcy's law, p^c-s^w and k_{rel}^α-s^w relationships, and appropriate boundary conditions provide closed sets of equations that describe the distribution of the velocities, pressures, and densities of the fluid phases.

In the present chapter, we are going to examine additional conditions that will enable us to also describe the distribution of chemical constituents within each phase. These equations are based on the mass conservation equations for each constituent and are referred to as the *mass transport equations*. The consideration of the equations will be informed by the fact that the description of the behavior of these constituents must be consistent with the description of the phase as a whole provided in the last chapter.

In one sense, the transport equations may be viewed as a refinement of the description of the flow of a phase. From another perspective, the transport equations might be considered as some conditions added to the flow equations. However, in cases where composition significantly impacts the density of the fluid phase, the description of the flow of a phase must account for this effect. No modification of the flow equations of the last chapter is required; however, the state equation for fluid phase α as given by $\rho^\alpha(p^\alpha)$ must be extended to a form $\rho^\alpha(p^\alpha, \omega^{i\alpha})$ for $i = 1, \ldots, N - 1$.[1]

---

[1] If we were to also consider heat effects, the dependence on temperature would have to be incorporated into the equation of state; and an energy equation that describes heat transport within the phase would also be needed. Our focus here is on isothermal systems, however, so the impact of temperature is not covered explicitly.

Essentials of Multiphase Flow and Transport in Porous Media, by George F. Pinder and William G. Gray
Copyright © 2008 by John Wiley & Sons, Inc.

In general, the problem of transport of chemical species begins with equation (2.109) from Chapter 2:

$$\frac{\partial(\varepsilon^\alpha \rho^\alpha \omega^{i\alpha})}{\partial t} + \nabla \cdot (\varepsilon^\alpha \rho^\alpha \omega^{i\alpha} \mathbf{v}^\alpha) + \nabla \cdot (\varepsilon^\alpha \mathbf{j}^{i\alpha}) = \varepsilon^\alpha r^{i\alpha} + \sum_{\beta \neq \alpha} e_{\alpha\beta}^{i\alpha} + \sum_{W=1}^{N_W} \rho_W^\alpha \omega_W^{i\alpha} Q_W^\alpha \delta(\mathbf{x} - \mathbf{x}^W)$$

$$\alpha = w, n, s; \quad i = 1, \ldots, N_\alpha \qquad (4.1)$$

It is possible to ignore Chapter 3 and develop a strategy for solution of this equation for each species in each phase. However, the approach we will adopt is to make use of the information from the last chapter that provides overall phase behavior and additionally solve for $N_\alpha - 1$ of the species equations for each phase. This is possible because the mass fractions in a phase are constrained by equation (2.65):

$$1 = \sum_{i=1}^{N_\alpha} \omega^{i\alpha} \quad \alpha = w, n, s \qquad (4.2)$$

and the dispersion vectors for the species that comprise a fluid phase, as defined in equation (2.72) satisfy the condition:

$$0 = \sum_{i=1}^{N_\alpha} \mathbf{j}^{i\alpha} \quad \alpha = w, n, s \qquad (4.3)$$

Equation (4.2) indicates that only $N_\alpha - 1$ of the species mass fractions are independent, because if $N_\alpha - 1$ are known, the N_αth may be calculated directly as:

$$\omega^{N_\alpha \alpha} = 1 - \sum_{i=1}^{N_\alpha - 1} \omega^{i\alpha} \quad \alpha = w, n, s \qquad (4.4)$$

Likewise, if $N_\alpha - 1$ of the species dispersion vectors are known, the dispersion vector of the N_αth species is obtained as:

$$\mathbf{j}^{N_\alpha \alpha} = - \sum_{i=1}^{N_\alpha - 1} \mathbf{j}^{i\alpha} \quad \alpha = w, n, s \qquad (4.5)$$

The overall mass exchange term between adjacent phases is equal to the sum of the exchange of species between the phases so that:

$$e_{\alpha\beta}^{N_\alpha \alpha} = e_{\alpha\beta}^\alpha - \sum_{i=1}^{N_\alpha - 1} e_{\alpha\beta}^{i\alpha} \quad \alpha = w, n, s; \quad \alpha\beta = wn, ws, ns \qquad (4.6)$$

Although the preceding makes the case for the number of species equations that need to be considered, there are four important issues that must be addressed before the description of species transport can be considered complete. First, equation (4.1) contains the phase velocity, $\mathbf{v}^\alpha$. To apply this equation to the fluid phases, a derivation must be performed that allows this velocity to be replaced with the Darcy velocity and terms that account for the solid phase deformation, as in the last chapter.

Second, although equations for flow of the solid phase were not developed in Chapter 3, the solid phase may exchange mass with the fluids. Therefore, species balance equations for the solid are needed that do not explicitly contain $\mathbf{v}^s$.

Third, the value of N_α may be different for each phase. Not all species will necessarily appear in each phase. Therefore, the N_αth species, the species not considered explicitly using a mass transport equation, will usually be different in each phase. In fact, it is typical to consider a fluid phase as being composed primarily of one chemical constituent and to work with the transport equations for the species dissolved in that constituent. For example, in modeling a wetting phase composed primarily of water with some organics dissolved in the water, the species transport equations would be applied to those dissolved constituents. The corresponding model for an organic phase would make use of a species transport equation for the small amount of water that might dissolve in that phase. This observation is not particularly complex, but it does point to the importance of keeping track of the chemical constituents being modeled.

Fourth, in solving the species transport equations, we are faced with the need to have the same number of equations as unknowns. This means that some constitutive forms will have to be developed for the *dispersion vectors*, $\mathbf{j}^{i\alpha}$, the *chemical reaction* expressions within a phase, $r^{i\alpha}$, and the *interphase transfer terms*, $e_{\alpha\beta}^{i\alpha}$. Additionally, all pumping rates, Q_W^α, will have to be specified along with the composition of any injected fluid.

The remainder of this chapter is devoted to addressing these issues, in turn, so that useful mass transport equations that complement the phase flow equations result. Because the presence of two fluid phases, as opposed to a single fluid phase, does not significantly complicate the derivation of the transport equations, the derivations will be performed for a three-phase w, n, s system. Simplification to a single-fluid system follows directly from the two-fluid equations.

4.2   VELOCITY IN THE SPECIES TRANSPORT EQUATIONS

For equation (4.1) applied to each species to be solvable in conjunction with the flow equations, the total number of unknowns must be equal to the number of equations. Each species equation may therefore introduce one unknown to the set of unknowns beyond those accounted for by other conservation, constitutive, and state equations. This additional unknown is the *mass fraction*, $\omega^{i\alpha}$. Therefore, we need to perform some manipulations to express other quantities in this equation, which have not appeared in the flow equations, in terms of solution variables (e.g., $\mathbf{q}^\alpha$, s^w, p^α, etc.) and state variables discussed in the last chapter. In this section, we will concentrate on the *advective term* of the transport equation, $\varepsilon^\alpha \rho^\alpha \omega^{i\alpha} \mathbf{v}^\alpha$, with the goal of eliminating the phase velocity, $\mathbf{v}^\alpha$, in terms of other variables. Although flow equations were developed in the last chapter, the flow quantity of interest was the Darcy velocity, which, strictly speaking, cannot be used to obtain the flow velocity without knowledge of the solid velocity because the Darcy velocity is defined as:

$$\mathbf{q}^\alpha = \varepsilon s^\alpha \left(\mathbf{v}^\alpha - \mathbf{v}^s \right) \quad \alpha = w, n \tag{4.7}$$

where $\varepsilon s^\alpha = \varepsilon^\alpha$. Thus, the problem of replacing the fluid velocity with the Darcy velocity is complicated when $\mathbf{v}^s$ is nonzero. In the previous chapter, the need to solve directly for $\mathbf{v}^s$ was replaced by the incorporation of the effects of solid deformation in the specific storage coefficient. With $\mathbf{v}^s$ unavailable, we need strategies for accounting for its impact on the transport equations for the fluid phases. Also, since species may be exchanged between the solid and fluids, the solid phase must be considered explicitly in transport problems, and the velocity of the solid phase will thus appear (i.e., $\alpha = s$ in equation (4.1)).

The three subsections that follow provide alternative approaches for incorporating the Darcy velocity into the transport equations without creating the need to calculate the solid phase velocity. The three approaches are called, respectively, the direct approach, the rigorous approach, and the species distribution approach. These approaches to the velocity are the first step in reducing the number of unknowns in the transport equations. Subsequently, we will also provide expressions for the dispersion vector, the interphase exchange terms, and the chemical reactions within the phase. The way these latter expressions are handled is independent of the approach used to modify the velocity term.

4.2.1 Direct Approach

The essence of this approach is that the velocity of the solid is deemed to be negligible when considering the transport equation. Thus from equation (4.7) for the purpose of use in the transport equation:

$$\mathbf{v}^\alpha \approx \frac{\mathbf{q}^\alpha}{\varepsilon s^\alpha} \quad \alpha = w, n \tag{4.8}$$

For the fluid phases, the equations formulated for the mass fraction are thus obtained from equation (4.1) in light of equation (4.8) as:

$$\frac{\partial \left(\varepsilon^\alpha \rho^\alpha \omega^{i\alpha} \right)}{\partial t} + \nabla \cdot \left(\varepsilon^\alpha \omega^{i\alpha} \mathbf{q}^\alpha \right) + \nabla \cdot \left(\varepsilon^\alpha \mathbf{j}^{i\alpha} \right) = \varepsilon^\alpha r^{i\alpha} + \sum_{\beta \neq \alpha} e^{i\alpha}_{\alpha\beta} + \sum_{W=1}^{N_W} \rho^\alpha_W \omega^{i\alpha}_W Q^\alpha_W \delta \left(\mathbf{x} - \mathbf{x}^W \right)$$

$$\alpha = w, n; \quad \beta = w, n, s; \quad i = 1, \ldots, N_\alpha \tag{4.9}$$

For the solid phase transport equation, both the solid velocity and dispersion within the solid phase are considered unimportant in the direct approach. Additionally, pumping of the solid phase is negligible so that the transport equations are:

$$\frac{\partial \left[(1 - \varepsilon) \rho^s \omega^{is} \right]}{\partial t} = (1 - \varepsilon) r^{is} + e^{is}_{ws} + e^{is}_{ns} \quad i = 1, \ldots, N_s \tag{4.10}$$

where use has been made of the fact that $\varepsilon^s = 1 - \varepsilon$. Although we have set the solid phase velocity to zero such that the solid matrix does not deform in response to mechanical stress, this equation still accounts for changes in porosity due to dissolution of, or deposition on, the surface of the solid.

Equations (4.9) and (4.10) are general forms of the direct method that can be solved for the mass fraction $\omega^{i\alpha}$ after additional closure conditions are developed. In practice, we would solve these for all but one of the species in the phase, and couple these equations to the phase flow equation. We note here that in some applications, the solution may be conveniently developed in terms of the mass of species i in phase α per unit volume of phase α, $\rho^{i\alpha} = \rho^{\alpha}\omega^{i\alpha}$.

Also, the concentration of i is alternatively expressed as moles of species i in phase α per unit volume of phase α as $c^{i\alpha} = \rho^{i\alpha}/M^i$, where M^i is the molecular weight of i. For example, to obtain the form of the transport equation in terms of molecular concentration, divide equations (4.9) and (4.10) by M^i and make use of the definition of molar concentration to obtain, respectively:

$$\frac{\partial\left(\varepsilon^{\alpha}c^{i\alpha}\right)}{\partial t} + \nabla\cdot\left(c^{i\alpha}\mathbf{q}^{\alpha}\right) + \nabla\cdot\left(\frac{\varepsilon^{\alpha}\mathbf{j}^{i\alpha}}{M^i}\right) = \frac{\varepsilon^{\alpha}r^{i\alpha}}{M^i} + \sum_{\beta\neq\alpha}\frac{e^{i\alpha}_{\alpha\beta}}{M^i} + \sum_{W=1}^{N_W}c^{i\alpha}_W Q^{\alpha}_W \delta\left(\mathbf{x}-\mathbf{x}^W\right)$$
$$\alpha = w, n; \quad i = 1,\ldots,N_{\alpha} \qquad (4.11)$$

and

$$\frac{\partial\left[(1-\varepsilon)c^{is}\right]}{\partial t} = \frac{(1-\varepsilon)r^{is}}{M^i} + \frac{e^{is}_{ws}}{M^i} + \frac{e^{is}_{ns}}{M^i} \quad i = 1,\ldots,N_s \qquad (4.12)$$

At the conclusion of the chapter, after all the needed constitutive forms have been presented, the various forms of the direct transport equations, as well as of the other forms derived, will be collected for easy comparison. For now, we simply note that the direct forms of the equations follow as a result of neglecting the solid phase deformation. The main problem with the direct forms is that when the transport equations for all species are added together, the equation that describes the phase behavior as a whole, as derived in Chapter 3, should be recovered. However, since the solid deformation has been neglected, the equations will not be the same as those of the last chapter which included the storage effects due to both solid compressibility and matrix deformation. Despite this drawback, the direct equations can be very useful in many situations. For example, if a w phase is contaminated with a small amount of a pollutant, the direct transport equation for the pollutant will be appropriate since most of the effects of solid deformation will apply to the chemical constituent present in the far greater amount. In other words, when the amount of contaminant is small, the contaminant transport will not affect the behavior of the phase as a whole, described by the phase flow equations, and will also not be affected significantly by the small solid phase deformations that contribute to phase storage. Some careful thought will lead to the conclusion that this is reasonable. Additionally, this observation can be confirmed by doing a more rigorous mathematical analysis, as in the next subsection.

4.2.2 Rigorous Approach

With this approach, we do not neglect the solid phase velocity but incorporate it into the equation of species transport in much the same way it was incorporated into the phase equation of the last chapter. The objective of this approach is to incorporate the Darcy velocity into the species transport equations in the fluid phases while taking into account the deformation of the solid. Within this approach, we do not explicitly derive a simplified balance equation for a species in the solid phase. Rather, the equation used for the solid phase is either that obtained from the direct method (which can be used when the species being modeled by the transport equation are at small concentration such that the changes in concentration are more important than solid motion) or using the distribution method of the next subsection (which takes account of the solid motion).

If we use the definition of the Darcy velocity as given by equation (4.7) to eliminate $\mathbf{v}^\alpha$ from equation (4.1), we obtain:

$$
\frac{\partial\left(\varepsilon^\alpha \rho^\alpha \omega^{i\alpha}\right)}{\partial t} + \nabla\cdot\left(\rho^\alpha \omega^{i\alpha}\mathbf{q}^\alpha\right) + \nabla\cdot\left(\varepsilon^\alpha \mathbf{j}^{i\alpha}\right) + \nabla\cdot\left(\varepsilon^\alpha \rho^\alpha \omega^{i\alpha}\mathbf{v}^s\right)
$$
$$
= \varepsilon^\alpha r^{i\alpha} + e_{wn}^{i\alpha} + e_{\alpha s}^{i\alpha} + \sum_{W=1}^{N_W} \rho_W^\alpha \omega_W^{i\alpha} Q_W^\alpha \delta\left(\mathbf{x}-\mathbf{x}^W\right)
$$
$$
\alpha = w, n; \quad i = 1,\dots,N_\alpha \qquad (4.13)
$$

The objective is to eliminate $\mathbf{v}^s$ by combining this equation with that for the solid phase. We can apply the product rule to the fourth term and make use of the definition of the *material derivative* with respect to the s phase. This allows for rearrangement of the equation to:

$$
\frac{D^s\left(\varepsilon^\alpha \rho^\alpha \omega^{i\alpha}\right)}{Dt} + \nabla\cdot\left(\rho^\alpha \omega^{i\alpha}\mathbf{q}^\alpha\right) + \nabla\cdot\left(\varepsilon^\alpha \mathbf{j}^{i\alpha}\right) + \varepsilon^\alpha \rho^\alpha \omega^{i\alpha}\nabla\cdot\mathbf{v}^s
$$
$$
= \varepsilon^\alpha r^{i\alpha} + e_{wn}^{i\alpha} + e_{\alpha s}^{i\alpha} + \sum_{W=1}^{N_W} \rho_W^\alpha \omega_W^{i\alpha} Q_W^\alpha \delta\left(\mathbf{x}-\mathbf{x}^W\right)
$$
$$
\alpha = w, n; \quad i = 1,\dots,N_\alpha \qquad (4.14)
$$

Equation (3.121) describes the *mass conservation of the solid phase* and may be rearranged to:

$$
\frac{1}{(1-\varepsilon)\rho^s}\frac{D^s\left[(1-\varepsilon)\rho^s\right]}{Dt} + \nabla\cdot\mathbf{v}^s = \frac{e_{ws}^s}{(1-\varepsilon)\rho^s} + \frac{e_{ns}^s}{(1-\varepsilon)\rho^s} \qquad (4.15)
$$

Elimination of $\nabla\cdot\mathbf{v}^s$ between these last two equations yields:

$$
\omega^{i\alpha}\left\{-\frac{\varepsilon^\alpha \rho^\alpha}{(1-\varepsilon)\rho^s}\frac{D^s\left[(1-\varepsilon)\rho^s\right]}{Dt} + \frac{\varepsilon^\alpha \rho^\alpha}{(1-\varepsilon)\rho^s}e_{ws}^s + \frac{\varepsilon^\alpha \rho^\alpha}{(1-\varepsilon)\rho^s}e_{ns}^s\right\}
$$
$$
+ \frac{D^s\left(\varepsilon^\alpha \rho^\alpha \omega^{i\alpha}\right)}{Dt} + \nabla\cdot\left(\rho^\alpha \omega^{i\alpha}\mathbf{q}^\alpha\right) + \nabla\cdot\left(\varepsilon^\alpha \mathbf{j}^{i\alpha}\right)
$$
$$
= \varepsilon^\alpha r^{i\alpha} + e_{wn}^{i\alpha} + e_{\alpha s}^{i\alpha} + \sum_{W=1}^{N_W} \rho_W^\alpha \omega_W^{i\alpha} Q_W^\alpha \delta\left(\mathbf{x}-\mathbf{x}^W\right)
$$
$$
\alpha = w, n; \quad i = 1,\dots,N_\alpha \qquad (4.16)
$$

The fraction of the total mass at a point associated with a phase is designated x_{mass}^α for $\alpha = w, n, s$. Also, consistent with the previous chapter, we make the assumption that the advective part of the material derivative associated with the s phase is so small that $D^s/Dt \approx \partial/\partial t$. Therefore, equation (4.16) simplifies to:

$$
\omega^{i\alpha} \left\{ -\frac{\varepsilon^\alpha \rho^\alpha}{(1-\varepsilon)\rho^s} \frac{\partial[(1-\varepsilon)\rho^s]}{\partial t} + \frac{x_{\text{mass}}^\alpha}{x_{\text{mass}}^s} e_{ws}^s + \frac{x_{\text{mass}}^\alpha}{x_{\text{mass}}^s} e_{ns}^s \right\}
$$
$$
+ \frac{\partial(\varepsilon^\alpha \rho^\alpha \omega^{i\alpha})}{\partial t} + \nabla \cdot (\rho^\alpha \omega^{i\alpha} \mathbf{q}^\alpha) + \nabla \cdot (\varepsilon^\alpha \mathbf{j}^{i\alpha})
$$
$$
= \varepsilon^\alpha r^{i\alpha} + e_{wn}^{i\alpha} + e_{\alpha s}^{i\alpha} + \sum_{W=1}^{N_W} \rho_W^\alpha \omega_W^{i\alpha} Q_W^\alpha \delta(\mathbf{x} - \mathbf{x}^W)
$$
$$
\alpha = w, n; \quad i = 1, \dots, N_\alpha \tag{4.17}
$$

The expression in braces is the only difference between this equation and equation (4.9). Therefore, this term must account for the solid phase deformation that was ignored in the direct approach. The fact that this term is multiplied by the mass fraction confirms the observation that when the mass fraction of a chemical species is small, the solid phase deformation has negligible impact on the transport equation for that species.[2]

For the moment, we will continue on with the rigorous derivation by introducing the fluid saturations into equation (4.17) via the identity $\varepsilon^\alpha = s^\alpha \varepsilon$ and also applying the product rule to the partial time derivative after the braces. With a bit of algebra, we can obtain:

$$
\frac{\partial(\varepsilon^\alpha \rho^\alpha \omega^{i\alpha})}{\partial t} = \varepsilon s^\alpha \frac{\partial(\rho^\alpha \omega^{i\alpha})}{\partial t} + \omega^{i\alpha} \left\{ \varepsilon \rho^\alpha \frac{\partial s^\alpha}{\partial t} - \frac{\rho^\alpha s^\alpha}{\rho^s} \frac{\partial[(1-\varepsilon)\rho^s]}{\partial t} \right.
$$
$$
\left. + \frac{\rho^\alpha s^\alpha (1-\varepsilon)}{\rho^s} \frac{\partial \rho^s}{\partial t} \right\} \tag{4.18}
$$

Substitution of this identity into equation (4.17) and then collection of terms multiplied by $\omega^{i\alpha}$ yield:

$$
\omega^{i\alpha} \left\{ \varepsilon \rho^\alpha \frac{\partial s^\alpha}{\partial t} - \frac{s^\alpha \rho^\alpha}{(1-\varepsilon)\rho^s} \frac{\partial[(1-\varepsilon)\rho^s]}{\partial t} + \frac{\rho^\alpha s^\alpha (1-\varepsilon)}{\rho^s} \frac{\partial \rho^s}{\partial t} + \frac{x_{\text{mass}}^\alpha}{x_{\text{mass}}^s} e_{ws}^s + \frac{x_{\text{mass}}^\alpha}{x_{\text{mass}}^s} e_{ns}^s \right\}
$$
$$
+ \varepsilon s^\alpha \frac{\partial(\rho^\alpha \omega^{i\alpha})}{\partial t} + \nabla \cdot (\rho^\alpha \omega^{i\alpha} \mathbf{q}^\alpha) + \nabla \cdot (\varepsilon^\alpha \mathbf{j}^{i\alpha})
$$
$$
= \varepsilon^\alpha r^{i\alpha} + e_{wn}^{i\alpha} + e_{\alpha s}^{i\alpha} + \sum_{W=1}^{N_W} \rho_W^\alpha \omega_W^{i\alpha} Q_W^\alpha \delta(\mathbf{x} - \mathbf{x}^W)
$$
$$
\alpha = w, n; \quad i = 1, \dots, N_\alpha \tag{4.19}
$$

[2] Some transport problems of importance, for instance of chlorinated hydrocarbons dissolved in water, involve mass fractions of a few parts per billion. Transport equations for these species can be specified just as well with the direct and rigorous methods.

We can make use of the relations for the solid and matrix compressibility given, respectively, by equations (3.128) and (3.129) so that the equation reduces further to:

$$
\omega^{i\alpha} \left\{ \varepsilon \rho^\alpha \frac{\partial s^\alpha}{\partial t} + s^\alpha \rho^\alpha [\alpha^b + (1 - \varepsilon) \beta^s] \frac{\partial p^s}{\partial t} + \frac{x_{\mathrm{mass}}^\alpha}{x_{\mathrm{mass}}^s} e_{ws}^s + \frac{x_{\mathrm{mass}}^\alpha}{x_{\mathrm{mass}}^s} e_{ns}^s \right\}
$$
$$
+ \varepsilon s^\alpha \frac{\partial (\rho^\alpha \omega^{i\alpha})}{\partial t} + \nabla \cdot (\rho^\alpha \omega^{i\alpha} \mathbf{q}^\alpha) + \nabla \cdot (\varepsilon^\alpha \mathbf{j}^{i\alpha})
$$
$$
= \varepsilon^\alpha r^{i\alpha} + e_{wn}^{i\alpha} + e_{\alpha s}^{i\alpha} + \sum_{W=1}^{N_W} \rho_W^\alpha \omega_W^{i\alpha} Q_W^\alpha \delta(\mathbf{x} - \mathbf{x}^W)
$$
$$
\alpha = w, n; \quad i = 1, \ldots, N_\alpha \qquad (4.20)
$$

This equation accounts for the effects of solid matrix deformation and compression in the transport process for each species i in a fluid phase α. Summation of this equation over the N_α species in the phase gives equation (3.130) without use of the fluid phase compressibility definition. Usually, equation (4.20) is simplified by using the observation that:

$$
\left| \varepsilon \rho^\alpha \frac{\partial s^\alpha}{\partial t} \right| \gg \left| s^\alpha \rho^\alpha [\alpha^b + (1 - \varepsilon) \beta^s] \frac{\partial p^s}{\partial t} + \frac{x_{\mathrm{mass}}^\alpha}{x_{\mathrm{mass}}^s} e_{ws}^e + \frac{x_{\mathrm{mass}}^\alpha}{x_{\mathrm{mass}}^s} e_{ns}^e \right| \qquad (4.21)
$$

This approximation applies except at the high and low ends of the saturation range (s_i^w and s_r^n). At these end locations, the change in pressure required to produce a change in saturation can be large. We thus obtain the form of the rigorous mass conservation commonly employed:

$$
\varepsilon \rho^\alpha \omega^{i\alpha} \frac{\partial s^\alpha}{\partial t} + \varepsilon s^\alpha \frac{\partial (\rho^\alpha \omega^{i\alpha})}{\partial t} + \nabla \cdot (\rho^\alpha \omega^{i\alpha} \mathbf{q}^\alpha) + \nabla \cdot (\varepsilon^\alpha \mathbf{j}^{i\alpha})
$$
$$
= \varepsilon^\alpha r^{i\alpha} + e_{wn}^{i\alpha} + e_{\alpha s}^{i\alpha} + \sum_{W=1}^{N_W} \rho_W^\alpha \omega_W^{i\alpha} Q_W^\alpha \delta(\mathbf{x} - \mathbf{x}^W)
$$
$$
\alpha = w, n; \quad i = 1, \ldots, N_\alpha \qquad (4.22)
$$

As with the direct method equation, this equation is also sometimes expressed in terms of the mass of species i per volume of α phase, $\rho^{i\alpha} = \rho^\alpha \omega^{i\alpha}$, or the moles of species i per volume of α phase, $c^{i\alpha} = \rho^\alpha \omega^{i\alpha}/\mathrm{M}^i$. For example, in terms of moles per volume, this equation is:

$$
\varepsilon c^{i\alpha} \frac{\partial s^\alpha}{\partial t} + \varepsilon s^\alpha \frac{\partial c^{i\alpha}}{\partial t} + \nabla \cdot (c^{i\alpha} \mathbf{q}^\alpha) + \nabla \cdot \left(\frac{\varepsilon^\alpha \mathbf{j}^{i\alpha}}{\mathrm{M}^i} \right)
$$
$$
= \frac{\varepsilon^\alpha r^{i\alpha}}{\mathrm{M}^i} + \frac{e_{wn}^{i\alpha}}{\mathrm{M}^i} + \frac{e_{\alpha s}^{i\alpha}}{\mathrm{M}^i} + \sum_{W=1}^{N_W} c_W^{i\alpha} Q_W^\alpha \delta(\mathbf{x} - \mathbf{x}^W)
$$
$$
\alpha = w, n; \quad i = 1, \ldots, N_\alpha \qquad (4.23)
$$

4.2.3 Distribution Approach

The objective of the direct and rigorous approaches is to obtain mass conservation equations for the chemical species in a phase with the Darcy velocity appearing in the equations. If the approximations made in the derivation are perfectly satisfied, the sum of the derived equations will describe the mass conservation for the phase as a whole. The distribution approach, on the other hand, leads to an equation that describes the distribution of mass fractions within the phase. Summation of the distribution equations over all species provides no information in addition to the fact that the sum of the species mass fractions must be 1. From a mathematical perspective, the distribution and conservation equations are identical. However, implementation of approximate solution procedures on the computer introduces different errors into the concentration fields obtained.

The derivation of the *distribution form of the species conservation equations* proceeds from the rigrous method equation (4.19). The product rule is applied to each of the first two terms following the braces to obtain:

$$\varepsilon s^{\alpha} \frac{\partial \left(\rho^{\alpha} \omega^{i\alpha}\right)}{\partial t} = \varepsilon \rho^{\alpha} s^{\alpha} \frac{\partial \omega^{i\alpha}}{\partial t} + \omega^{i\alpha} \varepsilon s^{\alpha} \frac{\partial \rho^{\alpha}}{\partial t} \tag{4.24}$$

and:

$$\nabla \cdot \left(\rho^{\alpha} \omega^{i\alpha} \mathbf{q}^{\alpha}\right) = \rho^{\alpha} \mathbf{q}^{\alpha} \cdot \nabla \omega^{i\alpha} + \omega^{i\alpha} \nabla \cdot \left(\rho^{\alpha} \mathbf{q}^{\alpha}\right) \tag{4.25}$$

Substitution back into equation (4.19) and rearrangement of terms yields:

$$\omega^{i\alpha} \left\{ \varepsilon \rho^{\alpha} \frac{\partial s^{\alpha}}{\partial t} - \frac{s^{\alpha} \rho^{\alpha}}{(1-\varepsilon)\rho^{s}} \frac{\partial \left[(1-\varepsilon)\rho^{s}\right]}{\partial t} + \frac{\rho^{\alpha} s^{\alpha}(1-\varepsilon)}{\rho^{s}} \frac{\partial \rho^{s}}{\partial t} + \varepsilon s^{\alpha} \frac{\partial \rho^{\alpha}}{\partial t} \right.$$
$$\left. + \nabla \cdot \left(\rho^{\alpha} \mathbf{q}^{\alpha}\right) \frac{x_{\text{mass}}^{\alpha}}{x_{\text{mass}}^{s}} e_{ws}^{s} + \frac{x_{\text{mass}}^{\alpha}}{x_{\text{mass}}^{s}} e_{ns}^{s} \right\}$$
$$+ \varepsilon \rho^{\alpha} s^{\alpha} \frac{\partial \omega^{i\alpha}}{\partial t} + \rho^{\alpha} \mathbf{q}^{\alpha} \cdot \nabla \omega^{i\alpha} + \nabla \cdot \left(\varepsilon^{\alpha} \mathbf{j}^{i\alpha}\right)$$
$$= \varepsilon^{\alpha} r^{i\alpha} + e_{wn}^{i\alpha} + e_{\alpha s}^{i\alpha} + \sum_{W=1}^{N_{W}} \rho_{W}^{\alpha} \omega_{W}^{i\alpha} Q_{W}^{\alpha} \delta\left(\mathbf{x} - \mathbf{x}^{W}\right)$$
$$\alpha = w, n; \quad i = 1, \dots, N_{\alpha} \tag{4.26}$$

Comparison of the term in braces with equation (3.126) in light of the usual assumption that the advective part of the solid phase material derivative is negligible (i.e., $\mathrm{D}^{s}/\mathrm{D}t \approx \partial/\partial t$) provides the identity that allows the terms in braces to be replaced by phase exchange terms and a pumping term so that we have:

$$\omega^{i\alpha} \left\{ e_{\alpha s}^{\alpha} + e_{wn}^{\alpha} + \sum_{W=1}^{N_{W}} \rho_{W}^{\alpha} Q_{W}^{\alpha} \delta\left(\mathbf{x} - \mathbf{x}^{W}\right) \right\}$$
$$+ \varepsilon \rho^{\alpha} s^{\alpha} \frac{\partial \omega^{i\alpha}}{\partial t} + \rho^{\alpha} \mathbf{q}^{\alpha} \cdot \nabla \omega^{i\alpha} + \nabla \cdot \left(\varepsilon^{\alpha} \mathbf{j}^{i\alpha}\right)$$
$$= \varepsilon^{\alpha} r^{i\alpha} + e_{wn}^{i\alpha} + e_{\alpha s}^{i\alpha} + \sum_{W=1}^{N_{W}} \rho_{W}^{\alpha} \omega_{W}^{i\alpha} Q_{W}^{\alpha} \delta\left(\mathbf{x} - \mathbf{x}^{W}\right)$$
$$\alpha = w, n; \quad i = 1, \dots, N_{\alpha} \tag{4.27}$$

Lastly, the mass exchange and pumping terms are collected so that the final form of the distribution equation is:

$$
\varepsilon \rho^\alpha s^\alpha \frac{\partial \omega^{i\alpha}}{\partial t} + \rho^\alpha \mathbf{q}^\alpha \cdot \nabla \omega^{i\alpha} + \nabla \cdot \left(\varepsilon^\alpha \mathbf{j}^{i\alpha} \right)
$$
$$
= \varepsilon^\alpha r^{i\alpha} + \left(e^{i\alpha}_{wn} - \omega^{i\alpha} e^\alpha_{wn} \right) + \left(e^{i\alpha}_{\alpha s} - \omega^{i\alpha} e^\alpha_{\alpha s} \right)
$$
$$
+ \sum_{W=1}^{N_W} \rho^\alpha_W \left(\omega^{i\alpha}_W - \omega^{i\alpha} \right) Q^\alpha_W \delta \left(\mathbf{x} - \mathbf{x}^W \right)
$$
$$
\alpha = w, n; \quad i = 1, \ldots, N_\alpha \tag{4.28}
$$

Note that the sum of each term in this equation over all species i in fluid phase α is zero. This characteristic makes the distribution form distinct from the previous forms in that it is not a statement of conservation of species mass which, when summed over all species, would give a statement of *total mass conservation*.

An alternative procedure for obtaining the distribution equations for the fluid phases and for the solid is relatively straightforward. Start with equation (4.1) and apply the product rule to the first two terms in a fashion analogous to equations (4.24) and (4.25). Then regrouping of terms provides:

$$
\omega^{i\alpha} \left\{ \frac{\partial \left(\varepsilon^\alpha \rho^\alpha \right)}{\partial t} + \nabla \cdot \left(\varepsilon^\alpha \rho^\alpha \mathbf{v}^\alpha \right) \right\} + \varepsilon^\alpha \rho^\alpha \frac{\partial \omega^{i\alpha}}{\partial t} + \varepsilon^\alpha \rho^\alpha \mathbf{v}^\alpha \cdot \nabla \omega^{i\alpha} + \nabla \cdot \left(\varepsilon^\alpha \mathbf{j}^{i\alpha} \right)
$$
$$
= \varepsilon^\alpha r^{i\alpha} + \sum_{\beta \neq \alpha} e^{i\alpha}_{\alpha\beta} + \sum_{W=1}^{N_W} \rho^\alpha_W \omega^{i\alpha}_W Q^\alpha_W \delta \left(\mathbf{x} - \mathbf{x}^W \right)
$$
$$
\alpha = w, n, s; \quad i = 1, \ldots, N_\alpha \tag{4.29}
$$

We can invoke mass conservation equation (2.116) for a phase to replace the term in brackets with mass exchange and pumping terms:

$$
\omega^{i\alpha} \left[\sum_{\beta \neq \alpha} e^\alpha_{\alpha\beta} + \sum_{W=1}^{N_W} \rho^\alpha_W Q^\alpha_W \delta \left(\mathbf{x} - \mathbf{x}^W \right) \right] + \varepsilon^\alpha \rho^\alpha \frac{\partial \omega^{i\alpha}}{\partial t} + \varepsilon^\alpha \rho^\alpha \mathbf{v}^\alpha \cdot \nabla \omega^{i\alpha}
$$
$$
+ \nabla \cdot \left(\varepsilon^\alpha \mathbf{j}^{i\alpha} \right) = \varepsilon^\alpha r^{i\alpha} + \sum_{\beta \neq \alpha} e^{i\alpha}_{\alpha\beta} + \sum_{W=1}^{N_W} \rho^\alpha_W \omega^{i\alpha}_W Q^\alpha_W \delta \left(\mathbf{x} - \mathbf{x}^W \right)
$$
$$
\alpha = w, n, s; \quad i = 1, \ldots, N_\alpha \tag{4.30}
$$

Terms in this equation may be gathered together so that the result is the same as equation (4.28) with the exception of the appearance of $\varepsilon^\alpha \mathbf{v}^\alpha$ instead of $\mathbf{q}^\alpha$ and the fact that the equation applies to the solid phase as well as the two fluid phases:

$$
\varepsilon^\alpha \rho^\alpha \frac{\partial \omega^{i\alpha}}{\partial t} + \varepsilon^\alpha \rho^\alpha \mathbf{v}^\alpha \cdot \nabla \omega^{i\alpha} + \nabla \cdot \left(\varepsilon^\alpha \mathbf{j}^{i\alpha} \right)
$$
$$
= \varepsilon^\alpha r^{i\alpha} + \sum_{\beta \neq \alpha} \left(e^{i\alpha}_{\alpha\beta} - \omega^{i\alpha} e^\alpha_{\alpha\beta} \right) + \sum_{W=1}^{N_W} \rho^\alpha_W \left(\omega^{i\alpha}_W - \omega^{i\alpha} \right) Q^\alpha_W \delta \left(\mathbf{x} - \mathbf{x}^W \right)
$$
$$
\alpha = w, n, s; \quad i = 1, \ldots, N_\alpha \tag{4.31}
$$

Because $\mathbf{v}^s \cdot \nabla \omega^{j\alpha}$ is negligible compared to $\mathbf{v}^\alpha \cdot \nabla \omega^{j\alpha}$ for the fluid phases,[3] $\varepsilon^\alpha \mathbf{v}^\alpha$ may be replaced with $\mathbf{q}^\alpha$ for $\alpha = w, n$ with negligible error to reproduce equation (4.28).

For the solid phase, we neglect the advective, diffusion, and pumping terms in equation (4.31) to obtain:

$$(1-\varepsilon)\rho^s \frac{\partial \omega^{is}}{\partial t} = (1-\varepsilon)r^{is} + \left(e_{ws}^{is} - \omega^{is}e_{ws}^s\right) + \left(e_{ns}^{is} - \omega^{is}e_{ns}^e\right)$$

$$i = 1, \ldots, N_s \qquad (4.32)$$

This last equation is the distribution form of the species equation for the solid phase and is an alternative to the conservation form given by equation (4.10) derived using the direct approach. Consistent with the property of the distribution result for the fluid phases, the sum of any of the four terms in the distribution equation for the s phase over all species in that phase is zero.

4.2.4 Summary

This section has been concerned with obtaining forms for the transport equation that make use of the Darcy velocity. Even in working through to some particular forms using three different approaches, we still have not provided simplifications to the forms that apply under some particular limiting cases. Some of the important and commonly encountered limiting cases include constant phase density ρ^α; no mass exchange between phases such that $e_{\alpha\beta}^{i\alpha} = 0$; moles per volume of phase, c^α, constant; and no chemical reactions within phase α such that $r^{i\alpha} = 0$. Additionally, cases occur where only some of the species transfer between phases or participate in intraphase reactions. It is therefore important to understand the general assumptions that have been applied in the direct, rigorous, and distributional methods and to be able to apply additional restrictions. The efforts in this section have focused exclusively on the terms involving phase velocities. The equations developed still need to be supplemented with information for modeling dispersion, interphase transport, and intraphase reactions. In this summary, we provide a synopsis of the features of each of the three derivations.

All of the derivations share the approximation that the movement of the solid phase is so slow that terms involving $\mathbf{v}^s \cdot \nabla$ are negligibly small. The direct approach involves the assumption that all movement of the solid phase, including deformation, can be neglected. The direct approach results are equations (4.9) for the fluid phases and (4.10) for the solid phase. The rigorous approach incorporates the deformation of the solid phase into the derivation by making use of the mass conservation equation for the solid and expressions for the solid and matrix compressibilities. The result of this approach is equation (4.20) for the fluid phases. No explicit equation for the solid phase is obtained.

The distribution form of the equation follows from use of the mass conservation equation for the fluid phases in conjunction with the rigorous equation result. The distributional form for fluid phase transport of a chemical species is equation (4.28),

---

[3] This may alternatively be justified by noting that $\partial \omega^{j\alpha}/\partial t \approx D^s \omega^{j\alpha}/Dt$.

which may also be obtained from a simpler path starting with the species conservation equations without passing through the rigorous form. This latter approach also provides the distributional form for a chemical species in the solid phase as equation (4.32). We comment that both the direct and distributional forms of the solid phase species equations assume that dispersion and pumping are negligible in addition to solid phase advection. The distributional form is mathematically as complete an expression of species transport as the rigorous form. However, because it describes how species distribute within and between phases while the rigorous form describes how mass is conserved, they may provide different solutions when subjected to additional assumptions or to the errors that are introduced by computational efforts to solve the equations.

A large array of assumptions can be applied that result in the alternative forms reducing to a single form. Many of these assumptions are reasonable descriptions of actual behavior encountered and should be applied when appropriate. By employing these approximations and studying their impact on the equation forms, one can gain insights into the relative importance of the various term sand make a prudent selection of the actual equation employed in a numerical solver.

In Figure 4.1 we provide a summary of the various strategies that can be used to derive the transport equation.

4.3 CLOSURE RELATIONS FOR THE DISPERSION VECTOR

The vector $\mathbf{j}^{i\alpha}$, referred to as the *mass dispersion vector*, actually accounts for both *diffusion* and *dispersion*. Thus, to aid our discussion, we write the overall dispersion vector in terms of its two mechanisms as:

$$\varepsilon^\alpha \mathbf{j}^{i\alpha} = \varepsilon^\alpha \mathbf{j}^{i\alpha}_{\text{dif}} + \varepsilon^\alpha \mathbf{j}^{i\alpha}_{\text{dis}} \quad \alpha = w, n, s; \quad i = 1, \dots, N_\alpha \tag{4.33}$$

where $\mathbf{j}^{i\alpha}_{\text{dif}}$ is the diffusion of species i in phase α and $\mathbf{j}^{i\alpha}_{\text{dis}}$ is the dispersion of species i in the α phase.

Diffusion is typically described as occurring when a dissolved chemical species moves within a phase from a region of higher concentration to a region of lower concentration due to random motion of molecules. Although diffusion within solids does occur, it is a very slow process and will not be considered here. We will thus employ the constitutive assumption:

$$\varepsilon^s \mathbf{j}^{is} = 0 \quad i = 1, \dots, N_s \tag{4.34}$$

Additionally, diffusion of mass can be enhanced or induced by large temperature or pressure gradients. These effects will be neglected here. The sum of all the diffusion vectors within a phase is zero such that the spreading of a species within a phase is balanced by the spreading of other species. The theory of diffusion in a fluid phase modeled at the microscale is based on *Fick's law*. The microscale generalized form of this law [1] indicates that diffusion depends on the gradients of chemical potential of each of the species in the phase. For simplicity, and consistent with the form commonly used in studying multiphase flows, we will replace the chemical potential with the mass fraction and assume that the macroscale diffusion of species

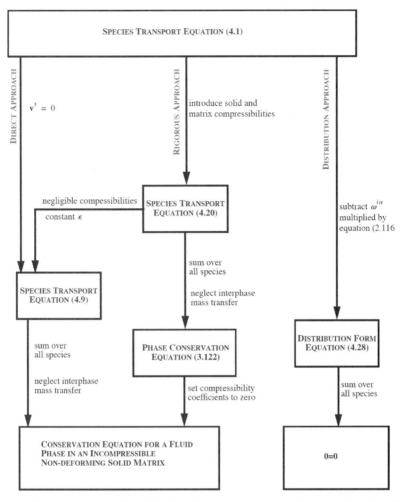

Figure 4.1: Alternative methods for obtaining the transport equation form used in simulations. All methods make use of the assumption that terms involving $\mathbf{v}^s \cdot \nabla$ are negligible.

i depends only on the gradient of the mass fraction of this species in the phase of interest. By analogy with the microscale situation, we write:

$$\varepsilon^\alpha \mathbf{j}_{\mathrm{dif}}^{i\alpha} = -\varepsilon^\alpha \rho^\alpha D_{\mathrm{dif}}^{i\alpha} \nabla \omega^{i\alpha} \quad \alpha = w, n; \quad i = 1, \ldots, N_\alpha \tag{4.35}$$

where $D_{\mathrm{dif}}^{i\alpha}$ is the *diffusion coefficient* of species i in the α phase and the negative sign is included to reflect the fact that diffusion will occur from regions of high mass fraction to low mass fraction.

When this equation is written for a microscale system, the diffusion coefficient is on the order of $10^{-5}\,\mathrm{cm}^2/\mathrm{sec}$ in a liquid and $10^{-2}\,\mathrm{cm}^2/\mathrm{sec}$ in a gas. However, at the macroscale, two features serve to decrease this coefficient. First, diffusion is represented as being driven by gradients of macroscale mass fraction to achieve macroscale homogeneity of concentration. This does not account for some microscale variations which would increase the mathematical driving force (i.e., a system can have macroscopically uniform concentration while having a microscopically non-uniform concentration). Second, the presence of the solid grains restricts the spreading of the chemical such that it must follow the tortuous path provided by the pore structure. This slows the apparent movement of the material between two macroscopic points. In fact, for an anisotropic pore space distribution, the solid phase may inhibit the spreading of a chemical species differently in different directions. This situation would require that the diffusion coefficient be replaced by a diffusion tensor so that:

$$\varepsilon^\alpha \mathbf{j}_{\text{dif}}^{i\alpha} = -\varepsilon^\alpha \rho^\alpha \mathbf{D}_{\text{dif}}^{i\alpha} \cdot \nabla \omega^{i\alpha} \quad \alpha = w, n; \quad i = 1, \ldots, N_\alpha \tag{4.36}$$

where $\mathbf{D}_{\text{dif}}^{i\alpha}$ is a 3×3 symmetric tensor. Since anisotropy of the diffusion tensor is caused by the solid structure, it seems reasonable that the directional properties of the diffusion tensor would be the same as for the intrinsic permeability such that $\mathbf{D}_{\text{dif}}^{i\alpha}$ would be proportional to $\mathbf{k}^s$. At the microscale, the diffusion coefficient depends only on the chemical species present. At the macroscale, it depends on the chemical species as well as the macroscale length and the structure of the porous medium. Modeling diffusion in a porous medium is not an easy problem.

However, in comparison to modeling dispersion, diffusion is straightforward. Whereas diffusion is due to movement of molecules relative to each other, dispersion is caused by nonuniformities in flow that cause mixing of the chemical species in a fluid. In a porous medium, the magnitude and distribution of fluid velocity, concentration inhomogeneities, and pore size distribution contribute to phase motion that causes the chemical species to mix. Dispersion is sometimes referred to as *mechanical dispersion* because it is the mechanical interaction of the fluid with the porous medium that causes mixing to occur.

The relative importance of diffusion and dispersion is related to the velocity of the flow. When the fluid is at rest, microscopically, no dispersion occurs and diffusion is the dominant cause of spreading of a contaminant. When a phase is flowing, the relative contribution of dispersion to mixing can render diffusion to be unimportant.

Dispersion is a mixing process. Thus, it seems reasonable to assume that when no concentration gradient exists, a chemical species will not mix further such that dispersion, like diffusion, is dependent on the concentration gradient.

These considerations suggest that from a mathematical perspective dispersion is similar to diffusion, except that the coefficient that accounts for the mixing is different. Thus, by analogy with equation (4.36) we propose:

$$\varepsilon^\alpha \mathbf{j}_{\text{dis}}^{i\alpha} = -\varepsilon^\alpha \rho^\alpha \mathbf{D}_{\text{dis}}^{i\alpha} \cdot \nabla \omega^{i\alpha} \quad \alpha = w, n; \quad i = 1, \ldots, N_\alpha \tag{4.37}$$

We expect, also, that $\mathbf{D}_{\text{dis}}^{i\alpha}$ will depend on the velocity of flow, with dispersion being zero when there is no flow. The dispersion coefficient $\mathbf{D}_{\text{dis}}^{i\alpha}$ is given in tensor form because dispersive mixing can depend on the direction of flow, even in an isotropic medium.

Based on some theoretical reasoning and experimental work, the dispersion coefficient is usually given the form:

$$\varepsilon^{\alpha}\mathbf{D}_{dis}^{i\alpha} = \alpha_{T}|\mathbf{q}^{\alpha}|\mathbf{I} + (\alpha_{L} - \alpha_{T})\frac{\mathbf{q}^{\alpha}\mathbf{q}^{\alpha}}{|\mathbf{q}^{\alpha}|} \quad \alpha = w, n; \quad i = 1, \ldots, N_{\alpha} \tag{4.38}$$

where α_{T} and α_{L} are called the *transverse* and *longitudinal* dispersivity, respectively, because they account for mixing in the directions normal to and aligned with the direction of flow, respectively. The longitudinal dispersivity is usually from 3 to about 80 times the size of the transverse dispersivity indicating that mixing in the direction of flow is greater than mixing orthogonal to the flow. Also note that the proposed functional form for $\varepsilon^{\alpha}\mathbf{D}_{dis}^{i\alpha}$ is the same for all chemical species. This implies that the mechanical mixing process impacts all species similarly. In fact, acquisition of data to support a more detailed model with species- or phase-dependent coefficients α_{L} and α_{T} cannot be justified based on any expected improved accuracy of this model.

If we examine equation (4.38) carefully, we see that the tensorial nature of $\mathbf{D}_{dis}^{i\alpha}$ is obtained solely due to the fact that the Darcy velocity can be different in each of the flow directions. The form proposed does not take into account anisotropy of the medium. The problem of modeling dispersion in an anisotropic medium presents serious theoretical and experimental challenges. We note that the anisotropy will be due to both the structure of the solid medium and the properties of the flow field. Here, we will not consider a dispersion coefficient more complex than that presented in equation (4.38).

The difficulties inherent in parameterizing a more complex correlation and determining an appropriate dependence on the velocity field have precluded wide spread use of such a representation. Usually, then, the dispersion vector employed is that for an isotropic medium obtained as the sum of equations (4.35) and (4.37) such that:

$$\varepsilon^{\alpha}\mathbf{j}^{i\alpha} = -\rho^{\alpha}\varepsilon^{\alpha}\mathbf{D}^{i\alpha} \cdot \nabla \omega^{i\alpha} \quad \alpha = w, n; \quad i = 1, \ldots, N_{\alpha} \tag{4.39}$$

where the diffusion/dispersion coefficient, $\mathbf{D}^{i\alpha}$, is approximated as:

$$\varepsilon^{\alpha}\mathbf{D}_{dis}^{\alpha} = \varepsilon^{\alpha}D_{dif}^{i\alpha}\mathbf{I} + \alpha_{T}|\mathbf{q}^{\alpha}|\mathbf{I} + (\alpha_{L} - \alpha_{T})\frac{\mathbf{q}^{\alpha}\mathbf{q}^{\alpha}}{|\mathbf{q}^{\alpha}|}$$
$$\alpha = w, n; \quad i = 1, \ldots, N_{\alpha} \tag{4.40}$$

The requirement that the dispersion vectors in a phase sum to zero is usually imposed by default. Typically, a phase is modeled using $N_{\alpha} - 1$ of the species balance equations (therefore requiring only $N_{\alpha} - 1$ values of $D_{dif}^{i\alpha}$) in conjunction with the mass balance equation for the entire phase. This latter equation, developed in the previous chapter, inherently requires that the sum of the dispersion vectors be zero. Therefore, the constitutive form of the dispersion vector for species N_{α} is not obtained from equation (4.39) with $i = N_{\alpha}$ but is, by default, obtained from equation (4.5):

$$\varepsilon^{\alpha}\mathbf{j}^{N\alpha\alpha} = -\sum_{i=1}^{N_{\alpha}-1}\varepsilon^{\alpha}\mathbf{j}^{i\alpha} = \sum_{i=1}^{N_{\alpha}-1}\rho^{\alpha}\varepsilon^{\alpha}\mathbf{D}^{i\alpha} \cdot \nabla \omega^{i\alpha} \tag{4.41}$$

The different character of this equation reinforces the idea that the $N_\alpha - 1$ species to be modeled using the species balance equation are those present in the smallest amounts. The mass fraction of the dominant phase is then obtained from equation (4.4). When the diffusive part of the diffusion/dispersion coefficient is negligible in comparison to the dispersive part, the diffusion/dispersion coefficient is independent of the chemical species and the dispersion model for the N_αth species is the same as for all other species in the α phase.

When the diffusive mixing is much less than that of dispersion, the diffusion coefficient, $D_{\mathrm{dif}}^{i\alpha}$, may be neglected without significant error. The model of the dispersion vector still requires specification of the two parameters α_{T} and α_{L}. The values of these parameters for field simulations (e.g., α_{L} typically on the order of tens of meters and $\alpha_{\mathrm{T}}/\alpha_{\mathrm{L}}$ typically on the order of 0.1) are actually dependent on the scale of the problem. Thus these coefficients can be approximated by fits of measured data to simulated concentration profiles. Also, perhaps as an indication that the theoretical models of dispersion are not particularly reliable or robust, some simulations are performed using a constant, scalar value for the diffusion/dispersion coefficient to simulate spreading.

We recall, also, that $\rho^\alpha \omega^{i\alpha} = c^{i\alpha} M^i$ where $c^{i\alpha}$ is the moles of species i pervolume of α phase and M^i is the molecular weight of i. When the spatial dependence of ρ^α can be neglected, the dispersion vector may be written in terms of molar concentrations for use in equations (4.11) or (4.23) as:

$$\frac{\varepsilon^\alpha \mathbf{j}^{i\alpha}}{M^i} = -\varepsilon^\alpha \mathbf{D}^{i\alpha} \cdot \nabla c^{i\alpha} \quad \alpha = w, n; \quad i = 1, \ldots, N_\alpha \tag{4.42}$$

4.4 CHEMICAL REACTION RATES

The quantity $r^{i\alpha}$ that appears in the species transport equation is the rate of production of mass of chemical species i due to chemical reactions occurring within phase α. Because the reactions considered involve only species within a single phase and not exchanges between phases or processes occurring at the interface between phases, $r^{i\alpha}$ is sometimes called the *homogeneous reaction rate*. Because these reactions cannot create mass, they are governed by the constraint:

$$\sum_{i=1}^{N_\alpha} r^{i\alpha} = 0 \quad \alpha = w, n, s \tag{4.43}$$

In many instances, it is convenient to work with chemical reactions expressed as rate of production of moles of a chemical species. The *molar rate of production*, $R^{i\alpha}$, is related to the *mass rate of production* by:

$$R^{i\alpha} = \frac{r^{i\alpha}}{M^i} \quad \alpha = w, n, s \tag{4.44}$$

Because chemical reactions do not necessarily conserve the number of moles in the system, the sum of the molar reaction rates is not zero. Rather, the sum of these reactions weighted by the molecular weight of each species will be zero, as can be seen by substitution of equation (4.44) into equation (4.43):

$$\sum_{i=1}^{N_\alpha} M^i R^{i\alpha} = 0 \quad \alpha = w, n, s \tag{4.45}$$

Chemical reaction rate expressions are most often determined from experimental studies and involve a rate parameter, or parameters, and the concentrations of the chemical constituents involved in the reaction. The experimental studies are typically done on well-mixed systems such that the chemical reaction expressions are obtained in terms of microscale quantities and are most often expressed in terms of molar concentrations. For example, a first order decay reaction of a chemical species has the form:

$$R_{i\alpha} = -k_{i\alpha} c_{i\alpha} \tag{4.46}$$

which can be transformed directly to the mass rate form by multiplying by M^i to obtain:

$$r_{i\alpha} = -k_{i\alpha} \rho_\alpha \omega_{i\alpha} \tag{4.47}$$

If the temperature is essentially constant within an averaging region such that $k_{i\alpha}$ may be treated as a constant, this microscale reaction rate expression can be integrated directly to:

$$r^{i\alpha} = -k_{i\alpha} \rho^\alpha \omega^{i\alpha} \tag{4.48}$$

The scripting on k is left as a subscript to emphasize that this coefficient is obtained from experiments that provide a microscale value. For a first order reaction, the expression of the reaction is such that the microscale coeffecient applies directly. We note that the decay rate indicated here depends on the phase being considered, as the decay rate in phase w may be different from that in phase n, although radioactive decay has a rate constant that is independent of the phase.

If all rate equations were first order (i.e., the product of a constant coefficient and a chemical concentration), the challenge of modeling chemical reactions in porous media would be greatly simplifed. However, this is not the case. Consider, as a simple example, a reaction that is found to obey second order kinetics by laboratory experiments such that:

$$R_{i\alpha} = -k_{i\alpha} c_{i\alpha}^{\,2} \tag{4.49}$$

Conversion of this expression to the mass based form yields:

$$r_{i\alpha} = -\frac{k_{i\alpha}}{M^i} (\rho_\alpha \omega_{i\alpha})^2 \tag{4.50}$$

Although $k_{i\alpha}/M^i$ is constant, the mass fractions may not be constant in the REV. When the mass fraction is not constant within the REV, the average of the mass fraction squared will be less than the square of the average mass fraction because:

$$\left(\rho^\alpha \omega^{j\alpha} \right)^2 \le \frac{1}{\delta V^\alpha} \int_{\delta V^\alpha} \left(\rho_\alpha \omega_{i\alpha} \right)^2 dv_\xi \tag{4.51}$$

where the equality applies only if $\rho^\alpha \omega_{i\alpha}$ is uniform within the integration region. Therefore, if one assumes the macroscopic version of equation (4.50) has the form:

$$r^{i\alpha} = -\frac{k_{i\alpha}}{M^i} \left(\rho^\alpha \omega^{i\alpha} \right)^2 \tag{4.52}$$

the predicted rate of decay will actually be less than what would be observed if the microscale field is not uniform.

Despite this fact, options for improved representation of nonlinear reactions are difficult to prescribe, for the most part because they require understanding of the state of mixing at the microscale. When the REV size is large, errors introduced by not accounting for variations in the concentration within the REV can be significant. In some instances, a high value of pH at a microscopic point could cause precipitation of a solid phase. However, this pH spike could be missed if one describes the system using only macroscale, spatially averaged, values of pH. At the very least, one needs to be alert to errors that can arise when working with macroscale expressions for chemical reactions and cognizant of the fact that sub-scale variations can dramatically alter system behavior.

The purpose of this section has not been to identify the range of possibilities of forms of chemical reactions. Besides decay reactions, one needs to consider generation of species. Also, the forms of the reaction rate expressions may depend on more than one species such that species transport equations become coupled (i.e., mass fractions of more than one species appear in the equation) and must be solved as a set rather than independently. The large number of chemical species that can be present in a subsurface system, the inherent difficulties in modeling reactions, issues of scale that impact the closure relation for dispersion, and the coupling of species equations are all issues that lead one to describe modeling of subsurface transport with confidence as either a nearly impossible exercise or as an area ripe for an influx of creative research ideas. We, of course, support the latter view.

4.5 INTERPHASE TRANSFER TERMS

The last process described in the species transport equation that we have to parameterize is the transfer of material between phases. Even when the phases are modeled as "immiscible" some small exchange of chemical species between the phases occurs. Thus, an oil spill contaminates water, and some water molecules will be transferred into the oil even if an interface between these phases can be identified. Transfer of mass between the solid and fluid can take place by dissolution of the solid, adsorption of a layer of chemical species from the fluid onto the solid surface, or desorption of species from the solid into the fluid.

These processes are all accounted for in the transport equations by terms of the form $e_{\alpha\beta}^{i\alpha}$ which describe the *transfer of species i in phase β into phase α*, or more precisely transfer from the $\alpha\beta$ interface. We will model systems here characterized by transfer between phases with no accumulation of material at the interface such that $e_{\alpha\beta}^{i\alpha} + e_{\alpha\beta}^{i\beta} = 0$. This indicates that the flux of species i into the α phase is exactly equal to its flux out of the β phase. To model this flux, we will make use of approximations that relate fluxes to concentrations in the phases.

There are two different conceptual models that can be used to model the interphase transfer. The first, the more complicated model, is the *kinetic formulation* or the *rate-limited model*. In this approach, the transfer of chemical species between phases is modeled as a rate process with the driving force for the process being the difference between the actual concentration and some equilibrium concentration. The system behaves in such a way that it moves toward equilibrium concentrations in each phase, but the rate at which it approaches equilibrium must be accounted for. The second model is the *equilibrium formulation*. This method considers the rate of transfer between phases within an REV to be so fast that at any point the concentrations of a chemical species are distributed between phases according to some equilibrium relation. Although the concentration need not be uniform within such a system, such that it can vary from location to location, the dynamics of interphase transfer are modeled as being instantaneous so that local equilibrium of concentration between phases exists. The advantage of this second approximation, from a computational perspective, is that if a concentration is solved for in one phase, the corresponding concentration in an adjacent phase is immediately known without having to solve a transport equation for concentration in that phase. The disadvantage is that equilibrium between phases is not achieved in many cases.

Both the kinetic and equilibrium models require that information be available concerning the *partitioning* of a chemical species among phases at equilibrium. In complicated mixtures of phases, the concentrations of each species in a phase impact the equilibrium concentrations of all species in that phase. Thus the equilibrium relations could be very complex functions. Additionally, equilibrium information is needed about each species in each phase. The development of equations for transport between phases is an active research area as efforts continue to account for the rate of interphase transfer as it is impacted by features such as the amount of interfacial area between phases, the coefficient describing the rate of transfer, dispersion processes that move species to an interface so that they are available for transfer to an adjacent phase, and surfactants that impact the surface tension between phases. We will not develop such complex functions, as the approach to modeling interphase transfer can be illustrated in the context of a simpler model.

4.5.1 Kinetic Formulation

The objective of this subsection is to provide a constitutive form for $e_{\alpha\beta}^{i\alpha}$ that can be used in modeling interphase transport. This quantity accounts for transport at the macroscale; thus it incorporates the impact of two microscale processes: (1) transport of material from the phase to the interphase between phases, and (2) transport across the interface. At the interface, the chemical species concentration can be considered to be at equilibrium with its concentration in the adjacent phase at the interface. However, the macroscale perspective is not able to describe the microscale interface concentrations. Thus, the transfer expression must be based on concentrations within the adjacent phases and equilibrium expressions involving those bulk concentrations.

The exchange term $e_{\alpha\beta}^{i\alpha}$ is positive when transfer is from the β phase to the α phase. In fact, such a transfer will take place when the chemical potential of species i is higher in the β phase than in the α phase. However, because modeling of chemical potential at the macroscale is complex and would require thermodynamic closure relations, concentration is typically used as a surrogate. Let us define $\omega_{eq\beta}^{i\alpha}$ as the concentration of species i in the α phase that would be in equilibrium with the actual concentration, $\omega^{i\beta}$, of i in the β phase. If we consider this concentration to be independent of other chemical species present in the system, the equilibrium relation may be expressed in terms of a function, $f_{eq\beta}^{i\alpha}(\omega^{i\beta})$ as:

$$\omega_{eq\beta}^{i\alpha} = f_{eq\beta}^{i\alpha}(\omega^{i\beta}) \tag{4.53}$$

Then the exchange term is approximated as:

$$e_{\alpha\beta}^{i\alpha} = -\kappa_{\alpha\beta}^{i\alpha}[\omega^{i\alpha} - f_{eq\beta}^{i\alpha}(\omega^{i\beta})] = -\kappa_{\alpha\beta}^{i\alpha}[\omega^{i\alpha} - \omega_{eq\beta}^{i\alpha}] \tag{4.54}$$

where the *coefficient of mass transfer*, $\kappa_{\alpha\beta}^{i\alpha}$, will, in general, be a function of the Darcy velocity, dispersion coefficient, species concentrations, temperature, and amount of interfacial area between the α and β phases per REV [2]. If this functional dependence is very complex, the use of equation (4.54) simply replaces one complicated function, $e_{\alpha\beta}^{i\alpha}$, with two others, $f_{eq\beta}^{i\alpha}(\omega^{i\beta})$ and $\kappa_{\alpha\beta}^{i\alpha}$. This situation could not be described as progress. However, the form provided by equation (4.54) has the property that the exchange term goes to zero when the concentrations in the phases are in equilibrium with respect to each other. Also, the coefficient of transfer can be approximated as a constant for some applications when the interfacial area in the system does not vary significantly with space and the distance from equilibrium of the concentration field is small.

In the preceding, transfer of the chemical species is viewed from the perspective of the α phase. We can also write corresponding formulas based on mass transfer behavior in the β phase. Equations similar to equations (4.53) and (4.54) with the superscripts and subscripts modified to describe the β phase are, respectively:

$$\omega_{eq\alpha}^{i\beta} = f_{eq\alpha}^{i\beta}(\omega^{i\alpha}) \tag{4.55}$$

and:

$$e_{\alpha\beta}^{i\beta} = -\kappa_{\alpha\beta}^{i\beta}[\omega^{i\beta} - f_{eq\alpha}^{i\beta}(\omega^{i\alpha})] = -\kappa_{\alpha\beta}^{i\beta}[\omega^{i\beta} - \omega_{eq\alpha}^{i\beta}] \tag{4.56}$$

Because no mass accumulates at the interface, the flux across the interface must be the same, regardless of which phase is referenced to describe the process. Therefore:

$$e_{\alpha\beta}^{i\alpha} = -e_{\alpha\beta}^{i\beta} = -\kappa_{\alpha\beta}^{i\alpha}[\omega^{i\alpha} - f_{eq\beta}^{i\alpha}(\omega^{i\beta})] = \kappa_{\alpha\beta}^{i\beta}[\omega^{i\beta} - f_{eq\alpha}^{i\beta}(\omega^{i\alpha})] \tag{4.57}$$

Utilization of the preceding equation requires that a form of the equilibrium function be obtained. If we consider the case of unsaturated flow where one phase is water and the nonwetting phase is air, a relation such as Henry's law may used as the equilibrium relation for the oxygen species with:

$$\omega_{\text{eqn}}^{iw} = f_{\text{eqn}}^{iw}(\omega^{in}) = \frac{H^{iwn}\rho^{n}}{\rho^{w}}\omega^{in} \tag{4.58}$$

where H^{iwn} is the Henry's law constant and is approximately 2×10^{-2} for oxygen in an aqueous system. Thus if we model the w phase, the exchange term is:

$$e_{wn}^{iw} = -\kappa_{wn}^{iw}\left[\omega^{iw} - \frac{H^{iwn}\rho^{n}}{\rho^{w}}\omega^{in}\right] \tag{4.59}$$

A linear expression of this form is often employed when modeling the mass exchange between two fluids where the parameter H^{iwn} is used to describe a linear equilibrium relation. More complex equilibrium relations may also be employed that fit data obtained over a range of values of mass fractions of interest.

Modeling the exchange of a chemical species i between the solid and one of the fluids, for example the w phase, typically makes use of:

$$e_{ws}^{is} = -e_{ws}^{iw} = -\kappa_{ws}^{is}\left[\omega^{is} - f_{\text{eqw}}^{is}(\omega^{iw})\right] = -\kappa_{ws}^{is}(\omega^{is} - \omega_{\text{eqw}}^{is}) \tag{4.60}$$

Then an *adsorption isotherm* is employed to relate the equilibrium values of ω^{is} and ω^{iw}. Here we will mention three isotherms that are in common use. We note that the forms of the isotherms, and particularly the values of the coefficients in the isotherms, take on different values depending on whether one expresses them in terms of mass fractions, mole fractions, mass per volume, or moles per volume. It is important that the values of parameters be adjusted between the different formulations to obtain consistency. Here, we will formulate all the isotherms in terms of mass fractions.

A *linear isotherm* is expressed:

$$\omega_{\text{eqw}}^{is} = K^{iws}\omega^{iw} \tag{4.61}$$

where K^{iws} is a dimensionless constant and is the constant slope of a plot of ω^{is} vs. ω^{iw}. One standard notation for this expression involves replacing K^{iws} by $K_{d}^{iws}\rho^{w}$, where K_{d}^{iws} is called the *distribution coefficient*. Over a range of values of mass fraction where this isotherm is applied, ρ^{w} is usually considered to be constant. If concentrations vary over a wide range, it may not be possible to use a single value of K^{iws} (or K_{d}^{iws}) to fit the data. In that case a piecewise linear fit may be used for a more complex isotherm. The linear isotherm is attractive because of its particularly simple functional form.

The *Freundlich isotherm* is more general than the linear isotherm and is given by:

$$\omega_{\text{eqw}}^{is} = K^{iws}(\omega^{iw})^{1/n_{\text{F}}} \tag{4.62}$$

where K^{iws} and n_F are constants. When $n_F = 1$, the Freundlich isotherm reduces to the linear isotherm. This isotherm typically applies over a range of values of mass fraction of interest to the problem at hand. One must be careful not to apply it beyond the range of measured data used to parameterize the expression.

The *Langmuir isotherm* is a relation describing the adsorption process that is physically based, rather than just a curve fit. It relies on the idea that chemical species adsorb in a single layer onto available sites on the solid. At low concentrations, the adsorption process is linear as there is no competition for adsorption sites. However, when the concentration of a chemical species becomes large, the sites on the solid become filled and species must compete for these locations. Thus the amount of adsorbed species approaches a constant limit. These ideas are incorporated into the Langmuir isotherm by the relation:

$$\omega_{eqw}^{is} = \frac{K^{iws}\omega_{max}^{is}\omega^{iw}}{1 + K^{iws}\omega^{iw}} \tag{4.63}$$

where K^{iws} and ω_{max}^{is} are constants. When the mass fraction in the fluid is small such that $K^{iws}\omega^{iw} \ll 1$, the Langmuir isotherm simplifies to the linear isotherm:

$$\omega_{eqw}^{is} = K^{iws}\omega_{max}^{is}\omega^{iw} \tag{4.64}$$

When ω^{iw} is large such that $K^{iws}\omega^{iw} \gg 1$, the Langmuir isotherm indicates that the amount of species i adsorbed does not depend on ω^{iw} but approaches a constant value with:

$$\omega_{eqw}^{is} = \omega_{max}^{is} \tag{4.65}$$

Although the Langmuir isotherm is attractive because it incorporates the fact that there is a limit to how much chemical species can be adsorbed on a solid, it does have some shortcomings. First, as it has been expressed, if there are multiple species dissolved in the w phase they can compete for available adsorption sites. Solutes that do not adsorb or react are called *conservative species* because their amount in the fluid phase changes only due to mass flux into and out of a region. Solutes that react or adsorb are referred to as *nonconservative species*, and it is the competition among the nonconservative species that changes the value of ω_{max}^{is} for each species. As long as the concentrations of all nonconservative species are small such that the Langmuir isotherm behaves linearly for each solute, this issue does not arise.

A second problem in modeling adsorption on a solid in multiphase flow is that the Langmuir isotherm does not reflect the fact that all fluid phases can be sources of constituents that can adsorb on the solid. Thus the equilibrium amount of i adsorbed on the solid could depend on the mass fractions of species i in all fluid phases. If we are studying a situation where the wetting phase completely coats the solid such that no other fluid phases are in contact with the solid, this problem will not arise. However, it will be important to model the mass exchange between the fluids accurately.

Although the discussion of shortcomings of the Langmuir isotherm is more extensive than that for the Freundlich isotherm, this does not necessarily imply that the Langmuir isotherm is inferior. Because it is based on assumptions about the physical and chemical behavior of the solid, we can identify situations when those assumptions break down such that the Langmuir isotherm may not model equilibrium adsorption well. On the other hand, the coefficients of the Freundlich isotherm are developed by curve fitting data. The isotherm applies only for the conditions under which the data was collected. It is not possible to speculate about the limits of validity of the fit since no physical insight is involved. When one tries to incorporate physical understanding into the development of coefficients for constitutive equations, the opportunity arises to consider limits of the equations when other physical situations exist. Thus the fact that we can identify situations in which the Langmuir isotherm may break down is helpful in that it identifies the need for research into other situations where more general isotherms need to be developed.

Modeling of the exchange of multiple species between phases is complicated when a large number of species is involved and relations are needed for equilibrium distributions of species among phases. Additionally, when modeling the kinetic transfer between phases, coefficients of mass transfer, $\kappa_{\alpha\beta}^{i\alpha}$, are also needed for each species. We note, however, that the condition $\kappa_{\alpha\beta}^{i\alpha} = 0$ implies that there is no transfer of species i between the α and β phases such that there is no need for an equilibrium function $f_{\text{eq}\beta}^{i\alpha}$. We can simply impose $e_{\alpha\beta}^{i\alpha} = e_{\alpha\beta}^{i\beta} = 0$. At the other extreme, when $\kappa_{\alpha\beta}^{i\alpha}$ is very large, the mass transfer rate will be finite only if $\omega_{\text{eq}\beta}^{i\alpha} \approx \omega^{i\alpha}$. This indicates that the concentrations in adjacent phases are essentially at equilibrium and equation (4.53) reduces to:

$$\omega^{i\alpha} = f_{\text{eq}\beta}^{i\alpha}(\omega^{i\beta}) \qquad (4.66)$$

This simplification defines the equilibrium formulation as the special case of the kinetic formulation that forms the topic of the next subsection.

4.5.2 Equilibrium Formulation

Besides being a special case of the kinetic formulation, the equilibrium formulation leads to a simplifed set of equations that have to be solved. Rather than solving the transport equation for species i in each phase, we only have to solve the transport equation for that species in one phase. Then the mass fraction in the adjacent phase is obtained from equilibrium equation (4.66).

The simplifications of the equilibrium formulation are not only convenient, they are also sometimes necessary. The necessity follows from the fact that modeling the rate of transfer $e_{\alpha\beta}^{i\alpha}$ as a product of a very large mass transfer coefficient and a very small difference between the actual and equilibrium mass fractions in the phase is difficult. The equilibrium formulation considers the mass transfer coefficient to be essentially infinite, and the difference between the actual and equilibrium concentrations in equation (4.54) is essentially zero. To manage this situation, we make use of the constraint that $e_{\alpha\beta}^{i\alpha} + e_{\alpha\beta}^{i\beta} = 0$ rather than treating each of the exchange terms in this sum separately. We also use the fact that the sum of the exchange of species is equal to the total mass exchange between phases as expressed in equation (4.6).

Complicating the situation is the fact that although one species in adjacent phases may partition between the phases according to the equilibrium relation, that species may distribute according to a kinetic formulation with another phase. Thus a truly comprehensive formulation of a multispecies, multiphase system would involve identifying each interphase exchange as kinetic or equilibrium and formulating that exchange appropriately. We have discussed formulation of kinetic exchanges in the last subsection; here we focus on equilibrium exchanges.

The basic maneuver in formulating transport equations to accommodate the equilibrium formulation for a species i in phases α and β is to add the transport equations for that species in those two phases together so that the mass exchange terms sum to zero. This can be employed using equations based on the direct, rigorous, or distribution forms of the transport equations. Here, by way of example, we will use the distribution form. We will consider the case where one species, species k, distributes itself between the w and n fluid phases according to the equilibrium model. We will also assume that no other species are exchanged between the w and n phases, that adsorption of species k onto the s phase, if any, must be modeled using a kinetic formulation, and that other species may be exchanged by the fluids and the solid. Thus we can write equation (4.28) for the w and n phases, respectively as:

$$
\begin{aligned}
&\varepsilon^w \rho^w \frac{\partial \omega^{kw}}{\partial t} + \rho^w \mathbf{q}^w \cdot \nabla \omega^{kw} + \nabla \cdot \left(\varepsilon^w \mathbf{j}^{kw} \right) \\
&= \varepsilon^w r^{kw} + \left(e^{kw}_{wn} - \omega^{kw} e^w_{wn} \right) + \left(e^{kw}_{ws} - \omega^{kw} e^w_{ws} \right) \\
&\quad + \sum_{W=1}^{N_W} \rho^w_W \left(\omega^{kw}_W - \omega^{kw} \right) Q^w_W \delta \left(\mathbf{x} - \mathbf{x}^W \right)
\end{aligned}
\tag{4.67}
$$

and:

$$
\begin{aligned}
&\varepsilon^n \rho^n \frac{\partial \omega^{kn}}{\partial t} + \rho^n \mathbf{q}^n \cdot \nabla \omega^{kn} + \nabla \cdot \left(\varepsilon^w \mathbf{j}^{kn} \right) \\
&= \varepsilon^n r^{kn} + \left(e^{kn}_{wn} - \omega^{kn} e^n_{wn} \right) + \left(e^{kn}_{ns} - \omega^{kn} e^n_{ns} \right) \\
&\quad + \sum_{W=1}^{N_W} \rho^n_W \left(\omega^{kn}_W - \omega^{kn} \right) Q^n_W \delta \left(\mathbf{x} - \mathbf{x}^W \right)
\end{aligned}
\tag{4.68}
$$

Since we know that the sum of the mass exchange terms between phases is zero and have specified that the only mass exchanged between the fluid phases is species k, the mass exchange terms in this example are related by:

$$
e^n_{wn} = e^{kn}_{wn} = -e^w_{wn} = -e^{kw}_{wn}
\tag{4.69}
$$

These conditions can be used in conjunction with equations (4.67) and (4.68) to eliminate all the terms involving mass exchange across the wn interface and obtain the single transport equation for species k:

$$
\begin{aligned}
&\varepsilon^w \rho^w \frac{\partial \omega^{kw}}{\partial t} + \rho^w \mathbf{q}^w \cdot \nabla \omega^{kw} + \nabla \cdot \left(\varepsilon^w \mathbf{j}^{kw} \right) \\
&+ \frac{1 - \omega^{kw}}{1 - \dot{\omega}^{kn}} \left[\varepsilon^n \rho^n \frac{\partial \omega^{kn}}{\partial t} + \rho^n \mathbf{q}^n \cdot \nabla \omega^{kn} + \nabla \cdot \left(\varepsilon^w \mathbf{j}^{kn} \right) \right] \\
&= \varepsilon^w r^{kw} + \left(e^{kw}_{ws} - \omega^{kw} e^w_{ws} \right) + \sum_{W=1}^{N_W} \rho^w_W \left(\omega^{kw}_W - \omega^{kw} \right) Q^w_W \delta \left(\mathbf{x} - \mathbf{x}^W \right) \\
&+ \frac{1 - \omega^{kw}}{1 - \omega^{kn}} \left[\varepsilon^n r^{kn} + \left(e^{kn}_{ns} - \omega^{kn} e^n_{ns} \right) + \sum_{W=1}^{N_W} \rho^n_W \left(\omega^{kn}_W - \omega^{kn} \right) Q^n_W \delta \left(\mathbf{x} - \mathbf{x}^W \right) \right]
\end{aligned}
\tag{4.70}
$$

The equilibrium relation that describes the distribution of species k between the w and n phases is the general form of equation (4.66):

$$\omega^{kn} = f_{\text{eqw}}^{kn}(\omega^{kw}) \tag{4.71}$$

Differentiation to obtain the time derivative and gradient of ω^{kn} provides:

$$\frac{\partial \omega^{kn}}{\partial t} = \left(\frac{\mathrm{d}f_{\text{eqw}}^{kn}}{\mathrm{d}\omega^{kw}}\right)\frac{\partial \omega^{kw}}{\partial t} \tag{4.72}$$

and:

$$\nabla \omega^{kn} = \left(\frac{\mathrm{d}f_{\text{eqw}}^{kn}}{\mathrm{d}\omega^{kw}}\right)\nabla \omega^{kw} \tag{4.73}$$

The last three equations can be used to eliminate ω^{kn} from equation (4.70):

$$\varepsilon^w \rho^w \left[1 + \frac{1-\omega^{kw}}{1-f_{\text{eqw}}^{kn}}\frac{\varepsilon^n \rho^n}{\varepsilon^w \rho^w}\left(\frac{\mathrm{d}f_{\text{eqw}}^{kn}}{\mathrm{d}\omega^{kw}}\right)\right]\frac{\partial \omega^{kw}}{\partial t}$$

$$+ \rho^w \left[\mathbf{q}^w + \frac{1-\omega^{kw}}{1-f_{\text{eqw}}^{kn}}\frac{\rho^n}{\rho^w}\left(\frac{\mathrm{d}f_{\text{eqw}}^{kn}}{\mathrm{d}\omega^{kw}}\right)\mathbf{q}^n\right]\cdot\nabla \omega^{kw}$$

$$+ \nabla\cdot\left(\varepsilon^w \overset{\star}{\mathbf{j}}{}^{kw}\right) + \frac{1-\omega^{kw}}{1-f_{\text{eqw}}^{kn}}\nabla\cdot\left(\varepsilon^n \overset{\star}{\mathbf{j}}{}^{kn}\right)$$

$$= \varepsilon^w r^{kw} + \left(e_{ws}^{kw} - \omega^{kw}e_{ws}^w\right) + \sum_{W=1}^{N_W}\rho_W^w(\omega_W^{kw} - \omega^{kw})Q_W^w\delta\left(\mathbf{x} - \mathbf{x}^W\right)$$

$$+ \frac{1-\omega^{kw}}{1-f_{\text{eqw}}^{kn}}\left[\varepsilon^n r^{kn} + \left(e_{ns}^{kn} - f_{\text{eqw}}^{kn}e_{ns}^n\right) + \sum_{W=1}^{N_W}\rho_W^n(\omega_W^{kn} - f_{\text{eqw}}^{kn})Q_W^n\delta\left(\mathbf{x} - \mathbf{x}^W\right)\right] \tag{4.74}$$

In the interest of having a complete formulation, let us substitute the constitutive forms for the dispersion vector given by equation (4.39) and again make use of equation (4.73) to eliminate $\nabla \omega^{kn}$ from the dispersion equation for k in phase n to obtain:

$$\varepsilon^w \rho^w \left[1 + \frac{1-\omega^{kw}}{1-f_{\text{eqw}}^{kn}}\frac{\varepsilon^n \rho^n}{\varepsilon^w \rho^w}\left(\frac{\mathrm{d}f_{\text{eqw}}^{kn}}{\mathrm{d}\omega^{kw}}\right)\right]\frac{\partial \omega^{kw}}{\partial t}$$

$$+ \rho^w \left[\mathbf{q}^w + \frac{1-\omega^{kw}}{1-f_{\text{eqw}}^{kn}}\frac{\rho^n}{\rho^w}\left(\frac{\mathrm{d}f_{\text{eqw}}^{kn}}{\mathrm{d}\omega^{kw}}\right)\mathbf{q}^n\right]\cdot\nabla \omega^{kw} - \nabla\cdot\left(\varepsilon^w \rho^w \mathbf{D}^{kw}\cdot\nabla \omega^{kw}\right)$$

$$- \frac{1-\omega^{kw}}{1-f_{\text{eqw}}^{kn}}\nabla\cdot\left[\varepsilon^n \rho^n \left(\frac{\mathrm{d}f_{\text{eqw}}^{kn}}{\mathrm{d}\omega^{kw}}\right)\mathbf{D}^{kn}\cdot\nabla \omega^{kw}\right]$$

$$= \varepsilon^w \left(r^{kw} + \frac{1-\omega^{kw}}{1-f_{\text{eqw}}^{kn}}\frac{\varepsilon^n}{\varepsilon^w}r^{kn}\right)$$

$$+ \left(e_{ws}^{kw} - \omega^{kw}e_{ws}^w\right) + \sum_{W=1}^{N_W}\rho_W^w(\omega_W^{kw} - \omega^{kw})Q_W^w\delta\left(\mathbf{x} - \mathbf{x}^W\right)$$

$$+ \frac{1-\omega^{kw}}{1-f_{\text{eqw}}^{kn}}\left[\left(e_{ns}^{kn} - f_{\text{eqw}}^{kn}e_{ns}^n\right) + \sum_{W=1}^{N_W}\rho_W^n(\omega_W^{kn} - f_{\text{eqw}}^{kn})Q_W^n\delta\left(\mathbf{x} - \mathbf{x}^W\right)\right] \tag{4.75}$$

Equation (4.75) can be solved for the distribution of species k in the w phase. Then the distribution in the n phase can be obtained from the equilibrium relation. In cases where the chemical reaction terms are nonzero and where exchanges with the solid phase are important, it may be necessary to solve this equation in conjunction with the transport equations for the species that are involved in chemical reactions that consume or produce k and with adsorption equation (4.32) for species k in the solid.

Despite its rather complex appearance, equation (4.75) can be explored to give some insight into the impact of the equilibrium exchange of mass on perceived system performance. This will be an approximate analysis motivated by the desire for insight rather than rigor, although the main approximation relates to the form of the equilibrium function f_{eqw}^{kn}. We will consider the case where ρ^n and ρ^w may be considered constant. Also, use a linear distribution relation so that $f_{eqw}^{kn} = K^{kwn}\omega^{kw}$. We will neglect the pumping terms since these do not contribute to the observations we will be making. Implementation of these stipulations in equation (4.75) yields:

$$
\varepsilon^w \rho^w \left[1 + \left(\frac{1-\omega^{kw}}{1-K^{kwn}\omega^{kw}} \right) \frac{\varepsilon^n \rho^n}{\varepsilon^w \rho^w} K^{kwn} \right] \frac{\partial \omega^{kw}}{\partial t}
$$

$$
+ \rho^w \left(\mathbf{q}^w + \left(\frac{1-\omega^{kw}}{1-K^{kwn}\omega^{kw}} \right) \frac{\rho^n}{\rho^w} K^{kwn} \mathbf{q}^n \right) \cdot \nabla \omega^{kw} - \nabla \cdot \left(\varepsilon^w \rho^w \mathbf{D}^{kw} \cdot \nabla \omega^{kw} \right)
$$

$$
- \left(\frac{1-\omega^{kw}}{1-K^{kwn}\omega^{kw}} \right) \nabla \cdot \left(\varepsilon^n \rho^n K^{kwn} \mathbf{D}^{kn} \cdot \nabla \omega^{kw} \right)
$$

$$
= \varepsilon^w \left[r^{kw} + \left(\frac{1-\omega^{kw}}{1-K^{kwn}\omega^{kw}} \right) \frac{\varepsilon^n}{\varepsilon^w} r^{kn} \right]
$$

$$
+ \left(e_{ws}^{kw} - \omega^{kw} e_{ws}^w \right) + \left(\frac{1-\omega^{kw}}{1-K^{kwn}\omega^{kw}} \right) \left(e_{ns}^{kn} - K^{kwn}\omega^{kw} e_{ns}^n \right) \tag{4.76}
$$

In the interest of our exploratory study,[4] we will consider the situation that the mass fractions are small so that:

$$
\frac{1-\omega^{kw}}{1-K^{kwn}\omega^{kw}} \approx 1 \tag{4.77}
$$

Thus the transport equation becomes:

---

[4] Remember that as long as we know what assumptions we are making and conditions we are imposing, we can go back and re-examine our system under less stringent assumptions in order to gain more insight or generality.

$$\varepsilon^w \rho^w \left(1 + \frac{\varepsilon^n \rho^n}{\varepsilon^w \rho^w} K^{kwn}\right) \frac{\partial \omega^{kw}}{\partial t} + \rho^w \left(\mathbf{q}^w + \frac{\rho^n}{\rho^w} K^{kwn} \mathbf{q}^n\right) \cdot \nabla \omega^{kw}$$

$$- \nabla \cdot \left[\varepsilon^w \rho^w \left(\mathbf{D}^{kw} + \frac{\varepsilon^n \rho^n}{\varepsilon^w \rho^w} K^{kwn} \mathbf{D}^{kn}\right) \cdot \nabla \omega^{kw}\right]$$

$$= \varepsilon^w \left(r^{kw} + \frac{\varepsilon^n}{\varepsilon^w} r^{kn}\right) + \left(e_{ws}^{kw} - \omega^{kw} e_{ws}^w\right) + \left(e_{ns}^{kn} - K^{kwn} \omega^{kw} e_{ns}^n\right) \qquad (4.78)$$

We can divide by the term in parentheses that multiplies the time derivative so that the time derivative term is identical to that in the original transport equation (4.67) for ω^{kw} obtained prior to applying the equilibrium formulation assumption. The resulting equation suggests that if we run an experiment to observe the movement of the chemical species, a conservative species that moves without exchange of mass will seem to advect with Darcy velocity $\mathbf{q}^w$, while species k will apparently advect at Darcy velocity $\mathbf{q}_{\mathrm{app}}^{kw}$ where:

$$\mathbf{q}_{\mathrm{app}}^{kw} = \left(1 + \frac{\varepsilon^n \rho^n}{\varepsilon^w \rho^w} K^{kwn}\right)^{-1} \left(\mathbf{q}^w + \frac{\rho^n}{\rho^w} K^{kwn} \mathbf{q}^n\right) \qquad (4.79)$$

Therefore, the apparent flow velocity of a species in the w phase will depend on both the magnitude of the equilibrium partitioning coefficient and the magnitude and direction of flow of the n phase. We can obtain an approximate expression for the effective dispersion coefficient, $\mathbf{D}_{\mathrm{app}}^{kw}$, by comparison of the dispersion terms in equations (4.67) and (4.78) of the form:

$$\mathbf{D}_{\mathrm{app}}^{kw} \approx \left(1 + \frac{\varepsilon^n \rho^n}{\varepsilon^w \rho^w} K^{kwn}\right)^{-1} \left(\mathbf{D}^{kw} + \frac{\varepsilon^n \rho^n}{\varepsilon^w \rho^w} K^{kwn} \mathbf{D}^{kn}\right) \qquad (4.80)$$

where the approximation comes in because we have considered spatial variation of the first term on the right to be small. When $|\mathbf{D}^{kn}| < |\mathbf{D}^{kw}|$, the effects of dispersion will be decreased so that spreading of species k will be decreased from that obtained in the absence of equilibrium exchange. Spreading will be enhanced when $|\mathbf{D}^{kn}| > |\mathbf{D}^{kw}|$.

As an additional study, let us consider the case where the transfer of species k among all three phases is an equilibrium process. We will only need one transport equation for this species along with two equilibrium expressions. To obtain the required form, we will combine equation (4.75) with equation (4.32) to eliminate the exchange term for species k between the w and s phases. For simplicity, we will consider that k is the only species that is exchanged among the phases. An equilibrium isotherm describing the partitioning of species k between the w and s phases is employed in the general form:

$$\omega^{ks} = f_{\mathrm{eqw}}^{ks}(\omega^{kw}) \qquad (4.81)$$

Typically this function would describe a linear, Freundlich, or Langmuir isotherm. The algebra required to complete the derivation is a straightforward extension of the manipulations employed in obtaining equation (4.75). The additional algebra leads to:

$$\varepsilon^w \rho^w \left[1 + \frac{1-\omega^{kw}}{1-f_{eqw}^{kn}} \frac{\varepsilon^n \rho^n}{\varepsilon^w \rho^w} \left(\frac{df_{eqw}^{kn}}{d\omega^{kw}} \right) + \frac{1-\omega^{kw}}{1-f_{eqw}^{ks}} \frac{(1-\varepsilon)\rho^s}{\varepsilon^w \rho^w} \left(\frac{df_{eqw}^{ks}}{d\omega^{kw}} \right) \right] \frac{\partial \omega^{kw}}{\partial t}$$

$$+ \rho^w \left[\mathbf{q}^w + \frac{1-\omega^{kw}}{1-f_{eqw}^{kn}} \frac{\rho^n}{\rho^w} \left(\frac{df_{eqw}^{kn}}{d\omega^{kw}} \right) \mathbf{q}^n \right] \cdot \nabla \omega^{kw}$$

$$- \nabla \cdot (\varepsilon^w \rho^w \mathbf{D}^{kw} \cdot \nabla \omega^{kw}) - \frac{1-\omega^{kw}}{1-f_{eqw}^{kn}} \nabla \cdot \left[\varepsilon^n \rho^n \left(\frac{df_{eqw}^{kn}}{d\omega^{kw}} \right) \mathbf{D}^{kn} \cdot \nabla \omega^{kw} \right]$$

$$= \varepsilon^w \left[r^{kw} + \frac{1-\omega^{kw}}{1-f_{eqw}^{kn}} \frac{\varepsilon^n}{\varepsilon^w} r^{kn} + \frac{1-\omega^{kw}}{1-f_{eqw}^{ks}} \frac{(1-\varepsilon)}{\varepsilon^w} r^{ks} \right]$$

$$+ \sum_{W=1}^{N_W} \rho_W^w (\omega_W^{kw} - \omega^{kw}) Q_W^w \delta(\mathbf{x} - \mathbf{x}^W)$$

$$+ \frac{1-\omega^{kw}}{1-f_{eqw}^{kn}} \left[\sum_{W=1}^{N_W} \rho_W^n (\omega_W^{kn} - f_{eqw}^{kn}) Q_W^w \delta(\mathbf{x} - \mathbf{x}^W) \right] \qquad (4.82)$$

Examination of this equation indicates that there are no changes to the coefficients of the terms involving advection, dispersion, and pumping in comparison to equation (4.75). This should not be surprising as these processes are considered unimportant to modeling the species transport in a solid. The coefficients of the time derivative and the chemical reaction have been altered by using equilibrium partitioning with the solid phase. We will examine these quantities.

We identify the quantity in brackets multiplying the time derivative in equation (4.82) as the *total retardation factor* for species k in the w phase, R_T^{kw}, so that:

$$R_T^{kw} = 1 + \frac{1-\omega^{kw}}{1-f_{eqw}^{kn}} \frac{\varepsilon^n \rho^n}{\varepsilon^w \rho^w} \left(\frac{df_{eqw}^{kn}}{d\omega^{kw}} \right) + \frac{1-\omega^{kw}}{1-f_{eqw}^{ks}} \frac{(1-\varepsilon)\rho^s}{\varepsilon^w \rho^w} \left(\frac{df_{eqw}^{ks}}{d\omega^{kw}} \right) \qquad (4.83)$$

This is referred to as a "retardation" factor be cause it acts as a factor that impacts the apparent velocity (i.e., it retards the apparent velocity). Retardation is most commonly considered in the context of a single fluid and the interaction of the species in that fluid with the solid. The key element in reducing the total retardation factor to the *standard retardation factor*, R_S^{kw}, is the assumption that species k is either insoluble in the n phase or must be modeled using the kinetic approximation in its exchanges from the w or s phases with the n phase. In either case, the standard retardation factor is obtained as:

$$R_S^{kw} = 1 + \frac{1-\omega^{kw}}{1-f_{eqw}^{ks}} \frac{(1-\varepsilon)\rho^s}{\varepsilon^w \rho^w} \left(\frac{df_{eqw}^{ks}}{d\omega^{kw}} \right) \qquad (4.84)$$

Also, processes associated with the n phase will not appear in the combined transport equation for species k obtained from the w and s transport equations except for some terms involving exchanges with the n phase. These terms will be zero in some instances (e.g., no n phase is present, or mass exchanges of all species with the

n phase are unimportant) but otherwise will have to be modeled using the kinetic formulation. Thus, equation (4.82) simplifies to:

$$
\begin{aligned}
\varepsilon^w \rho^w R_S^{kw} \frac{\partial \omega^{kw}}{\partial t} + \rho^w \mathbf{q}^w \cdot \nabla \omega^{kw} - \nabla \cdot \left(\varepsilon^w \rho^w \mathbf{D}^{kw} \cdot \nabla \omega^{kw} \right) \\
= \varepsilon^w \left[r^{kw} + \frac{1-\omega^{kw}}{1-f_{eqw}^{ks}} \frac{(1-\varepsilon)}{\varepsilon^w} r^{ks} \right] \\
+ \left(e_{wn}^{kw} - \omega^{kw} e_{wn}^w \right) + \left(\frac{1-\omega^{kw}}{1-K^{kws}\omega^{kw}} \right) \left(e_{ns}^{ks} - K^{kws}\omega^{kw} e_{ns}^s \right) \\
+ \sum_{W=1}^{N_W} \rho_W^w (\omega_W^{kw} - \omega^{kw}) Q_W^w \delta(\mathbf{x} - \mathbf{x}^W)
\end{aligned}
\tag{4.85}
$$

The retardation factor accounts for the fact that if the Darcy velocity of the flow is $\mathbf{q}^w$, the adsorption and desorption of species k on the solid behave as if it is actually traveling only at a Darcy velocity of $\mathbf{q}^w / R_S^{kw}$.[5] Interaction with only the essentially immobile solid retards the flow, whereas interaction with the n phase when the equilibrium formulation for mass exchange is appropriate indicates that movement of that phase may either retard or advance the flow.

In the most common form of R_S^{kw} a linear isotherm, as in equation (4.61), is employed with:

$$
f_{eqw}^{ks} = K^{kws} \omega^{kw}
\tag{4.86}
$$

Then the standard retardation factor becomes:

$$
R_S^{kw} = 1 + \left(\frac{1-\omega^{kw}}{1-K^{kws}\omega^{kw}} \right) \frac{(1-\varepsilon)\rho^s}{\varepsilon^w \rho^w} K^{kws}
\tag{4.87}
$$

Usually, the factor $(1-\omega^{kw})/(1-K^{kws}\omega^{kw})$ is overlooked in formulating the standard retardation factor. However, the fact that the mass fractions studied are typically very small means that the factor is close to 1 in any event. Introducing the distribution coefficient K_d^{kws} by setting $K^{kws} = K_d^{kws} \rho^w$, we then obtain:

$$
R_S^{kw} = 1 + \frac{(1-\varepsilon)\rho^s}{\varepsilon^w} K_d^{kws}
\tag{4.88}
$$

More complex forms of the retardation coefficient result when the Freundlich and Langmuir isotherms are employed. With these nonlinear forms, the expression for the retardation factor will be dependent on the species mass fraction so that the magnitude of the retardation will depend on the concentration.

The fact that this section on the equilibrium formulation made use of the distribution forms of the species conservation equations is not an essential part of the

---

[5] This behavior is the basis for chromatography whereby a solvent containing a mixture of solutes flows through a column and the various species separate along the column because they have different apparent velocities of flow along the column due to differences in their respective retardation coefficients.

derivations. The same conclusions are reached using the direct and rigorous forms of the equations. The defining feature of the equilibrium formulation is the use of partitioning relations between fluids or isotherms for fluid-solid interactions to eliminate the need to solve a transport equation for a species in each phase. When distributional equilibrium is quickly achieved, knowledge of concentration in one phase provides knowledge of the concentration in another phase through the equilibrium relation. The method simplifies computation; however if an averaging region is large, the equilibrium assumption may not apply so that simulation must employ the kinetic approach. Systems composed of many species can make use of the kinetic approach for some and the equilibrium approach for others. The issues of scale, the need for appropriate kinetic and equilibrium constitutive expressions, and the need to support whichever model is chosen with accurate coefficients remains a challenging area of research.

4.5.3 Summary: Kinetic vs. Equilibrium Formulations

The mathematics in the preceding subsections indicates the quite different implications arising from the application of the kinetic and equilibrium formulations. The mathematics is important. However, the mathematics is employed to describe what is observed physically. The mathematical descriptions will improve as the observations improve, and vice-versa. Here, we present some of the physical observations regarding the behavior of nonaqueous phase liquid (NAPL) in the subsurface that inform the selection and use of the alternative mass exchange formulations.

The subsurface environment of NAPLs involves four phases: solid, water, NAPL, and a gas phase. The addition of a third fluid phase complicates the simulation problem because an additional set of conservation equations is needed, and the closure relations tend to become more complex.

When modeling NAPL in near surface granular soils, the following interphase exchange mechanisms must be considered:

- dissolution from the NAPL phase into the water phase;
- evaporation from the NAPL phase into the gas phase;
- evaporation of NAPL that has dissolved in the water phase into the gas phase;
- adsorption of the NAPL dissolved in the water phase onto the solid.

The assumption is also made that primarily the water phase makes direct contact with the solid so that adsorption directly from the NAPL phase or from the gas phase can be neglected.

Choices must be made concerning modeling of the transfer processes kinetically or using the equilibrium formulation. The following considerations are useful:

- The kinetic formulation of mass transfer provides predictive flexibility, as local equilibrium conditions can be simulated by increasing the mass transfer coefficient.
- The kinetic formulation is more accurate than the equilibrium formulation for heterogeneous soils, inhomogeneous residual NAPL distribution, inhomogeneous blob size distribution, and high fluid flow rates.

- Modeling pump-and-treat remediation of NAPL-contaminated soils must be based on the kinetic formulation of mass transfer to mimic both bench-scale experimental and field-scale data, specifically effluent concentration tailing.
- When the desorption transport path of NAPL is from soil to water and then from water to gas, data suggests that only one of the mass transfer processes needs to be modeled kinetically as the slower kinetic process will allow for equilibrium to be achieved for the other transfer process.
- A linear equilibrium formulation making use of the Henry's law constant is usually used to define NAPL transport between the gas and water phases.
- The maximum NAPL mass fraction that can be adsorbed onto the soil is usually defined using an organic carbon-based model.

Based on these factors, the interphase transfer relations for the organic NAPL are formulated. Transport equations are combined, as allowed, when the mass exchange is modeled as an equilibrium process. When the NAPL is present as a mixture of organic chemicals, decisions have to be made as to whether an accurate simulation can be achieved without modeling each organic, but rather, for example, by grouping the chemicals by molecular weight. We emphasize that the preceding items are observations that help inform the modeling process. The actual process involves making decisions and selections based on available data, computational power, the simulation code being employed, and the questions to be answered by the simulation. Theory, experiments, field observation, and computational power are best employed synergistically to obtain useful simulations.

4.6 INITIAL AND BOUNDARY CONDITIONS

After the species transport equation is closed through use of constitutive equations for the dispersion vector, homogeneous chemical reactions, and interphase transfer, initial and boundary conditions on the mass fractions must be specified. The governing species transport equation is actually similar in form to the groundwater flow equation, and thus considerations of the auxiliary conditions are similar to those in Subsection 3.6.5.

If the distributions of the contaminants are varying with time such that they are described by the species transport equation with the time derivative included, the distribution at the beginning time of the modeling process, t^0, must be specified. This condition is:

$$\omega^{i\alpha}(t^0, \mathbf{x}) = \Omega^{i\alpha}(\mathbf{x}) \quad \alpha = w, n, s; \quad i = 1, \ldots, N_\alpha \qquad (4.89)$$

where $\Omega^{i\alpha}$ is the specified distribution. For modeling movement of contamination in the past, such as the previous evolution of a gasoline spill at a service station, this condition requires that an initial distribution of contamination can be specified. Thus, for example, if a contaminant plume is detected today and distribution of the contamination in the subsurface is to be simulated for the last five years, it is necessary to know the contaminant distribution existent five years ago. If this data is not available, as it typically would not be, it is necessary to try to estimate a possible initial distribution from some earlier time. This is not easy, as such estimates are often unreliable. Errors in the initial condition can seriously degrade the accuracy of the simulation of plume movement.

At all points on the boundary of the domain, one condition on the mass transfer must be specified for each species in each phase for which a transport equation must be solved. Recall that if an equilibrium distribution between phases for a species exists, a single transport equation for that species is formulated that takes into account the interaction. Alternatives for the conditions to be specified at a point on the boundary, $\mathbf{x}_b$, are a first type condition:

$$\omega^{i\alpha}(t, \mathbf{x}_b) = \Omega^{i\alpha}(t, \mathbf{x}_b) \quad \alpha = w, n, s; \quad i = 1, \ldots, N_\alpha - 1 \tag{4.90}$$

a second type condition:

$$\left(\mathbf{n} \cdot \nabla \omega^{i\alpha}\right)\big|_{\mathbf{x}_b} = \Omega^{i\alpha\prime}(t, \mathbf{x}_b) \quad \alpha = w, n, s; \quad i = 1, \ldots, N_\alpha - 1 \tag{4.91}$$

or a third type condition:

$$\mathbf{n} \cdot \left[\left(\rho^\alpha \omega^{i\alpha i\alpha} \mathbf{q}^\alpha\right) - \varepsilon^\alpha \rho^\alpha \mathbf{D}^{i\alpha} \cdot \nabla \omega^{i\alpha}\right]\big|_{\mathbf{x}_b} = F^{i\alpha}(t, \mathbf{x}_b)$$
$$\alpha = w, n, s; \quad i = 1, \ldots, N_\alpha - 1 \tag{4.92}$$

where the quantities on the right side of each of these conditions are specified.

The first type condition is the specification of the species concentration at the boundary of the domain. The second type condition is the normal gradient at the boundary. For a situation where no flow occurs at the boundary and the species is not reacting at that boundary, the normal gradient will be zero (i.e., $\Omega^{i\alpha\prime}(t, \mathbf{x}_b) = 0$). The third type condition is the specification of the normal flux of concentration at the boundary due to advection and dispersion. The total flux is equal to the specified quantity $F^{i\alpha}(t, \mathbf{x}_b)$.

For a transient problem, initial conditions must be specified throughout the domain, and one of the three optional boundary conditions must be specified at each point on the boundary. For a steady state problem, no initial condition is needed; any one of the three boundary condition types may be specified at each point on the boundary although a first or third type condition must be specified at least at one point.

4.7 CONCLUSION

For simulation of a past field contamination problem, the specification of boundary conditions that are consistent with what were the actual boundary conditions impacting the system is difficult. Excellent historical records are needed, but are seldom available. It is uncommon for an illegal dumper of toxic waste to keep precise records of the times, locations, and magnitudes of fluxes of contamination applied at the boundary of an environmental system. Specification of exchanges of contamination with adjacent streams at the boundary of a domain of interest are also dependent on knowledge of historical additions of contamination to the stream. Simulations also require information on pumping rates and the addition and removal of chemical species through wells. Uncertainties in all these important elements coupled with uncertainties in specification of expressions for dispersion fluxes, interphase mass exchange, and chemical reaction rates suggest that simulation of chemical transport is a difficult business.

We would also be remiss if we did not note that in the case of density dependent flow, changes in fluid density due to changes in fluid chemical composition can impact the flow field of the phase. In those instances, one cannot solve the flow equations and then examine the distribution of chemicals in the phase. Transport feeds back to the flow field. Therefore an approach to solving the flow and transport equations simultaneously must be employed.

4.8 EXERCISES

1. The direct approach led to transport equation (4.9) for a chemical species in a fluid phase. The rigorous approach led to equation (4.22). By examining the assumptions involved in their derivations, indicate situations where one form might be preferable to the other for a simulation problem.

2. The rigorous approach led to transport equation (4.22) for a chemical species in a fluid phase. The distribution approach led to equation (4.28). Compare the assumptions that went into these alternative forms of the equation and identify situations where one might be theoretically preferable to the other for a simulation problem.

3. The closure equation for the dispersion vector is equation (4.39). Obtain the expressions for each component of the dispersion vector by expanding this expression so that the elements of the $\mathbf{D}^{i\alpha}$ and of $\nabla \omega^{i\alpha}$ appear explicitly.

4. Derive the standard retardation factor for the case where the equilibrium distribution between the fluid and solid follows a Langmuir isotherm.

5. Consider equation (4.85) which expresses the transport equation for use when exchange of mass between the w and s phases may be considered an equilibrium process. Assume the species of interest, k, is a radioactive isotope that decays according to:

$$r^{k\alpha} = -\lambda_{\mathrm{d}}^k \rho^{\alpha \omega k \alpha} \quad \alpha = w, s$$

where λ_{d}^k is a constant decay coefficient that is the same in both the w and s phases. We have seen that the standard retardation factor causes a decrease in the apparent velocity of species k. Does this factor also cause an apparent decrease in the reaction rate? Explain this answer on physical grounds.

6. Comment on or correct the following statement: The retardation factor multiplies the time derivative in a transient problem. Therefore, if one considers a steady state problem, the retardation factor does not appear; and the species of interest experiences no decrease in apparent velocity.

BIBLIOGRAPHY

[1] Bird, R.B., W.E. Stewart, and E.N. Lightfoot, *Transport Phenomena*, Second Edition, John Wiley & Sons, New York, 2002.

[2] Imhoff, P.T., P.R. Jaffe, and G.F. Pinder, An Experimental Study of the Dissolution of Trichloroethylene in Saturated Porous Media, Princeton University Water Resources Program Report: WR-92-1, 1992.

5

SIMULATION

In this chapter we will expand, through simulation, on various porous-medium flow and transport mathematical-physical concepts introduced in the previous chapters. The examples will illustrate porous medium flow systems of varying degrees of complexity. We begin, in Section 5.1 and Section 5.2 with one-dimensional, two-phase flow problems. In Section 5.3 we extend this work to two space dimensions. The concept of phase dissolution and mass transport is examined in Section 5.4. In Section 5.5 we look at the impact that seven different model attributes have on the behavior of flow and mass transport in a single-phase, water saturated system. In this section we also illustrate the importance of using three-space-dimensional models to simulate problems involving vertical flow. In the final section in this chapter, we extend our studies to include three-phase flow and also provide a comparison between simulated and physical model results.

5.1 1-D SIMULATION OF AIR-WATER FLOW

In this section we introduce *multiphase simulation* using a one-dimensional example first presented by Guarnaccia [1] depicted in Figure 5.1. The physical system consists of a column that is 67 cm long and 2 cm in diameter. The boundary at the top is open to the atmosphere. The usual procedure in modeling such a system is to treat the air phase as being passive and at a constant pressure. Here, we will model both the air and water phases.

For this simulation the column is initially fully saturated with water such that the initial *water saturation* is $s^w = 1.0$. The air pressure is set to a constant value of zero at the top with a no-flow condition on the water phase. At the beginning of the first simulation, which is described below in Subsection 5.1.1, the pressure of water at the bottom of the column is set to 31.5 cm of water so that the water begins to drain

Essentials of Multiphase Flow and Transport in Porous Media, by George F. Pinder and William G. Gray
Copyright © 2008 by John Wiley & Sons, Inc.

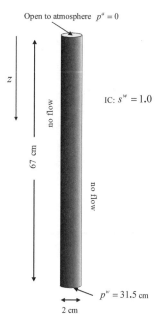

Figure 5.1: One-dimensional porous-medium filled column used in example problems discussed in this chapter. Note that the vertical axis is positive downward. Note that p^a and p^w are the pressure values of air and water, respectively.

from the column (drainage phase). A no-flow condition on the air phase exists at the bottom of the column. After a specified period of time water is again introduced at the top of the column, and we examine the saturation change behavior during secondary imbibition.

The conservation of mass equation governing flow in each fluid phase of the system is given by equation (3.122):

$$\frac{\partial(s^\alpha \varepsilon \rho^\alpha)}{\partial t} + \nabla \cdot (s^\alpha \varepsilon \rho^\alpha \mathbf{v}^\alpha) = e_{\alpha s}^\alpha + e_{wn}^\alpha + \sum_{W=1}^{N_W} \rho_W^\alpha Q_W^\alpha \delta(\mathbf{x} - \mathbf{x}^W) \quad \alpha = w, n \qquad (5.1)$$

where $\alpha = w$ for the water phase and $\alpha = n$ for the nonwetting gas phase, s^α is the *saturation* of the a phase, ρ^α is the *α phase density*, and ε is the *porosity*. If we assume that no mass is being exchanged between phases and note that there is no pumping, the terms on the right side are zero. Also, the only velocity component is in the z direction so the equation simplifies to:

$$\frac{\partial(s^\alpha \varepsilon \rho^\alpha)}{\partial t} + \frac{\partial(s^\alpha \varepsilon \rho^\alpha v_z^\alpha)}{\partial z} = 0 \quad \alpha = w, n \qquad (5.2)$$

The flow equation is based on equation (3.131):

$$\mathbf{q}^\alpha = -\frac{\mathbf{k}^{s\alpha}}{\mu^\alpha} \cdot (\nabla p^\alpha - \rho^\alpha g) \quad \alpha = w, n \qquad (5.3)$$

This equation is simplified to its one-dimensional form with the movement of solid grains deemed negligible and with the isotropic form of the intrinsic permeability as given by equation (3.132) so that $\mathbf{k}^{s\alpha} = k^s k_{\mathrm{rel}}^{\alpha} \mathbf{I}$. Therefore, we obtain:

$$s^\alpha \varepsilon v_z^\alpha = -\frac{k^s k_{\mathrm{rel}}^{\alpha}}{\mu^\alpha}\left(\frac{\partial p^\alpha}{\partial z} - \rho^\alpha g\right) \quad \alpha = w, n \tag{5.4}$$

The solution to equations (5.2) and (5.4) requires constitutive relationships for $k_{\mathrm{rel}}^{\alpha}$ and a relation for the pressure difference between phases as a function of saturation (as discussed in regard to equation (3.135) with $p^c(s^w) = p^n - p^w$). The model will be formulated in terms of the effective saturation of water, s_e^w, as introduced in equation (3.141) and the capillary head, h^c, as defined in equation (3.136). The p^c-s^w model to be employed is the van Genuchten form given by equation (3.146), which can be rearranged to give:

$$h^c = \frac{1}{\alpha}\left[\left(s_e^w\right)^{-1/M} - 1\right]^{1/N} \tag{5.5}$$

Here the parameter α is assigned a value α_d for the wetting phase drainage and α_i for wetting phase imbibition. The parameter N is related by curve fitting to the pore size distribution, and we will make use of the usual expression for M in terms of N, $M = 1 - 1/N$. Hysteresis exists in the p^c-s^w relation because of fluid entrapment and contact angle effects. Besides altering the value of α depending on whether drainage or imbibition is occurring, we make use of an effective, albeit complicated, strategy for moving along scanning curves. The details of this procedure are beyond the scope of this presentation. The interested reader can find the specifics of this protocol in [2].

The $k_{\mathrm{rel}}^{\alpha}$-$s^w$ model parameters are defined by equations (3.167) and (3.168) based on the Mualem porosity distribution function and the van Genuchten capillary pressure relationship:

$$k_{\mathrm{rel}}^w(s_e^w) = \left(s_e^w\right)^\kappa \left\{1 - \left[1 - \left(s_e^w\right)^{1/M}\right]^M\right\}^2 \tag{5.6}$$

$$k_{\mathrm{rel}}^n(s_e^w) = \left(1 - s_e^w\right)^\zeta \left[1 - \left(s_e^w\right)^{1/M}\right]^{2M} \tag{5.7}$$

Other physical constants that are required are found in Table 5.1. For convenience, plots of $k_{\mathrm{rel}}^{\alpha}$-$s_e^w$ and p^c-s_e^w based on the parameter values in the table are given in Figures 5.2, 5.3, and 5.4 respectively.

5.1.1 Drainage in a Homogeneous Soil

The first problem we consider is water drainage from the fully saturated column shown in Figure 5.1. As noted, the boundary conditions are zero air pressure (zero pressure head) and no flow of water at the top of the column and a water pressure of 31.5 cm of water and no flow of air at the lower end. Although the air-water system has been described mathematically, the complexity of the system requires that the governing equations be solved using numerical methods. In the following

Table 5.1: Parameter values used in drainage and imbibition of water example

Fluid Properties

$\rho^w = 0.9982\,\text{g/cm}^3$	$\rho^n = 0.000129\,\text{g/cm}^3$
$\mu^w = 0.01$ poise	$\mu^n = 0.0015$ poise

p^c-s^w Model Definition

$\alpha_d = 0.04\,\text{cm}^{-1}$	$\alpha_i = 0.06\,\text{cm}^{-1}$
$s_i^w = 0.12$	$s_r^n = 0.02$
	$N = 10 = 1/(1-M)$

k_{rel}^{α}-s^w Model Definition

$s_e^w = \left(s^w - s_i^w\right)/\left(1 - s_r^n - s_i^w\right)$	$\kappa = \zeta = 0.5$

Field Properties

$\varepsilon = 0.37$	$k^s = 5.0 \times 10^{-7}\,\text{cm}^2$

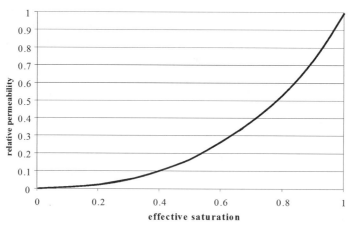

Figure 5.2: Relationship between relative permeability of water and effective water saturation (data from [2]).

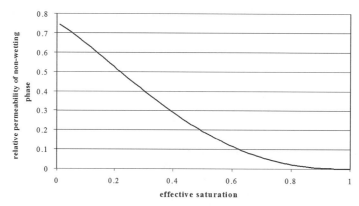

Figure 5.3: Relationship between relative permeability of nonwetting phase and effective water saturation (data from [2]).

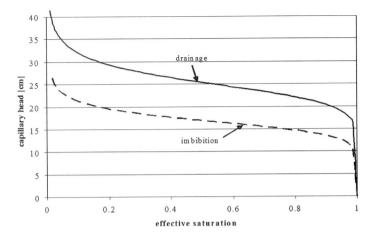

Figure 5.4: Saturation-capillary pressure function used in one-dimensional drainage simulation (data from [2])

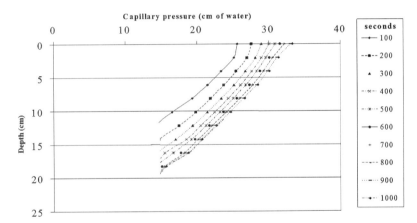

Figure 5.5 Capillary pressure expressed in terms of centimeters of water plotted agains depth for five time values expressed in seconds.

multiphase flow examples a *collocation finite element method* is employed. The comcomitant computer program that does the calculatons is called NAPL, is fully documented, and is available free of charge at *http://www.epa.gov/ada/csmos/models/ napl.html*.

The NAPL code models both the water and air phases, in contrast to a standard Richards' model that would treat the air aerostatically. Nevertheless, we will discuss the results of the computation in terms of the pressure difference between the phases, the capillary pressure. The simulated changes in capillary pressure and satu- ration as a function of time and of depth below the land surface are shown in Figures 5.5 and 5.6. Consider, for example, the curve denoted as 200 in Figure 5.5. The curve tells us that after 200 sec of simulated time the capillary pressure at the top of the column has increased from 0.0 to 27 cm of water (2.65 mbar). The water pressure has decreased by this amount since the air pressure remains near atmospheric. The water pressure has decreased the most at the top of the column. The magnitude of

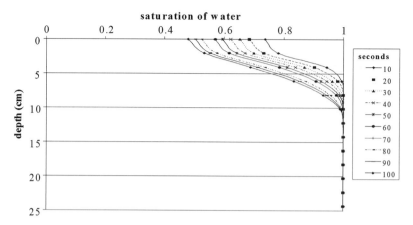

Figure 5.6: Water saturation plotted against depth for drainage over the period 0 to 100 sec.

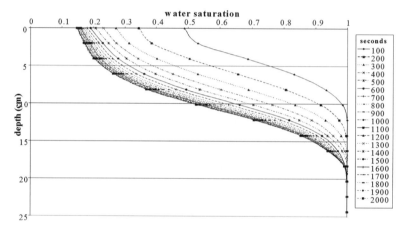

Figure 5.7: Water saturation plotted against depth for drainage over the period 0 to 2000 sec.

the vertical gradient of water pressure is smaller in the top 4 cm than it is in the next 10 cm. Below this depth, the pressure is essentially unchanged from its initial value.

As time proceeds up to 2000 sec, the capillary pressure increases at all spatial locations. At 1000 sec the capillary pressure head is 34 cm at the land surface and decreases with depth for a distance of about 20 cm. Under ideal conditions, at steady state, there should be no vertical gradient in total head, so the water pressure head increase with depth (capillary pressure head decrease) should balance the elevation head decrease. Thus the water pressure should increase linearly from the land surface to a depth of 35.0 cm, which is the location consistent with the boundary condition for water pressure of 31.5 cm at the base of the model. Note that the rate of change in capillary pressure is decreasing as the simulation proceeds. To the left of these curves the capillary pressure is not defined since $s^w = 1$.

Figure 5.6 shows the early-time change in saturation with depth at 10-sec intervals between 0 and 100 sec. Figure 5.7 is a plot of the saturation over the total 2000-sec simulation period. The saturation is initially at unity and drops to 0.74 at the end of 10 sec at the top of the column as can be seen in Figure 5.6. Since the saturation and the pressure are related via the pressure-saturation curve found in Figure 5.4, the saturation changes in lock-step with the pressure shown in Figure 5.5. Note that the relationship in Figure 5.4 shows that there will be pressure intervals where a considerable change in saturation for a relatively small change in capillary pressure will be observed.

5.1.2   Drainage in a Heterogeneous Soil

We now consider the complicating issue of a variable value of permeability. Using the preceding example we replace a layer at the base of the column with a layer of thickness approximately 12 cm which has a permeability that is two orders of magnitude greater ($k^s = 5.0 \times 10^{-5}\,\mathrm{cm}^2$) than the value ($k^s = 5.0 \times 10^{-7}\,\mathrm{cm}^2$) that exists in the rest of the column. The constitutive curves are the same as those used in the preceding example. With an initial saturation of unity, a specified air pressure of zero, no flow of water at the top, and a specified pressure of water of 31.5 cm and no flow of air at the bottom, we begin to drain the column of water. The resulting saturation profile is shown in Figure 5.8.

The behavior presented in this figure can be compared with that in Figure 5.6. Note that the time scale is different so that not all curves correspond directly. However, it is evident that the impact of the high permeability layer is not very significant because most of the dynamics of the system occur in the upper 18 cm of the column for the time duration considered.

We now reverse the situation and place the high permeability layer at the top. The low permeability layer at the bottom is about 12 cm. The results of this experiment are shown in Figure 5.9. Now the impact of the high permeability layer is quite

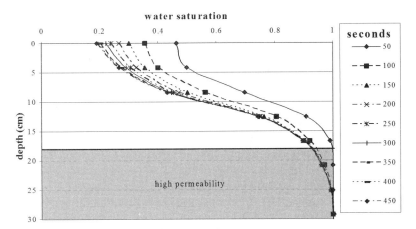

Figure 5.8: Saturation distribution when the patterned area below the heavy line has a permeability two orders of magnitude higher than the area above the heavy line. The series of curves represent elapsed time of drainage in seconds.

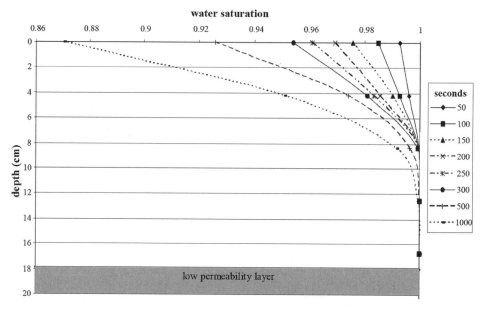

Figure 5.9: Saturation distribution when the patterned area below the heavy line has a permeability two orders of magnitude lower than the area above the heavy line. The series of curves represent elapsed time of drainage in seconds.

dramatic. When viewing this figure, be sure to notice the change in the depth and saturation scales as compared to those in previous figures. Additionally, a curve representing the state of the system at 1000 sec has been added. Overall, drainage is significantly retarded by the low permeability layer such that there is very little change in the saturation over the 500-sec, and indeed over the entire 1000-sec, modeling interval.

The conclusion that can be drawn from this experiment is the following: the fact that the high permeability layer exists at depth does not impact the behavior at shallow depths at early time. At shallow depths and early times the system behavior is relatively insensitive to the difference between the pressure at the transition depth of about 15 cm and that at the bottom of the column.

5.1.3   Imbibition in Homogeneous Soil

The counterpart of drainage is imbibition. Imbibition is the process of introducing wetting fluid into a multiphase system. Starting from the saturation state of the system after drainage has been ongoing for 50,000 sec, the water pressure at the top of the column is now increased to 35.5 cm such that the total head is higher than the total head at the base of the column (−35.0 cm). As seen in Figure 5.10, the water saturation at the top of the column immediately increases and continues to increase during the duration of the simulation. A saturation front moves downward causing the saturation to increase within the column as well. Simultaneously, drainage continues at the initial front (bottom of the profile), moving the initial saturation front further down the column. This is consistent with the imposed water head gradient (from 35.5 cm at the top of the column to −35 cm at the base). Although not

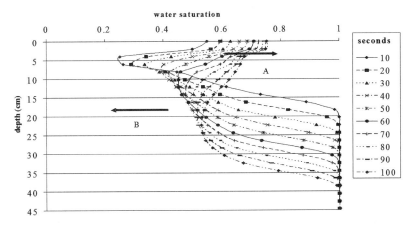

Figure 5.10: Water imbibing into a partially drained system. Time is in seconds. The arrow A indicates the curves are younger to older moving left to right, and arrow B indicates time evolution from right to left.

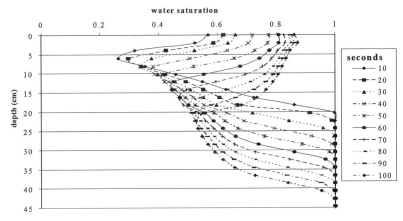

Figure 5.11: Water imbibing into a partially drained system. Time is in seconds. The simulation differs from that of Figure 5.10 only inasmuch as there is no hysteresis.

apparent from the figures, upward movement of air is also observed in the model output.

All of the examples to this point have assumed hysteretic behavior. To see the impact of this phenomenon on the solution, we show in Figure 5.11 the same problem presented in Figure 5.10, but without hysteresis. While the plots are similar, a difference does exist. Note that the maximum saturation in the "no hysteresis" example is greater at late times near the ground surface; the front has propagated further; but the satuations at the front are smaller.

5.2   1-D SIMULATION OF DNAPL-WATER FLOW

In this subsection we consider a second two-phase flow example; the case of *dense nonaqueous phase (DNAPL) flow*. The equations governing flow in this system are the same as equations (5.2) through (5.7) where the nonwetting phase is DNAPL rather than air. The DNAPL physical constants are $\mu^n = 0.09$ poise and $\rho^n = 1.5$ g/cm^3. The constitutive relationships used are the same as those introduced earlier; these are the p^c-s^w and k_{rel}^α-s^w curves. In general these curves are unique, depending upon the soil type and the nature of the wetting and nonwetting fluids. However, in this case the same k_{rel}^α-s^w curves were used for the water and DNAPL.

5.2.1 Primary DNAPL Imbibition in Homogeneous Soil

The boundary condition imposed on the bottom of the column is similar to that provided earlier, except that DNAPL, not air, is now the nonwetting fluid (i.e., a pressure of water of 31.5 cm of water and no DNAPL flow), but the top boundary condition is changed. In the case of primary imbibition into a water saturated column we use a constant DNAPL pressure of 0.0 and no water flow at the top. Thus movement is due solely to the gravitational effect.

The results of this simulation are seen in Figure 5.12. The curves represent saturation as a function of depth and time (the curves represent different elapsed times in seconds). The DNAPL saturation increases at the top of the column, and, due to its greater density, the DNAPL front moves down the column at a rate that decreases with time.

5.2.2 Density Effect

Now let us examine the impact of density on the rate of propogation of the DNAPL front. Holding all other aspects of the physical system as in the preceding example, we decrease the density from $\rho^n = 1.5$ g/cm^3 to $\rho^n = 1.2$ g/cm^3. Once again, we simulate

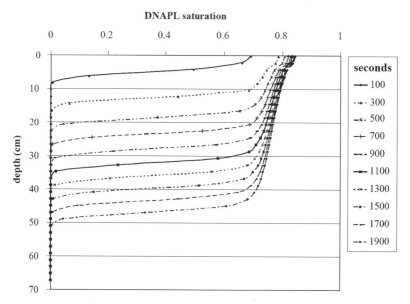

Figure 5.12: Primary imbibition of DNAPL into water saturated porous medium. Curves represent elapsed time since initiation of simulation.

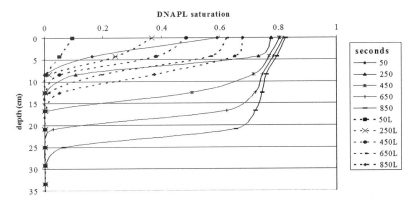

Figure 5.13: Primary imbibition of DNAPL into a water saturated porous medium. Curves represent elapsed time since initiation of simulation. The dashed curves are generated using $\rho^n = 1.2\,g/cm^3$ and the solid curves are those appearing in Figure 5.12 where a value of $\rho^n = 1.5\,g/cm^3$ is used.

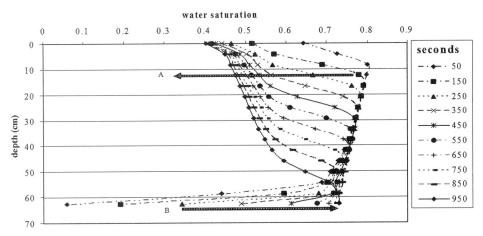

Figure 5.14: Primary drainage of DNAPL coexisting with water. Values associated with DNAPL curves indicate time since initiation of drainage cycle in seconds.

primary DNAPL imbibition. The dashed and solid curves in Figure 5.13 represent the time evolution of DNAPL saturation with ρ^n of $1.2\,g/cm^3$ and $1.5\,g/cm^3$, respectively. As one would anticipate, the DNAPL with the greater density moves downward faster due to the effects of gravity.

5.2.3   DNAPL Drainage in Homogeneous Soil

In the next simulation we consider the phenomenon of DNAPL drainage. The intial conditions in this example represent the state of the system at the end of the primary imbibition phase, that is, at an elapsed time of 1000 sec. The top boundary condition is changed from a zero pressure of DNAPL to no flow of DNAPL and the pressure of water is set to zero. The boundary condition on the bottom of the column remains as a water pressure of 31.5 cm of water.

The results of the simulation are found in Figure 5.14. At the top of the column the saturation of water increases as the DNAPL drains. As in the air-water example, the arrow denoted by the letter A indicates that the time evolution at the top of the column is from right to left; that is, the curve on the right-hand side represents the saturation profile at the earliest time. As time progresses the DNAPL saturation decreases along most of the length of the column. However, at the base of the column s^n begins to increase as the DNAPL accumulates due to the imposed water-pressure boundary condition. Time evolution of the curves at the base is the opposite to that at the top; that is, curves representing earlier DNAPL saturation profiles are on the left.

5.2.4 Secondary Imbibition of DNAPL in Homogeneous Soil

Using the DNAPL drainage saturation profile found in Figure 5.14 at 1000 sec, we now reverse the situation once again and imbibe DNAPL from the top of the column (*secondary imbibition*). To achieve this we replace the zero pressure of water boundary condition with a zero flux condition and change the DNAPL condition from zero flux to zero pressure. The water boundary condition is specified as no flow. The boundary condition on the bottom of the column remains as a water pressure of 31.5 cm of water and no flow of the DNAPL phase.

One would expect the saturation of DNAPL to increase, intially at the top, and then progressively downward as time increases. Such behavior is observed in Figure 5.15. Once again the arrows identified as A and B indicate the direction of advancing time at the top and bottom of the column, respectively. We observe that, over the period of analysis, the DNAPL saturation increases at the top of the column while drainage occurring at the base causes the DNAPL saturation to decrease there.

It is interesting to compare the state of the system in Figures 5.12 and 5.15 at the same point in simulated time (in the sense of time since each simulation began). Consider, for example the DNAPL saturation profiles provided in Figure 5.16. The solid and dashed lines represent the time evolution of the primary and secondary imbibition saturation fronts. The impact of the pre-existing DNAPL saturation in the case of secondary imbibition is dramatic. Observe, for example, the location of the saturation front at a time of 950 sec. It is evident that the secondary imbibition

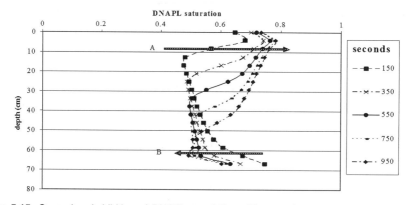

Figure 5.15: Secondary imbibition of DNAPL coexisting with water in porous medium. Curves are time since secondary imbibition started, in seconds.

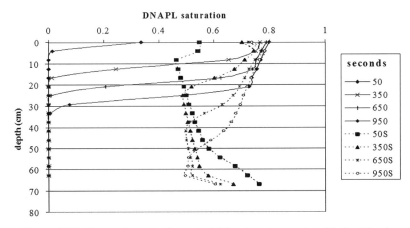

Figure 5.16: Comparison of primary (solid lines) and secondary (dashed lines).

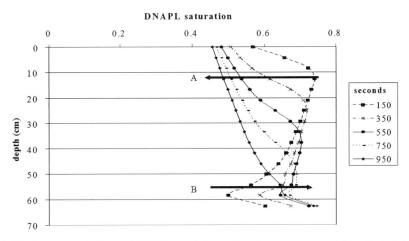

Figure 5.17: Secondary drainage of DNAPL coexisting with water in a porous medium. Curves denote time since secondary drainage started, in seconds.

front moves much faster than the primary imbibition front due to the effect of the saturation distribution of the DNAPL in the drained profile.

5.2.5   Secondary Drainage in Homogeneous Soil

Now we turn our attention to secondary drainage, the final simulation in this sequence. The secondary imbibition profile observed at 1000 sec is employed as the initial condition for saturation. The constant DNAPL pressure boundary condition at the top is replaced with a no flux condition, and a zero water flux condition is replaced with a zero water pressure condition. The boundary condition on the bottom of the column remains as a water pressure of 31.5 cm of water and no flow of the DNAPL phase. Under these conditions we observe the behavior documented in Figure 5.17. The general behavior appears similar to that found in Figure 5.14 with DNAPL saturations decreasing at the top of the column and increasing at the base. However, it is evident that the dynamics of secondary drainage are different in detail from those of primary drainage.

5.2.6   Primary Imbibition in Heterogeneous Soil

We return now to a discussion of the impact of heterogeneity, more specifically layering, on DNAPL movement. We assume, as earlier, that the column has a low permeability layer about 17 cm from the column top. Holding all other conditions as in the previous primary imbibition calculation, namely that at the bottom of the column the water pressure is 31.5 cm and no flow of DNAPL occurs while at the top of the column, the DNAPL pressure is 0.0 and no water flow occurs, we simulate primary imbibition. The results are presented in Figure 5.18. When compared to primary imbibition into a homogeneous sand, the propogation of the saturation front is very slow. Compare, for example, the 950 sec profile in Figure 5.12 with the 1000 sec contour in Figure 5.18. It is evident that the lower permeability layer is impeding the downward movement of the water which, in turn, impedes the downward movement of the DNAPL. It also appears that there is some pooling on the low permeability interface.

It is helpful to compare directly the movement of the saturation fronts for the cases of homogeneous and layered material. Such a comparison is presented in Figure 5.19, where the solid and dashed lines represent the saturation profiles in homogeneous materials and layered materials, respectively. As expected, the saturation front moves more readily through the homogeneous material.

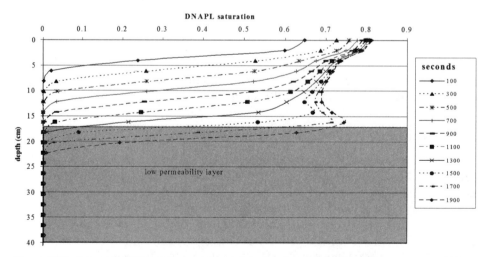

Figure 5.18: Primary imbibition into a column composed of materials of two different permeabilities. The upper portion of the column has a permeability of 5.0×10^{-7} cm^2 and the lower portion has a value of 5.0×10^{-8} cm^2. Notice that the elapsed time has increased to 1500 sec.

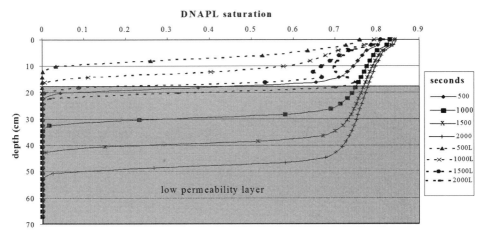

Figure 5.19: Imbibition fronts for the case of heterogeneous (dashed curve) and homogeneous (solid curve) soils.

5.3 2-D SIMULATION OF DNAPL-WATER FLOW

While a number of interesting multiphase phenomena were illustrated in the preceding section on one-dimensional flow, two-dimensional systems can be used to introduce additional ideas. In this section we consider several two-dimensional systems, each illustrating an aspect of multiphase-flow behavior.

5.3.1 DNAPL Descent into a Water-Saturated Reservoir

The physical system we will model in this subsection is illustrated in Figure 5.20. It consists of a cross section with a width of 50 cm and a depth of 37.5 cm. The mass conservation equations that describe this system are special cases of equation (5.1) written for each fluid phase. When mass exchange between phases is neglected and no pumping is occurring, the equations are

$$\frac{\partial(s^\alpha \varepsilon \rho^\alpha)}{\partial t} + \frac{\partial(s^\alpha \varepsilon \rho^\alpha v_x^\alpha)}{\partial x} + \frac{\partial(s^\alpha \varepsilon \rho^\alpha v_z^\alpha)}{\partial z} = 0 \quad \alpha = w, n \tag{5.8}$$

where v_x^α and v_z^α are the x and z velocities of the α phase, respectively. The flow equations follow from equation (5.3) where we will consider isotropic conditions and flow in the x and z directions only such that:

$$q_x^\alpha = s^\alpha \varepsilon v_x^\alpha = -\frac{k^s k_{rel}^\alpha}{\mu^\alpha}\left(\frac{\partial p^\alpha}{\partial x}\right) \quad \alpha = w, n \tag{5.9}$$

and

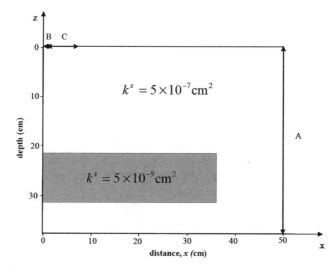

Figure 5.20: Definition sketch for the DNAPL vertical transport problem.

$$q_z^\alpha = s^\alpha \varepsilon v_z^\alpha = -\frac{k^s k_{rel}^\alpha}{\mu^\alpha}\left(\frac{\partial p^\alpha}{\partial z} + \rho^\alpha g\right) \quad \alpha = w, n \tag{5.10}$$

where the solid grain movement has been considered negligible.

The reservoir permeability is that of a sand with $k^s = 5 \times 10^{-7}\,\text{cm}^2$ except for a less permeable rectangular area of clay where $k^s = 5 \times 10^{-9}\,\text{cm}^2$. Initially the reservoir is fully saturated with water such that $s^w = 1$. Other parameters for this problem can be found in the following table:

Fluid Property	Water	DNAPL
μ^α [viscosity (poise)]	0.01	0.005
ρ^α [density (g/cm³)]	1.0	1.5
s_i^w/s_r^n [irred./resid. saturation]	0.1	0.15
ε [porosity]		0.34

The left side of the region is impermeable, as is the base. The head along the right-hand vertical side is set at hydrostatic (no vertical flow). Along the section marked as B on the top, DNAPL is injected at a rate of 0.02 l³/sec and along the segment marked as C the rate is 0.04 l³/sec. From the rightmost end of the region C to the right-hand side boundary the top is impermeable.

For this system, DNAPL will enter the reservoir in the top left-hand corner and move primarily vertically downward under the influence of gravity. It will then encounter the lower permeability layer and begin to pool since its vertical movement is retarded. When the saturation of DNAPL has reached a critical level, the DNAPL will move horizontally to the right along the top of the lower permeability layer. Thereafter it will again move more expeditiously downward under the influence of gravity.

The following three figures show the movement of DNAPL over a period of 1500 seconds. In Figure 5.21 the DNAPL saturation distribution after 300 seconds is

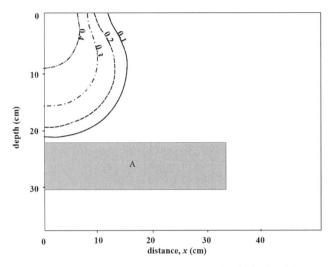

Figure 5.21: Saturation distribution of DNAPL after 300 seconds of injection. The contours are saturation distribution.

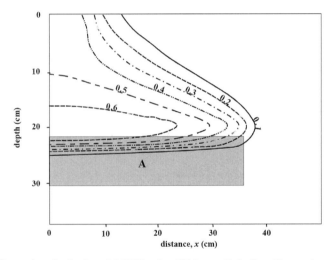

Figure 5.22: Saturation distribution of DNAPL after 1000 sec of injection. The contours are saturation distribution, and the lower permeability region is denoted by the letter A.

provided. At this point the influence of the underlying lower permeability layer marked A is not evident. However, after 1000 seconds the DNAPL has begun to pool on the lower permeability layer as seen by the increased saturation to 0.6 above the lower permeability layer A as illustrated in Figure 5.22. Finally, in Figure 5.23 we observe that the DNAPL has moved laterally and is cascading down through a breach in the low permeability layer B. It is interesting to note that from a practical perspective the downward movement may lead to DNAPL penetrating to significant depths from which it is difficult to remove.

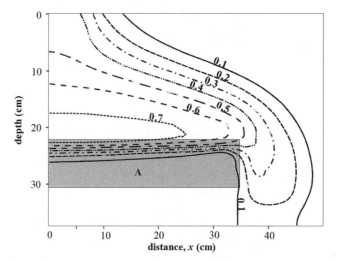

Figure 5.23: Saturation distribution of DNAPL after 1500 sec of injection. The contours are saturation distribution, and the lower permeability region is denoted by the letter A.

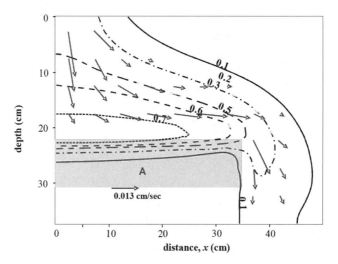

Figure 5.24: DNAPL velocity at an elapsed time of 1500 sec overlying the saturation profile of DNAPL. The scale for the velocity is found on the figure in centimeters per second.

In Figure 5.24 the DNAPL velocity is superimposed on the DNAPL flow field so that the DNAPL movement can be visualized; water velocity is shown in Figure 5.25.

5.4   SIMULATION OF MULTIPHASE FLOW AND TRANSPORT

In this section we present a couple of representative solutions to problems that are described by making use of both the flow and transport theory presented in the preceding chapters, especially Chapters 3 and 4. Here, again, we will describe the

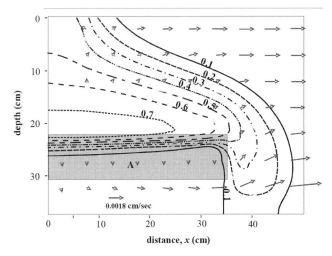

Figure 5.25: Water velocity at an elapsed time of 1500 sec overlying the saturation profile of DNAPL. The scale for the velocity is found on the figure in centimeters per second.

physical problem, obtain the simplified form of the governing equations, employ any needed constitutive relations, and provide solutions along with some observations.

5.4.1   1-D Two-Phase Flow and Transport

Consider the problem described in Subsection 5.1.3, imbibition of water into a homogeneous soil containing water and air. Let the water that is imbibing enter the system with a constant solute mass fraction for species i of $\omega^{iw} = \omega_0^{iw}$ at the top of the column. Assume that the chemical species is nonreacting, nonvolatile, and non-adsorbing onto the solid phase. To describe the transport process, we can use one of the forms obtained in Chapter 4. Here, we will select the distribution form, equation (4.31), which for species i in the w phase with the reaction, interphase transport, and pumping terms on the right side neglected, takes the form:

$$\varepsilon s^w \rho^w \frac{\partial \omega^{iw}}{\partial t} + \varepsilon s^w \rho^w \mathbf{v}^w \cdot \nabla \omega^{iw} + \nabla \cdot \left(\varepsilon s^w \mathbf{j}^{iw} \right) = 0 \qquad (5.11)$$

To close this equation, we need a constitutive form for the dispersion tensor. This is obtained from Chapter 4 as equation (4.39), which for the species of interest here in the w phase is:

$$\varepsilon s^w \mathbf{j}^{iw} = -\rho^w \varepsilon s^w \mathbf{D}^{iw} \cdot \nabla \omega^{iw} \qquad (5.12)$$

where the diffusion/dispersion coefficient, $\mathbf{D}^{iw}$, is approximated, according to equation (4.40), as:

$$\varepsilon s^w \mathbf{D}^{iw} = \varepsilon s^w D_{\text{dif}}^{iw} \mathbf{I} + \varepsilon s^w \alpha_{\mathrm{T}} |\mathbf{v}^w| \mathbf{I} + \varepsilon s^w (\alpha_{\mathrm{L}} - \alpha_{\mathrm{T}}) \frac{\mathbf{v}^w \mathbf{v}^w}{|\mathbf{v}^w|} \qquad (5.13)$$

where the solid phase deformation has been considered negligible such that $\mathbf{q}^w = \varepsilon s^w \mathbf{v}^w$.

We substitute equation (5.12) into equation (5.11) to obtain the closed equation:

$$\varepsilon s^w \rho^w \frac{\partial \omega^{iw}}{\partial t} + \varepsilon s^w \rho^w \mathbf{v}^w \cdot \nabla \omega^{iw} - \nabla \cdot (\rho^w \varepsilon s^w \mathbf{D}^{iw} \cdot \nabla \omega^{iw}) = 0 \qquad (5.14)$$

For the problem of interest, we note that the macroscale velocity has a component only in the z direction and that there are no gradients of ω^{iw} in the transverse directions. If we also consider that the concentrations of contaminant are small enough that the density of the flow may be considered constant, the preceding equation simplifies to:

$$\varepsilon s^w \frac{\partial \omega^{iw}}{\partial t} + \varepsilon s^w v_z^w \frac{\partial \omega^{iw}}{\partial z} - \frac{\partial}{\partial z}\left(\varepsilon s^w D_{zz}^{iw} \frac{\partial \omega^{iw}}{\partial z} \right) = 0 \qquad (5.15)$$

and:

$$D_{zz}^{iw} = D_{\text{dif}}^{iw} + \alpha_L |v_z^w| \qquad (5.16)$$

The value of v_z^w in these expressions is obtained as a solution to the flow equation (5.4). Because of the tortuous flow path, the macroscale diffusion coefficient, D_{dif}^{iw}, is typically reduced from its microscale antecedent by tortuosity. Here, we will use a value of $\alpha_L = 1\,\text{cm}$ and

$$D_{\text{dif}}^{iw} = (\varepsilon s^w)^{1/3} (s^w)^2 D_{iw_{\text{dif}}} \qquad (5.17)$$

where $D_{iw_{\text{dif}}}$ is the microscale diffusion coefficient. The value of the microscale diffusion coefficient is assumed to be $D_{iw_{\text{dif}}} = 1.0 \times 10^{-5}\,\text{cm}^2/\text{sec}$.

The propagation of the concentration front is seen in Figure 5.26, where $\omega^{i\alpha}/\omega_0^{i\alpha}$ is plotted as a function of depth. The interesting feature of this figure is in the comparison of the movement of the concentration front relative to the movement of the saturation front provided in Figure 5.11. At an elapsed time of 100 sec one can see that both the saturation and the concentration are increasing at a depth of 35 cm. However, the concentration increase is not attributable to convection at this depth but rather to dispersion and diffusion. The maximum value of v_z^w in this experiment was 0.04 cm/sec. Therefore after 100 sec, the concentration front would have convected no more than about 4 cm.

5.4.2 2-D Two-Phase Flow and Transport

In this subsection we will model solute transport as related to DNAPL flow. While DNAPL has slight solubility in water, this solubility can be important when the DNAPL solute is transported by the water phase. In general the DNAPL phase will dissolve, albeit slowly, into the aqueous phase and can also volatilize into a gaseous phase both from the DNAPL phase and the aqueous phase. In the following we add

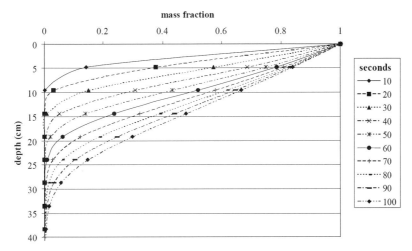

Figure 5.26: Propagation of the normalized concentration due to imbibed water into a drained column.

the complexity of interphase transport to the DNAPL imbibition problem defined in Subsection 5.3.1. Specifically we will consider the case when DNAPL exchange with the water phase is modeled as a kinetic process. Because water initially fills the pore space, we will assume that interaction of DNAPL with the solid occurs only through the wetting phase. This approximation suggests that the wetting phase remains in contact with the solid throughout the system. We will consider that the solid movement is negligible and that no gas phase is present. The model therefore requires that we consider flow of the two fluid phases and transport of DNAPL within the w phase.

As a starting point for specification of the transport equation employed, use equation (4.85) for species i (rather than k) with the reaction terms r^{iw} and r^{is} set to zero since no reactions are occurring and with the pumping term neglected:

$$\varepsilon^w \rho^w R_S^{iw} \frac{\partial \omega^{iw}}{\partial t} + \rho^w \mathbf{q}^w \cdot \nabla \omega^{iw} - \nabla \cdot \left(\varepsilon^w \rho^w \mathbf{D}^{iw} \cdot \nabla \omega^{iw} \right)$$
$$= \left(e_{wn}^{iw} - \omega^{iw} e_{wn}^w \right) + \left(\frac{1 - \omega^{iw}}{1 - K^{iws} \omega^{iw}} \right) \left(e_{ns}^{is} - K^{iws} \omega^{iw} e_{ns}^s \right) \qquad (5.18)$$

Since we are neglecting mass transfer between the n and s phases, both e_{ns}^{is} and e_{ns}^s are zero. Also, since DNAPL is the only species being transferred between the w and n phases, $e_{wn}^{iw} = e_{wn}^w$. With these stipulations imposed, equation (5.18) becomes:

$$\varepsilon^w \rho^w R_S^{iw} \frac{\partial \omega^{iw}}{\partial t} + \rho^w \mathbf{q}^w \cdot \nabla \omega^{iw} - \nabla \cdot \left(\varepsilon^w \rho^w \mathbf{D}^{iw} \cdot \nabla \omega^{iw} \right) = \left(1 - \omega^{iw} \right) e_{wn}^{iw} \qquad (5.19)$$

The transport between the w and n phases is modeled as a kinetic process using the linear closure relation of equation (4.54):

$$e_{wn}^{iw} = -\kappa_{wn}^{iw}\left(\omega^{iw} - \omega_{eqn}^{iw}\right) \tag{5.20}$$

where ω_{eqn}^{iw} is the *mass fraction of the solubility limit* of DNAPL in water. As was stated, the solid movement is being neglected; thus $\mathbf{q}^w = \varepsilon s^w \mathbf{v}^w$. The model to be employed is two-dimensional, so we can expand the vector notation of equation (5.18) to the explicit form:

$$
\begin{aligned}
&\varepsilon s^w \rho^w R_S^{iw} \frac{\partial \omega^{iw}}{\partial t} + \varepsilon s^w \rho^w v_x^w \frac{\partial \omega^{iw}}{\partial x} + \varepsilon s^w \rho^w v_z^w \frac{\partial \omega^{iw}}{\partial z} \\
&- \frac{\partial}{\partial x}\left[\varepsilon s^w \rho^w\left(D_{xx}^{iw}\frac{\partial \omega^{iw}}{\partial x} + D_{xz}^{iw}\frac{\partial \omega^{iw}}{\partial z}\right)\right] \\
&- \frac{\partial}{\partial z}\left[\varepsilon s^w \rho^w\left(D_{zx}^{iw}\frac{\partial \omega^{iw}}{\partial x} + D_{zz}^{iw}\frac{\partial \omega^{iw}}{\partial z}\right)\right] \\
&= -\left(1 - \omega^{iw}\right)\kappa_{wn}^{iw}\left(\omega^{iw} - \omega_{eqn}^{iw}\right)
\end{aligned}
\tag{5.21}
$$

To solve equation (5.21), the parameters must be specified. Adsorption of the DNAPL from the *w* phase onto the solid phase is modeled making use of a *linear adsorption isotherm* so that the retardation coefficient, R_S^{iw}, is given by equation (4.88):

$$R_S^{iw} = 1 + \frac{(1-\varepsilon)\rho^s}{\varepsilon^w}K_d^{iws} \tag{5.22}$$

with the *distribution coefficient*, introduced in equation (4.88) as:

$$K_d^{iws} = \frac{\omega^{is}}{\left(\rho^w \omega^{iw}\right)}$$

assigned the value $K_d^{iws} = 1.0 \times 10^{-6} \text{ cm}^3/\text{g}$.

The diffusion/dispersion tensor is again parameterized using the constitutive form of equation (4.40) where the elements of the tensor are:

$$
\begin{aligned}
D_{xx}^w &= \alpha_T|\mathbf{v}^w| + (\alpha_L - \alpha_T)(v_x^w)^2/|\mathbf{v}^w| + (\varepsilon s^w)^{1/3}(s^w)^2 D_{iw_{dif}} \\
D_{zz}^w &= \alpha_T|\mathbf{v}^w| + (\alpha_L - \alpha_T)(v_z^w)^2/|\mathbf{v}^w| + (\varepsilon s^w)^{1/3}(s^w)^2 D_{iw_{dif}} \\
D_{xz}^w &= D_{zx}^w = (\alpha_L - \alpha_T)v_x^w v_z^w/|\mathbf{v}^w|
\end{aligned}
\tag{5.23}
$$

For the physical system being considered here, we select $\alpha_L = 1.0\,\text{cm}$, $\alpha_T = 0.2\,\text{cm}$, and $D_{iw_{dif}} = 1.0 \times 10^{-5}\,\text{cm}^2/\text{sec}$.

The *kinetic rate coefficient*, κ_{wn}^{iw}, describes the rate at which DNAPL crosses the DNAPL-water interface and is parameterized according to the relationship [3]:

$$\kappa_{wn}^{iw} = \kappa_{wn0}^{iw}\left(\varepsilon s^n\right)^{1/2}|\mathbf{v}^w| \tag{5.24}$$

This form attempts to account for the influence of flow velocity and the amount of interfacial area present in the system on the rate at which mass is transferred. Here, we specify $\kappa_{wn0}^{iw} = 1.0 \times 10^1$ g/cm^3. The *mass fraction solubility limit* of DNAPL in water employed here is $\omega_{eqn}^{iw} = 1.1 \times 10^{-3}$.

In this example we are using the same flow model as provided in Subsection 5.3.1. Initially the mass fraction of DNAPL is zero in the water phase. The boundary conditions for the mass fraction are zero normal gradient (second type $\mathbf{n} \cdot \nabla \omega^{iw} = 0$) on all sides. The resulting system permits convective, but not dispersive, movement across the boundaries.

The concentration calculated after 300 sec is provided as the solid lines in Figure 5.27. Also shown on this figure are the DNAPL saturation contours. In this physical system all of the water movement is generated due to displacement of the water by the NAPL. Thus the concentration distribution is due to this convective movement, dissolution of the DNAPL, adsorption, and dispersion.

In Figure 5.28 the water velocity arrows have been added and only the 0.1 DNAPL saturation contour is provided. One observes that the water velocity is relatively small in the interior of the DNAPL and is higher to the right of the DNAPL due to continuing outward motion of the DNAPL saturation front. The water velocity is also responding to the lower permeability layer marked A. The state of the system after 600 sec of simulation is shown in Figure 5.29. Note that the movement of the DNAPL phase over the end of the low permeability layer is causing a corresponding response in the water-phase velocity field.

Since DNAPL movement in this example is primarily gravity driven, it is interesting to consider how a sloping aquifer would influence its movement. In Figure 5.30 the movement of a DNAPL phase along an inclined surface is illustrated. The problem is the same as presented earlier with two exceptions: (1) the low permeability layer is inclined and has a permeability value of 5.0×10^{-11} cm^2 and (2) the left vertical boundary has been changed from one allowing no dispersive mass transport to one where the DNAPL mass fraction in the water phase is held at zero. The resulting impact on concentration after 600 sec of simulation is evident from the

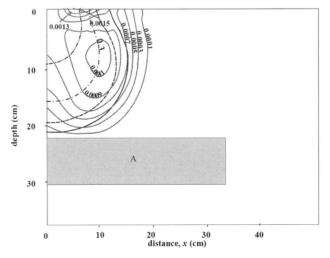

Figure 5.27: Concentration distribution of DNAPL associated with dissolution from the DNAPL plume shown in Figure 5.21 which documents the saturation distribution after 300 sec.

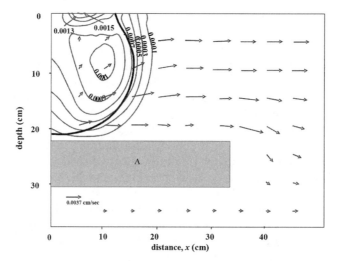

Figure 5.28: Concentration distribution of DNAPL after 300 sec associated with dissolution from the DNAPL plume shown in Figure 5.21. Included are the water velocity vectors.

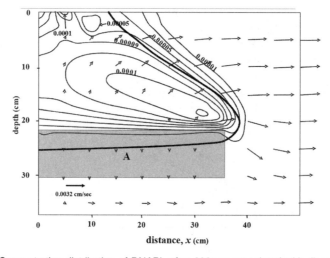

Figure 5.29: Concentration distribution of DNAPL after 600 sec associated with dissolution from the DNAPL plume. Included are the 0.1 saturation contour and water velocity vectors.

change in the concentration contours in that area. Note also that the steplike surface of the low permeability layer is employed due to rectangular grid-definition limitations of the model in describing a sloped surface.

We now consider the impact of imposing a groundwater flow velocity in a direction opposite to that in which the DNAPL tends to move due to the sloping low permeability layer. All other aspects of the model are the same as those that generated Figure 5.30. The water velocity is generated by specifying head on the left side of the model to be 5 cm lower than that on the right.

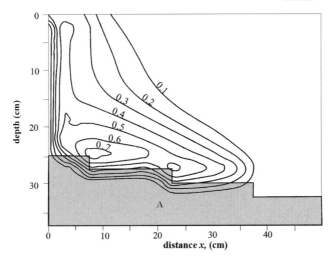

Figure 5.30: DNAPL saturation distribution when the low-permeability layer is sloping. Elapsed time 600 sec.

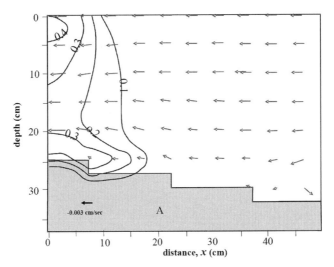

Figure 5.31: Saturation distribution of DNAPL after 600 sec when there is a pressure head gradient in the water phase due to a change in water phase pressure head of 5 cm from right to left. Also shown are the water velocity vectors.

Figure 5.31 reveals a consistent water flow from right to left. More interesting, however, is the fact that this flow is sufficiently strong to inhibit significantly the movement of the DNAPL down the slope. It is useful to compare Figure 5.30, which does not have the water flow from right to left, with Figure 5.31, which does. It is evident that the counterflow of water has a significant impact on the flow of DNAPL in this system.

5.5   2-D SINGLE-PHASE FLOW AND TRANSPORT

In this section we focus on two-dimensional saturated flow and mass transport where the dependence of the fluid density on the concentration of the dissolved solute is assumed negligible. The objective is to illustrate the influence of a range of physical processes on species transport behavior. The physical system is shown in Figure 5.32 and is a $600 \times 400\,$ft rectangular horizontal region for which all properties are considered to be uniform in the vertical.

For modeling the two-dimensional flow, we will make use of the vertically integrated flow equation (3.112):

$$S\frac{\partial h^w}{\partial t} - \nabla' \cdot \left(\mathbf{T}^{w''} \cdot \nabla' h^w \right) + \left(\mathbf{n} \cdot \mathbf{q}^w \right)\big|_{z_T} + \left(\mathbf{n} \cdot \mathbf{q}^w \right)\big|_{z_B} = \sum_{W=1}^{N_W} Q_W^{wt} \tag{5.25}$$

In developing this equation we have assumed that although transfer of chemical species between phases occurs, the amount of mass transferred is not significant enough that it impacts the flow equation for the w phase. We will consider the system to be isotropic such that the two-dimensional *transmissivity* is $\mathbf{T}^{w''} = T^w \mathbf{I}''$. Then expansion of the vector notation into component parts yields:

$$S\frac{\partial h^w}{\partial t} - \frac{\partial}{\partial x}\left(T^w \frac{\partial h^w}{\partial x} \right) - \frac{\partial}{\partial y}\left(T^w \frac{\partial h^w}{\partial y} \right) + \left(\mathbf{n} \cdot \mathbf{q}^w \right)\big|_{z_T} + \left(\mathbf{n} \cdot \mathbf{q}^w \right)\big|_{z_B} = \sum_{W=1}^{N_W} Q_W^{w'} \tag{5.26}$$

The vertically integrated transport equation for species i in the w phase is obtained using the same methods employed in Subsection 3.6.6 for the flow equation. We start from species transport equation (4.1) where the w phase is the only fluid phase present such that $\varepsilon^w = \varepsilon$. The result is provided here without the detailed derivation:

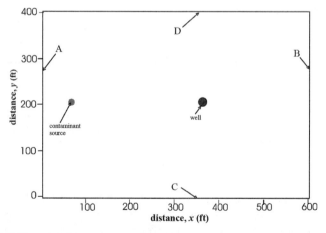

Figure 5.32: Definition of problem area considered in the two-dimensional saturated-flow and transport problems. The letters A–D define boundary conditions described in the text. The contaminant source and location of pumping well are labeled. The flow is uniform from side A to side B and the units of distance are feet.

$$\frac{\partial(\varepsilon\rho^{\omega}b\omega^{iw})}{\partial t} + \nabla' \cdot (\varepsilon\rho^w b\omega^{iw}\mathbf{v}^{w'}) + \nabla' \cdot (\varepsilon b\mathbf{j}^{iw'})$$
$$+ [\mathbf{n} \cdot (\rho^w\omega^{iw}\mathbf{q}^w + \varepsilon\mathbf{j}^{iw})]\big|_{z_T} + [\mathbf{n} \cdot (\rho^w\omega^{iw}\mathbf{q}^w + \varepsilon\mathbf{j}^{iw})]\big|_{z_B}$$
$$= \varepsilon b r^{iw} + b e_{ws}^{iw} + \sum_{W=1}^{N_W} \rho_W^w \omega_W^{iw} Q_W^{wt} \tag{5.27}$$

where the quantities other than those explicitly denoted as being evaluated at the top (z_T) or the bottom (z_B) of the study region are averages over the height (b) and the $'$ indicates a two-dimensional (in the lateral directions) vertically averaged vector. Because advection, dispersion, and pumping for the solid phase can be neglected, the vertically integrated transport equation for species i in the s phase is:

$$\frac{\partial[(1-\varepsilon)\rho^s b\omega^{is}]}{\partial t} = \varepsilon b r^{is} + b e_{ws}^{is} \tag{5.28}$$

For this study, we will make use of the equilibrium approach to transfer of species i between the s and w phases. Therefore, we make use of the fact that $e_{ws}^{iw} + e_{ws}^{is} = 0$ to eliminate this exchange term between equations (5.27) and (5.28) to obtain:

$$\frac{\partial(\varepsilon\rho^w b\omega^{iw})}{\partial t} + \frac{\partial[(1-\varepsilon)\rho^s b\omega^{is}]}{\partial t} + \nabla' \cdot (\varepsilon\rho^w b\omega^{iw}\mathbf{v}^{w'}) + \nabla' \cdot (\varepsilon b\mathbf{j}^{iw'})$$
$$+ [\mathbf{n} \cdot (\rho^w\omega^{iw}\mathbf{q}^w + \varepsilon\mathbf{j}^{iw})]\big|_{z_T} + [\mathbf{n} \cdot (\rho^w\omega^{iw}\mathbf{q}^w + \varepsilon\mathbf{j}^{iw})]\big|_{z_B}$$
$$= \varepsilon b r^{iw} + \varepsilon b r^{is} + \sum_{W=1}^{N_W} \rho_W^w \omega_W^{iw} Q_W^{w'} \tag{5.29}$$

The transport may be modeled using this form, or we can reduce the equation further by making use of the phase equation obtained by summing equation (5.27) over all species in the w phase:

$$\frac{\partial(\varepsilon\rho^w b)}{\partial t} + \nabla' \cdot (\varepsilon\rho^w b\mathbf{v}^{w'}) + [\mathbf{n} \cdot (\rho^w\mathbf{q}^w)]\big|_{z_T} + [\mathbf{n} \cdot (\rho^w\mathbf{q}^w)]\big|_{z_B} = b e_{ws}^w + \sum_{W=1}^{N_W} \rho_W^w Q_W^{w'} \tag{5.30}$$

and summing equation (5.28) over all species in the s phase:

$$\frac{\partial[(1-\varepsilon)\rho^s b]}{\partial t} = b e_{ws}^s \tag{5.31}$$

Then, since $e_{ws}^w + e_{ws}^s = 0$, these last two equations may be combined to:

$$\frac{\partial(\varepsilon\rho^w b)}{\partial t} + \frac{\partial[(1-\varepsilon)\rho^s b]}{\partial t} + \nabla' \cdot (\varepsilon\rho^w b\mathbf{v}^{w'})$$
$$+ [\mathbf{n} \cdot (\rho^w\mathbf{q}^w)]\big|_{z_T} + [\mathbf{n} \cdot (\rho^w\mathbf{q}^w)]\big|_{z_B} = \sum_{W=1}^{N_W} \rho_W^w Q_W^{w'} \tag{5.32}$$

Multiplication of equation (5.32) by ω^{iw} and subtraction of the result from equation (5.29) while making use of the product rule yields:

$$
\begin{aligned}
\varepsilon\rho^w b\frac{\partial\omega^{iw}}{\partial t} &+ (1-\varepsilon)\rho^s b\frac{\partial\omega^{is}}{\partial t} + (\omega^{is}-\omega^{iw})\frac{\partial[(1-\varepsilon)\rho^s b]}{\partial t} \\
&+ \varepsilon\rho^w b\mathbf{v}^{w'}\cdot\nabla'\omega^{iw} + \nabla'\cdot(\varepsilon b\mathbf{j}^{iw'}) \\
&+ \mathbf{n}\cdot(\rho^w\mathbf{q}^w)\big|_{z_T}(\omega^{iw}_{z_T}-\omega^{iw}) + \mathbf{n}\cdot(\varepsilon\mathbf{j}^{iw'})\big|_{z_T} \\
&+ \mathbf{n}\cdot(\rho^w\mathbf{q}^w)\big|_{z_B}(\omega^{iw}_{z_B}-\omega^{iw}) + \mathbf{n}\cdot(\varepsilon\mathbf{j}^{iw'})\big|_{z_B} \\
&= \varepsilon br^{iw} + \varepsilon br^{is} + \sum_{W=1}^{N_W}\rho^w_W(\omega^{iw}_W-\omega^{iw})Q^{w'}_W
\end{aligned}
\tag{5.33}
$$

The third term in this equation is smaller than the first two because the deformation of the solid is very small. We will therefore ignore this term. We will also assume that no reaction involving species i occurs in the solid phase (i.e., $r^{is} = 0$) and that the dispersion vectors normal to the top and bottom surfaces are zero (i.e., no dispersion into or out of the system occurs at the top and bottom of our study region). Additionally, the interphase mass transfer is considered to occur at equilibrium. We will make use of a linear adsorption isotherm as given by equation (4.66) written for the s and w phases as:

$$
\omega^{is} = \rho^w K_d^{iws}\omega^{iw}
\tag{5.34}
$$

We will also make use of the closure relation for the vertically averaged, two-dimensional dispersion vector analogous to equation (5.12):

$$
\mathbf{j}^{iw'} = -\rho^w\mathbf{D}^{iw''}\cdot\nabla'\omega^{iw}
\tag{5.35}
$$

With these conditions incorporated, equation (5.33) simplifies to:

$$
\begin{aligned}
\varepsilon\rho^w b&\left[1+\frac{(1-\varepsilon)\rho^s}{\varepsilon}K_d^{iws}\right]\frac{\partial\omega^{iw}}{\partial t} \\
&+ \varepsilon\rho^w b\mathbf{v}^{w'}\cdot\nabla'\omega^{iw} - \nabla'\cdot(\varepsilon\rho^w b\mathbf{D}^{iw''}\cdot\nabla'\omega^{iw}) \\
&+ \mathbf{n}\cdot(\rho^w\mathbf{q}^w)\big|_{z_T}(\omega^{iw}_{z_T}-\omega^{iw}) + \mathbf{n}\cdot(\rho^w\mathbf{q}^w)\big|_{z_B}(\omega^{iw}_{z_B}-\omega^{iw}) \\
&= \varepsilon br^{iw} + \sum_{W=1}^{N_W}\rho^w_W(\omega^{iw}_W-\omega^{iw})Q^{w'}_W
\end{aligned}
\tag{5.36}
$$

In the present example, we will make use of the common assumption that changes in density and porosity are important only insofar as they contribute to the storage coefficient in equation (5.26). Therefore, make the approximation:

$$
\nabla'\cdot(\varepsilon\rho^w b\mathbf{D}^{iw''}\cdot\nabla'\omega^{iw}) \approx \varepsilon\rho^w\nabla'\cdot(b\mathbf{D}^{iw''}\cdot\nabla'\omega^{iw})
\tag{5.37}
$$

Finally, we can expand equation (5.36) in terms of the elements of the vector components and divide by $\varepsilon\rho^w$ to obtain:

$$
bR_S^{iw}\frac{\partial\omega^{iw}}{\partial t}+bv_x^w\frac{\partial\omega^{iw}}{\partial x}+bv_y^w\frac{\partial\omega^{iw}}{\partial y}
$$

$$
-\frac{\partial}{\partial x}\left(bD_{xx}^{iw}\frac{\partial\omega^{iw}}{\partial x}+bD_{xy}^{iw}\frac{\partial\omega^{iw}}{\partial y}\right)-\frac{\partial}{\partial y}\left(bD_{yx}^{iw}\frac{\partial\omega^{iw}}{\partial x}+bD_{yy}^{iw}\frac{\partial\omega^{iw}}{\partial y}\right)
$$

$$
+\frac{1}{\varepsilon\rho^w}\mathbf{n}\cdot(\rho^w\mathbf{q}^w)\Big|_{z_T}(\omega_{z_T}^{iw}-\omega^{iw})+\frac{1}{\varepsilon\rho^w}\mathbf{n}\cdot(\rho^w\mathbf{q}^w)\Big|_{z_B}(\omega_{z_B}^{iw}-\omega^{iw})
$$

$$
=b\frac{r^{iw}}{\rho^w}+\frac{1}{\varepsilon\rho^w}\sum_{W=1}^{N_W}\rho_W^w(\omega_W^{iw}-\omega^{iw})Q_W^{w'} \tag{5.38}
$$

where R_S^{iw} is the standard retardation factor as defined in equation (4.88).

The specific problem we are going to model is vertically integrated flow and transport in a system where there is only one well ($N_W = 1$). Flow is out of the well with magnitude Q (volume per area per time) and therefore the concentration of species i in the well discharge is equal to the concentration in the system at that point ($Q_1^{w'}=-Q$ and $\omega_1^{iw}=\omega^{iw}$). The thickness of the system is constant (b = constant). There is no flow out the bottom of the system ($(\mathbf{n}\cdot\mathbf{q}^w)\big|_{z_B}=0$); the flow at the top could be from leakage from an overlying formation. It is given a positive value when flow enters the study region ($(\mathbf{n}\cdot\mathbf{q}^w)\big|_{z_T}=-I$); and the infiltrating fluid contains no species i ($\omega_{z_T}^{iw}=0$). The fluid density entering the system and porosity at z_T are equal to their corresponding vertical averages, and species i may decay by a *first order chemical reaction* ($r^{iw}=-k_{rxn}^{iw}\rho^w\omega^{iw}$). With these restrictions, equation (5.26) is expressed as:

$$
S\frac{\partial h^w}{\partial t}-\frac{\partial}{\partial x}\left(T^w\frac{\partial h^w}{\partial x}\right)-\frac{\partial}{\partial y}\left(T^w\frac{\partial h^w}{\partial y}\right)-I+Q=0 \tag{5.39}
$$

Species transport equation (5.38) simplifies to:

$$
bR_S^{iw}\frac{\partial\omega^{iw}}{\partial t}+bv_x^w\frac{\partial\omega^{iw}}{\partial x}+bv_y^w\frac{\partial\omega^{iw}}{\partial y}-\frac{\partial}{\partial x}\left(bD_{xx}^{iw}\frac{\partial\omega^{iw}}{\partial x}+bD_{xy}^{iw}\frac{\partial\omega^{iw}}{\partial y}\right)
$$

$$
-\frac{\partial}{\partial y}\left(bD_{yx}^{iw}\frac{\partial\omega^{iw}}{\partial x}+bD_{yy}^{iw}\frac{\partial\omega^{iw}}{\partial y}\right)+\frac{1}{\varepsilon}I\omega^{iw}+bk_{rxn}^{iw}\omega^{iw}=0 \tag{5.40}
$$

We will parameterize the dispersion coefficients using the usual form analogous to equations (5.23) with $s^w = 1$ and the y coordinate replacing the z coordinate in those expressions.

With reference to Figure 5.32, the left boundary (A) and the right boundary (B) of the study domain have specified head values of 120 and 112 ft respectively. The top (D) and bottom (C) boundaries are specified as no flow. The initial head is uniform with $h^w = 120$ ft, and initial mass fraction is $\omega^{iw} = 0.0$ throughout the domain. The contaminant mass fraction is modeled relative to the value at the contaminant source. When a pumping well is employed, it is located as indicated in the figure. Unless stated otherwise the transmissivity is $T^w = 900$ ft²/day, the storage coefficient is $S = 0.09$, the porosity is 0.2, and the aquifer thickness is $b = 90$ ft. The logitudinal and transverse dispersivities are $\alpha_L = 5$ ft and $\alpha_T = 0.1$ ft.

Table 5.2: Tabulation of attributes of figures depicting two-dimensional, single-phase transport

Figure	t days	T^w (inset) ft^2/day	$(1 - \varepsilon)\rho^s K_d$ –	Q ft^3/day/ft^2	I ft/day	k_{rxn}^{iw} day^{-1}
5.33	300	900.0	0.0	0.0	0.0023	0.0
5.34	600	900.0	0.0	0.0	0.0023	0.0
5.35	900	900.0	0.0	0.0	0.0023	0.0
5.36	900	900.0	0.0	0.0	**0.0**	0.0
5.37	900	900.0	0.0	0.0	**0.023**	0.0
5.38	900	900.0	0.0	**2000**	0.0023	0.0
5.39	900	900.0	**0.9**	0.0	0.0023	0.0
5.40	600	**180.0**	0.0	0.0	0.0023	0.0
5.41	1200	**180.0**	0.0	0.0	0.0023	0.0
5.42	1800	**180.0**	0.0	0.0	0.0023	0.0
5.43	300	**1800.0**	0.0	0.0	0.0023	0.0
5.44	600	**1800.0**	0.0	0.0	0.0023	0.0
5.45	900	**1800.0**	0.0	0.0	0.0023	0.0
5.46	900	900.0	0.0	0.0	0.0023	1.5×10^{-4}

In Table 5.2 we have summarized additional information regarding the examples to be presented below in this section. The first column indicates the figure number for the problem considered. The row identified with each figure number provides the information unique to that figure. The second column specifies the time elapsed since the beginning of the represented simulation. Column three provides the transmissivity values for the blocks of soil used to represent heterogeneous formations. The fourth column provides the coefficient associated with adsorption. Column five provides the pumping rate, column six the net inflow at z_T, and column seven the reaction rate coefficient.

5.5.1 Base Case

The base case, as described in the first row of Table 5.2, will be used for making comparisons to other examples. The base system consists of a homogeneous medium with inflow at z_T and no pumping, reactions, or adsorption. Figures 5.33 through 5.35, corresponding to the first three rows of the table, show the development of the contaminant plume emanating from its source and moving, generally, left to right across the model domain. Note that around the source area relative mass fraction contours between 0.6 and 1.0 have been omitted in the interest of clarity of the figure.

5.5.2 Effect of Inflow

In Figure 5.36 the inflow at z_T is removed. That is the flux attributable to this source is set to zero (i.e., $I = 0$). The figure documents the plume geometry after 900 days of simulation without inflow from an overlying formation (compare Figure 5.35 and Figure 5.36). The results are, perhaps, counter intuitive. One might expect, with the elimination of inflow, that the velocity would decrease and the plume would be smaller at any given time. The argument would be along the lines that when water

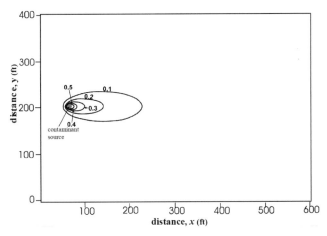

Figure 5.33: Concentration after 300 days. Contour values from 0.5 to 1.0 are omitted for clarity of presentation.

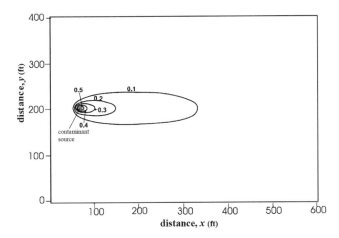

Figure 5.34: Concentration after 600 days. Contour values from 0.5 to 1.0 are omitted for clarity of presentation.

enters the system via vertical inflow, the amount leaving the system would increase, and therefore the velocity would also increase relative to the case of no inflow. If the velocity increases, then the plume should get larger. However, the plume, in fact, is larger when there is no inflow. Why this is the case is best illustrated by exaggerating the influence of inflow.

In Figure 5.37 we show the results of increasing the inflow by a factor of 10 relative to the base case from 0.0023 ft/day to 0.023 ft/day. In this instance, the plume is smaller than the base case with less inflow at a corresponding time of 300 days (see Figure 5.33). The reason for this behavior can be discerned from the groundwater velocity plot accompanying the concentration contours in Figure 5.37. The inflow provides water that, throughout the system, can be a source of outflow. Therefore, the flow from the left side (Boundary A) is diminished. Additionally, the inflow serves to dilute the contaminant.

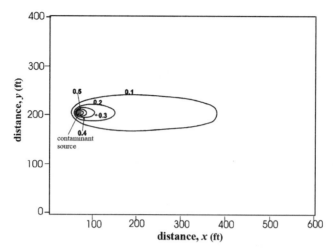

Figure 5.35: Concentration after 900 days. Contour values from 0.5 to 1.0 are omitted for clarity of presentation.

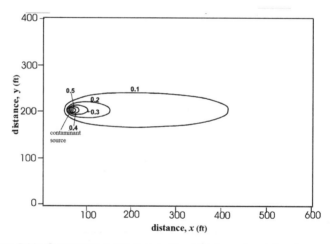

Figure 5.36: Concentration distribution after 900 days when inflow is removed.

5.5.3 Impact of Well Discharge

The use of discharging or recharging wells in the design of remediation systems for the containment and removal of subsurface contaminants is well established.

The concept is to change the groundwater flow path such that contaminants are "hydrodynamically herded" to contain the plume or to remove mass from it. In this example, we look at the impact of introducing a discharge well within the model area. As seen from Figure 5.38, after 900 days of simulation the perimeter of the plume has been modified from that appearing in the base case (see Figure 5.35). The velocity vector plot shows that the diversion of flow from a left-to-right trajec-

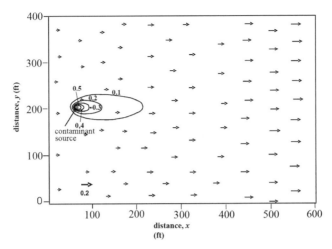

Figure 5.37: The results after 300 days of increasing the inflow by a factor of 10 relative to the base case from 0.0023 ft/day to 0.023 ft/day.

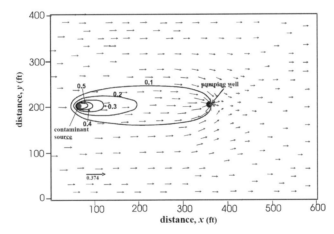

Figure 5.38: Concentration after 900 days with well pumping at 10 gal/min. Contour values from 0.5 to 1.0 are omitted for clarity of presentation.

tory to one that enters the influence of the well modifies the plume topology. In essence, the plume has been entirely captured by the well; as long as the well continues to pump, the plume is essentially contained.

5.5.4 Effect of Adsorption

The objective of this example is to illustrate the impact on species transport of adsorption of species i onto the solid phase. In this example, we make use of a linear isotherm approximation. If tetrachloroethylene were species i, $(1 - \varepsilon)\rho^s K_d \approx 0.9$. We will use this value for our simulation and note that, since $\varepsilon = 0.2$, $R_S^{iw} = 5.5$.

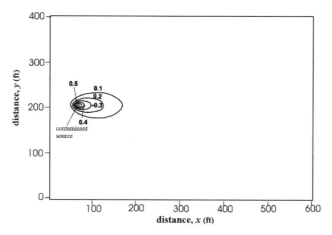

Figure 5.39: Concentration after 900 days assuming a linear adsorption isotherm with a K_d of 4.53 typical of tetrachloroethylene. Contour values from 0.5 to 1.0 are omitted for clarity of presentation.

Figure 5.39 illustrates the topology of the contaminant plume after 900 days of simulation. It is clear that the plume does not move as fast as in the base case. The adsorption onto the solid decreases the effective velocity of the flow, as was previously discussed in Chapter 4 following equation (4.85). The apparent advective velocity is inversely proportional to the retardation coefficient, R_S^{iw}; thus the effective velocity for the case depicted in Figure 5.39 is on the order of 20% of that with no adsorption depicted in Figure 5.35. The value of R_S^{iw} is very species-specific and takes on a wide range of values depending on what is being adsorbed and on the properties of the solid.

5.5.5   Effect of a Low Transmissivity Region

Subsurface reservoirs are seldom homogeneous. Typically geologic materials show a variety of grain sizes and often zones of high and low permeability materials are present within a single region of interest. In this example we embed a rectangular zone of low permeability material within the reservoir depicting the base case. More specifically, the low permeability zone has a transmissivity of $T^w = 180\,\text{ft}^2/\text{day}$ as compared to the surrounding $T^w = 900\,\text{ft}^2/\text{day}$ material.

The behavior of the contaminant plume for this scenario is shown in Figures 5.40 through 5.45. To facilitate an understanding of the resulting behavior, a vector plot of the groundwater velocity is provided in each figure. The velocity pattern confirms that water moves more slowly in the low transmissivity zone and tends to move in the more permeable material around this zone. We observe that the flow in the low permeability block is not exactly parallel to the x axis. Flow that enters the low permeability zone at the left side tends to move toward the upper and lower boundaries of the zone. As a consequence, the contaminant plume initially expands forming a bulbous shape within the low permeability block. Upon moving through the block, the plume contracts as would be expected based upon the convergence of the groundwater flow depicted by the velocity vectors. Thereafter the plume continues

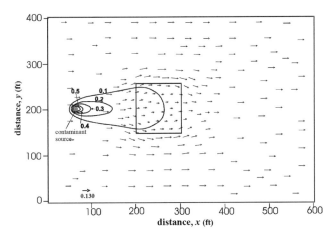

Figure 5.40: Concentration after 600 days. Rectangular area is of lower hydraulic conductivity (2 ft/day versus 10 ft/day for remaining area; velocity vector units are ft/day). Contour values from 0.5 to 1.0 are omitted for clarity of presentation.

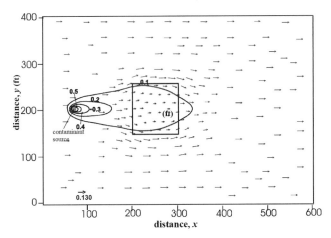

Figure 5.41: Concentration after 1200 days. Rectangular area is of lower hydraulic conductivity (2 ft/day versus 10 ft/day for remaining area). Contour values from 0.5 to 1.0 are omitted for clarity of presentation.

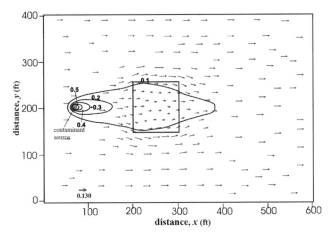

Figure 5.42: Concentration after 1800 days. Rectangular area is of lower hydraulic conductivity (2 ft/day versus 10 ft/day for remaining area). Contour values from 0.5 to 1.0 are omitted for clarity of presentation.

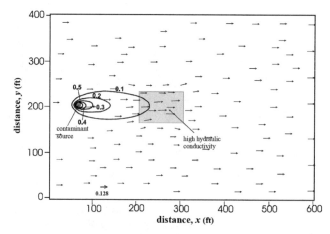

Figure 5.43: Concentration after 300 days. Rectangular area is of higher hydraulic conductivity (20 ft/day versus 10 ft/day for remaining area). Contour values from 0.5 to 1.0 are omitted for clarity of presentation.

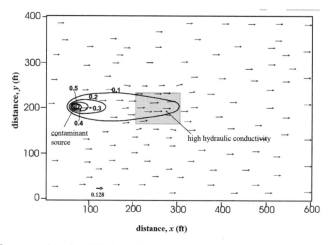

Figure 5.44: Concentration after 600 days. Rectangular area is of higher hydraulic conductivity (20 ft/day versus 10 ft/day for remaining area). Contour values from 0.5 to 1.0 are omitted for clarity of presentation.

its movement beyond the block with a topology similar to the base case. However, in viewing these figures it is important to note that the elapsed time has increased by a factor of 2 from the base case. These plots capture the system at 600, 1200, and 1800 days.

5.5.6 Effect of a High Transmissivity Region

In this example we look at the case when the imbedded block of soil has a higher transmissivity than the surrounding region. The transmissivity of block is $T^w = 1800$ ft^2/day while that of the surrounding soil remains at $T^w = 900$ ft^2/day. The evolution of the contaminant plume is presented in Figures 5.43 through 5.45. Once again,

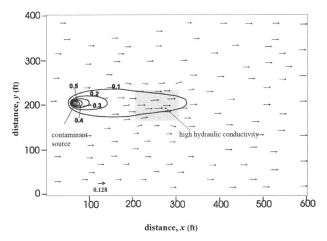

Figure 5.45: Concentration after 900 days. Rectangular area is of higher hydraulic conductivity (20 ft/day versus 10 ft/day for remaining area). Contour values from 0.5 to 1.0 are omitted for clarity of presentation.

examination of the groundwater velocity vectors provides insight into the mechanisms of plume spreading. Because the transmissivity contrast is relatively low (a factor of 2) the impact of the higher permeability layer is not as dramatic as presented in the example of Subsection 5.5.5 where the transmissivities differed by a factor of 5. Nevertheless, the velocity within the high permeability block is higher than that in the surrounding soil and the flow preferentially enters the high permeability region to find a "path of least resistance." After 900 days of simulation the lateral spreading of the plume front is less for this case than for the base case.

It is noteworthy that the plume has not moved further downstream than for the base case, given the existence of the high permeability area. Because the higher permeability block is isolated within a lower permeability soil it has little influence on long-term plume behavior. Although flow will move faster through the block than in the surrounding area, the contaminant carried by this flow at the smaller times will be met at the block exit by relatively fresh water and thus will be diluted. The 0.1 concentration front will extend approximately the same distance as for the base case, or even less depending on time of observation and the transmissivity ratio in the two regions.

5.5.7   Effect of Rate of Reaction

In this instance we consider the impact of a chemical reaction on the concentration distribution. We specifically consider the case when the species being transported is nonadsorbing and decays according to a first order reaction (i.e., $r^{iw} = -k_{rxn}^{iw}\rho^{w}\omega^{iw}$). If the *half-life*, $t_{1/2}$, is the time it takes for half the species present to react and the reaction is first order, the rate constant is obtained as:

$$k_{rxn}^{iw} = \frac{\ln 2}{t_{1/2}} \tag{5.41}$$

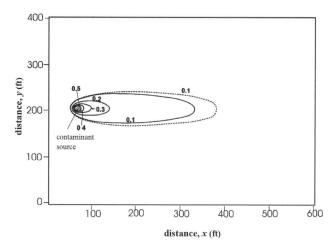

Figure 5.46: The base-case simulation (dashed line) compared with the same simulation but with the addition of a decay reaction (solid line) after 900 days.

For example, tritium is a species that decays in this manner with a half-life of 12.43 years such that the reaction rate constant is $k_{rxn}^{iw} = 1.5 \times 10^{-4}\,\mathrm{day}^{-1}$. We will use this coefficient in the present example while recognizing that for other species, rate constants will vary by orders of magnitude and that the forms of reaction rate expressions can also be very complex.

The plume is simulated using the base case situation with the addition of the decay reaction. After 900 days, the plume topology obtained is as shown in Figure 5.46. For reference, the base plume 0.1 contour is provided as the dashed line in this figure. As expected, the plume is smaller relative to the base case since species i has been depleted through the decay reaction.

5.6   3-D SINGLE-PHASE FLOW AND TRANSPORT

The straightforward single-phase flow examples presented in Section 5.5 were based on an areal two-dimensional model. The vertical dimension was formally accommodated through integration and definition of suitable vertically averaged parameters and state variables. The question naturally arises as to how important the vertical dimension is in addressing real world problems. In this section we briefly address this question through consideration of a three-dimensional extension of the base problem described in Subsection 5.5.1 augmented by inflow from above as presented in Subsection 5.5.2. We selected this problem as an example because it illustrates the "plunging plume" phenomenon very commonly encountered in the field.

The three-dimensional transient flow and transport model was developed using columnar finite elements with triangular cross-section with each of the five vertical layers in the model being discretized into the same triangular mesh while the meshes are connected using vertical lines. Thus, in cross section, each element projects as a rectangle. The significant difference between this model and that of Subsection 5.5.2

is that the aquifer with vertical thickness $b = 90$ ft is now subdivided into five layers. The top layer is 10 ft thick and the thicknesses of the other four are each 20 ft. A second experiment was also conducted which employed five layers for a system with $b = 50$ ft to demonstrate the importance of aquifer thickness on three-dimensional flow.

The concentration distribution in each of the layers after an elapsed period of 1000 days for the case of high inflow (0.023 ft/day) is provided in Figure 5.47. The sequence goes from top layer (A) to bottom layer (E). In this simulation, the concentration source is considered to be present only in the top layer.

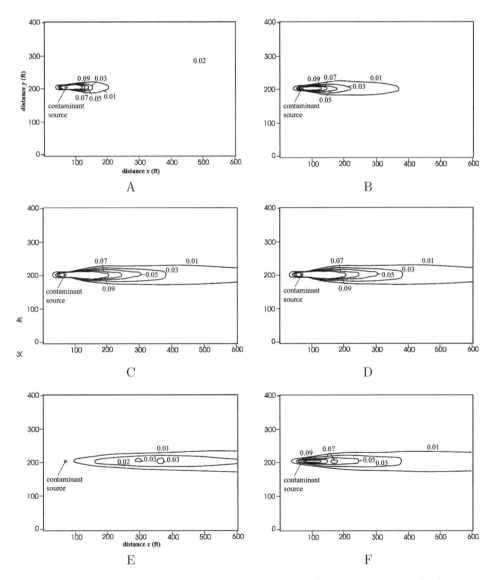

Figure 5.47: Plunging plume due to infow with representation of increasing area with depth as one moves through panels B, C, D, and E at time of 1000 days. Panel F is the behavior in the bottom layer of a five layer model where the thickness is 50 ft.

Examination of Figure 5.47 illustrates the phenomenon of a "plunging plume." The contamination emanating from the source moves, as would be expected, in the direction of groundwater flow. However, due to the vertical inflow, there is a significant downward velocity component. As a result the plume moves downward providing a projection of increasing area in panels B, C, D, and E. In panel F we present a simulation of the behavior of a plume in the bottom layer of a five layer model of an aquifer that has a thickness of 50 ft. Panels E and F can be directly compared. As one might expect, the concentration at the bottom of the 50 ft thick aquifer is higher than that at the bottom of the 90 ft thick aquifer because (1) the contaminant is spread over a smaller vertical interval and (2) less time is required for the contaminant to reach the bottom of the thinner region.

It is important to keep in mind that the very different results in the two-dimensional and three-dimensional models presented here are not due to heterogeneity. Both models are homogeneous. However, when the physical system is such that vertical variation of properties, such as flow or concentration, or vertical flow influences the solution, a two-dimensional model may be unable to capture the dynamics correctly. In such a case, a three-dimensional model is required.

In a second illustration of the significance of using a three-dimensional simulation in lieu of a two-dimensional surrogate, we consider the high inflow case discussed in the preceding example, but using a fully penetrating source; that is, the source is assumed to extend the entire depth of the model rather than just extending through the top layer. Figure 5.48 shows the concentration distribution after 1000 days, a slightly longer period than the 900 days represented in Figure 5.37, in the top 10 ft of the model. The results are comparable to panel A in Figure 5.47. The results for the middle layer, which extends from 30 to 50 ft below ground surface, are shown in Figure 5.49. This figure is comparable to panel C in Figure 5.47. While the results are similar, the fully penetrating well example generates a plume that is somewhat larger than that created using the source located only in the upper layer.

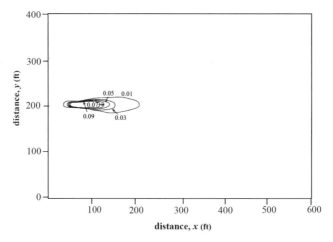

Figure 5.48: Concentration distribution in top 10 ft of the model using a fully penetrating source after 1000 days.

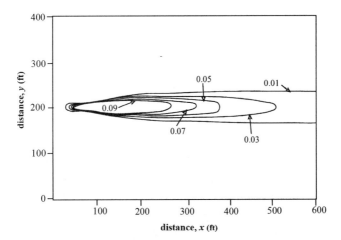

Figure 5.49: Concentration distribution in the middle layer of the model (30 to 50 feet below ground surface) using a fully penetrating source after 1000 days.

5.7 2-D THREE-PHASE FLOW

Although the derivations in this text have referred primarily to one- or two-phase flow, the extension to three phases is conceptually consistent with the approaches. The challenges in obtaining governing equations with appropriate closure relations are heightened for the case of two immiscible fluids plus a gas moving in a porous medium, but simulators have nevertheless been developed that attempt to model such systems and also provide insights into the mechanisms that require further attention if the models are to be useful predictive tools. Here we provide the results of such a model.

Perhaps the simplest physically meaningful three-phase flow problem is that which involves transient flow in a vertical plane. A physical experiment that satisfies these constraints was conducted by M. Fishman of EPA. Tetrachloroethylene (also called perchloroethylene PCE), a denser than water nonaqueous phase liquid (DNAPL) with a density of $\rho^n = 1.62\,\text{g/cm}^3$ and solubility limit of $\rho^w \omega^{iw}_{eqn} = 1.5 \times 10^{-4}\,\text{g/cm}^3$ (20 °C), was introduced into an artificial aquifer as illustrated in Figure 5.50. As described in [2];

The inside dimensions of the box (defining the volume of the sand) are 67 cm deep, 49 cm wide and 2 cm thick. The top boundary is open to the atmosphere and the bottom boundary is impermeable. The vertical sides are constructed such that water can flow across the boundary but not air. A constant phreatic surface is defined by specifying appropriate water source/sink ports along the vertical sides. The box is filled with a uniform medium-grained sand in a manner which is assumed to result in a homogeneous, isotropic porous medium. The soil properties are reported to be:

Permeability (k^s)	$3.5 \times 10^{-7}\text{cm}^2$
Porosity (ε)	0.37

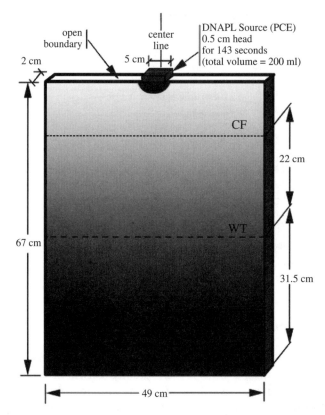

Figure 5.50: Experimental setup used by M. Fishman of EPA as reported in [2]. The acronym CF refers to capillary fringe and WT denotes watertable.

The sand-filled box is imbibed with water to the top and allowed to equilibrate to create an initial condition where the system is in static equilibrium and $s^w = 1$ throughout. The following fluid properties are provided:

	w (water)	n (DNAPL)	g (gas)
Density (g/cm³)	1.0	1.626	0.00129
Viscosity (poise)	0.01	0.0093	0.0002
Interfacial tension (dynes/cm)	γ^{wn}	39.5	
	γ^{gn}	31.74	
	γ^{wg}	72.75	

From this initial condition, three sequential displacement experiments are run:

1. The phreatic surface is lowered to elevation 35.5 cm from the top of the box and, the system is allowed to return to equilibrium conditions. . . .
2. Given the initial condition from Part 1, the PCE source is applied as shown in Figure 5.50. That is, a 0.5 cm head of PCE is applied uniformly over a 10 cm² surface at the center/top of the box until 200 cm³ enters the domain. Note that for the experiment this took 143 sec. . . .

3. Given the initial condition from Part 2, that is, the data at time = 143 sec, the PCE source is removed and the system is allowed to return to equilibrium for a period of 3,452 sec (total elapsed time since the DNAPL was applied is 3,595 sec). . . .

The results of the experiment are shown in Figures 5.51 through 5.54. The left panel presents the experimental observations, and is the relevant information

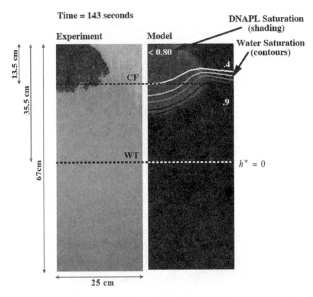

Figure 5.51: Distribution of DNAPL 143 sec after source was applied. At this time the source was removed [2]. The variable h^W is the point at which the pressure is atmospheric. The left panel presents the experimental results and the right panel the calculated saturations.

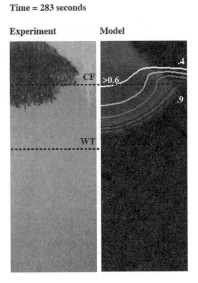

Figure 5.52: Distribution of DNAPL 283 sec after first introduction and 140 seconds after source was stopped [2].

Time = 1195 seconds

Experiment   Model

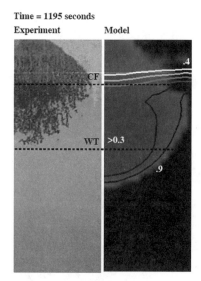

Figure 5.53: Distribution of DNAPL 1195 sec after the source was applied and 1052 sec after the source was removed [2].

Time = 3595 seconds

Experiment   Model

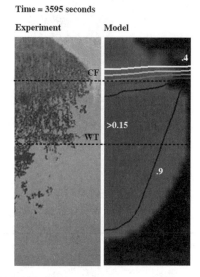

Figure 5.54: Distribution of DNAPL 3595 sec after the source was applied and 3452 sec after the source was removed [2].

needed from these figures at this time. We will address the panels on the right shortly. When viewed in combination with Figure 5.50 one can see from Figure 5.51 that the DNAPL moves through the unsaturated zone with limited lateral spread and begins to enter the *capillary fringe*. In Figure 5.52 we observe that although the DNAPL moves across the top of the capillary fringe, it appears to spread more horizontally. In Figure 5.53 one sees movement of the DNAPL through the capillary fringe and an apparent decrease in DNAPL saturation. Finally, in Figure 5.54, the DNAPL has passed below the water table and is moving through fully water-saturated porous media. Note that the saturation of the DNAPL, as indicated by the inequalities, is decreasing as time increases. This is to be expected since a fixed volume of DNAPL entered the system and is now spreading over an increasing volume of porous medium.

The mathematical representation of the DNAPL problem requires us to specify the initial and boundary conditions. The simulation actually requires the solution of three problems. They are described in [2] as follows:

1. Starting with the initial condition of full water phase saturation, drop the water table to match the experimental condition, and allow the system to approach steady-state conditions. . . .
2. Given the initial conditions from submodel 1, apply the DNAPL source for the specified time period (i.e., 143 sec).
3. Given the initial conditions from submodel 2, remove the DNAPL source and allow the system to re-equilibrate for 3,452 sec.

Boundary conditions for the three sequential simulations (A–C in Figure 5.55) are specified in this figure. In simulation A the objective is to provide appropriate initial conditions for simulation B. The DNAPL source is introduced in simulation B via the specification of a DNAPL pressure head of 0.5 cm as indicated. Simulation C describes the movement of the DNAPL subsequent to discontinuing the source.

The model used for this simulation is NAPL, an EPA supported collocation–finite element based computer code. The code documentation is found in [2]. The results of the simulation are provided in the right-hand panels of Figures 5.51 through 5.54. By comparing the left-hand panels with the right-hand side panels in each figure, a qualitative comparison of the computed and observed DNAPL saturation can be achieved. While the filamentous nature of the observed DNAPL plume cannot be

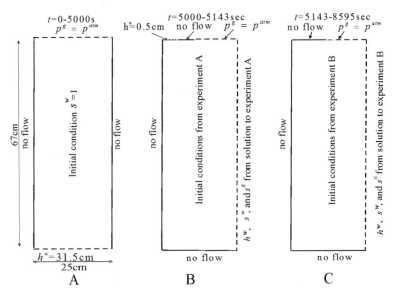

Figure 5.55: Identification of boundary conditions used in the three-phase simulation model (modified from [2]).

reproduced numerically using this continuum model, the general behavior of the DNAPL is appropriately simulated.

5.8   SUMMARY

The equations developed in earlier chapters have been used in this chapter in conjunction with numerical simulators to illustrate interesting features of porous medium flow and transport in some simple situations. Initially we considered one-dimensional multiphase flow examples; air-water and DNAPL-water systems were examined.

Drainage of water from a saturated homogeneous soil (Subsection 5.1.1) and, subsequently, from a heterogeneous layered soil (Subsection 5.1.2) were the first two problems to be considered. With the drained soil conditions used as the initial system state, water imbibition was examined for both soil systems (Subsection 5.1.3).

The second series of problems involved one-dimensional DNAPL-water flow. After considering the imbibition of DNAPL into a water-saturated homogeneous soil (Subsection 5.2.1), the impact of NAPL density was considered (Subsection 5.2.2). The contrast in behavior of two fluids with different densities was demonstrated.

Using the simulation results at the end of the primary DNAPL imbibition phase as the initial state of the system, DNAPL drainage was next considered (Subsection 5.2.3). Given the state of the system after this drainage phase, secondary imbibition of DNAPL was simulated (Subsection 5.2.4), followed by secondary drainage (Subsection 5.2.5). Primary imbibition into heterogeneous layered soils was also simulated (Subsection 5.2.6).

Next the sinking of DNAPL introduced at the top into a water-saturated reservoir. The downward motion was simulated as well as lateral motion induced by the presence of a lower permeability zone (Subsection 5.3.1).

In Section 5.4 we considered the case of one-dimensional two-phase flow and transport (Subsection 5.4.1) followed by two-dimensional two-phase flow and transport (Subsection 5.4.2). While one-dimensional problems are illustrative of many basic concepts, the extension to two space dimensions reveals behavior that cannot be captured in one-dimensional simulations. Movement around areas of lower hydraulic conductivity and the representation of irregular sharp-interface propagation, for example, cannot be viewed using one-dimensional models. The classic problem of DNAPL movement from a source of limited areal extent into a water-saturated reservoir was used as the physical system of interest in the two-dimensional simulations.

Section 5.5 focused on single-phase flow and transport in two space dimensions using a vertically integrated model. The objective was to show the impact of a number of factors on contaminant transport behavior. The impact of inflow from above (Subsection 5.5.2), well discharge (Subsection 5.5.3), adsorption (Subsection 5.5.4), heterogeneous hydraulic conductivity (Subsections 5.5.5 and 5.5.6), and a first order chemical decay reaction (Subsection 5.5.7) were examined. We note that these were simple examples aimed at demonstrating mechanisms rather than complex examples designed to demonstrate simulation capabilities.

Section 5.6 extends the single-phase flow and transport simulations to three dimensions. This section demonstrates the utility of modeling the vertical dimension, as well as the lateral dimensions, when vertical flow occurs or when gradients of concentration in the vertical are important. When the source of contamination is near the top of the region, flow from the top can cause the "plunging plume" phenomenon whereby the contaminant is transported toward the bottom of the reservoir. Such a situation cannot be simulated accurately unless variation in the vertical dimension is accounted for.

In Section 5.7, three-phase, three-dimensional flow and transport are presented for demonstration purposes. A qualitative comparison between computed and observed behavior suggests that at least some of the mechanisms that influence system dynamics are properly parameterized mathematically.

As a final disclaimer, it is important to state that the material presented here, though perhaps complex in some sense, is indeed introductory in nature. Additional work that integrates theory, experiments, and simulation is required to obtain increased understanding of the influence of heterogeneity, anisotropy, scale, phase interactions, chemical reactions, solid phase dynamics, and biological activity on mass, momentum, and energy transport in multiphase porous media systems. From this point, the research must continue.

BIBLIOGRAPHY

[1] Guarnaccia, J., course notes, 2003.

[2] Guarnaccia, J., G.F. Pinder, and M. Fishman, NAPL: Simulator Documentation, EPA/600/SR-97/102, 1997.

[3] Imhoff, P.T., P.R. Jaffe, and G.F. Pinder, An Experimental Study of the Dissolution of Trichloroethylene in Saturated Porous Media, Princeton University Water Resources Program Report: WR-92-1, 1992.

SELECT SYMBOLS

Roman Letters

A	cross-sectional area of column, L^2
a	air phase
a	fitting parameter for Brooks-Corey relationship, —
b	aquifer thickness$(= z_\text{T} - z_\text{B})$, L
C_c	coefficient of curvature, —
C_u	uniformity coefficient, —
C^w	water capacity function, $1/L$
$c^{i\alpha}$	macroscale moles of species i per volume of phase α, $moles/L^3$
D	particle diameter, L
D_i	geometric mean of the grain size of sieve i and sieve $i + 1$, L
$D_{iw\,\text{dif}}$	microscale diffusion coefficient for species i in phase w, L^2/T
$D^{i\alpha}$	dispersion coefficient of species i in phase α, L^2/T
$D^{i\alpha}_{\text{dif}}$	macroscale diffusion coefficient of species i in phase α, L^2/T
$\mathbf{D}^{i\alpha}_{\text{dif}}$	macroscale diffusion coefficient tensor for species i in phase α, L^2/T
$\mathbf{D}^{i\alpha}_{\text{dis}}$	macroscale dispersion coefficient tensor for species i in phase α, L^2/T
D_n	grain size greater than or equal to n% of the grains by weight, L
D_p	pore diameter, L
D^w	soil water diffusivity, L^2/T
d_p	tube diameter, L
e	void ratio, —
$e^{\alpha}_{\alpha\beta}$	rate of mass transfer into phase α at the $\alpha\beta$ interface, $M/(L^3T)$
$e^{i\alpha}_{\alpha\beta}$	rate of mass transfer of species i into the α phase at the $\alpha\beta$ interface, $M/(L^3T)$
$\mathbf{F}$	general vector function with continuous first spatial derivatives within the volume V
$F^{i\alpha}$	general macroscale scalar property of species i in phase α
$\mathbf{F}|_\alpha$	the microscale vector $\mathbf{F}$ evaluated in the α phase
f	an arbitrary scalar function
f	pore-size distribution function, —
f^w	fraction of the linear sum of the two phase velocities that is due to the water phase, —

Essentials of Multiphase Flow and Transport in Porous Media, by George F. Pinder and William G. Gray
Copyright © 2008 by John Wiley & Sons, Inc.

$f\vert_\alpha$	the microscale scalar f evaluated in the α phase
$\mathbf{g}$	gravity vector, L/T^2
g	magnitude of gravity vector, L/T^2
H	initial macroscale head distribution, L
H^b	macroscale head specified on the boundary of the domain, L
H_{ext}	measure of the head external to the region of study at the boundary that influences the flow, L
$H^{i\alpha\beta}$	Henry's law constant for species i in phase β in equilibrium with its mass fraction in phase α, —
h	hydraulic head, L
h^c	macroscale capillary head, L
h_i	microscale hydraulic head at location $\mathbf{x}_i$, L
h^α	macroscale head in the α phase, L
$\mathbf{I}$	unit tensor, —
$J(s^w)$	Leverett J-function, —
$\mathbf{j}_i$	microscale diffusive/dispersive transport flux of species i, $M/(L^2T)$
$\mathbf{j}^{i\alpha}$	macroscale dispersion vector of species i in phase α, $M/(L^2T)$
K	hydraulic conductivity, L/T
K^{iws}	constant used in constitutive equations for adsorption isotherms for species i in fluid phase w on solid s, —
K_d^{kws}	distribution coefficient for species k between the w and s phases, L^3/M
$\mathbf{K}^w$	hydraulic conductivity tensor for the w phase, L/T
K^w	hydraulic conductivity for the w phase, L/T
$\mathbf{K}^{w\prime\prime}$	two-dimensional hydraulic conductivity tensor in the lateral plane for the w phase, L/T
$\mathbf{k}$	unit vector in z-coordinate direction, —
k_{rxn}^{iw}	macroscale rate constant for first order chemical reaction of species i in phase w, $1/T$
$k_{i\alpha}$	microscale first order decay coefficient for species i in phase α, $1/T$
k_{maxm}	maximum value of relative permeability obtained with $s^w = s_i^w$ on main drainage curve, —
k_{maxp}	maximum value of relative permeability obtained with $s^w = s_i^w$ on primary drainage curve, —
$\mathbf{k}^s$	intrinsic permeability tensor, L^2
k^s	intrinsic permeability, L^2
k^{sw}	effective conductivity of the wetting phase, L^2
$\mathbf{k}^{s\alpha}$	effective conductivity tensor of the α phase $(= k_{rel}^\alpha \mathbf{k}^s)$, L^2
k_{rel}^α	relative permeability for the α phase, —
L	length of column, L
L^p	macroscale distance, L
M	fitting parameter in van Genuchten p^c-s^w relation, —
$\mathbf{M}^i$	molecular weight of chemical species i, $M/mole$
M^{iw}	mass of species i in the w phase, M
M_0^{iw}	initial total amount of species i in phase w, M
M^α	mass of the α phase, M
m	mass of solid particles in a volume of fluid in a hydrometer study, M

m	fitting parameter in wetting phase relative permeability relationship, —
m_m	fitting parameter in main nonwetting drainage curve relative permeability relationship, —
m_p	fitting parameter in primary nonwetting drainage curve relative permeability relationship, —
m_α	mass of α phase, M
N	number of species, —
N	fitting parameter in van Genuchten p^c-s^w relation, —
N_b	Bond number (ratio of gravitational to capillary forces), —
N_c	capillary number (ratio of viscous to capillary forces), —
N_{int}	number of intervals between sieves, —
N_W	number of wells in the system, —
N_α	number of species in the α phase, —
n	nonwetting phase
$\mathbf{n}$	unit normal vector outward from a volume at a point on the surface, —
$\mathbf{n}^{ext}$	outwardly directed unit normal to external boundary S of volume V, —
n_F	constant in Freundlich isotherm, —
$\mathbf{n}_\alpha$	outward-directed unit normal to surface of α phase, —
$\mathbf{n}_{\alpha\beta}$	outward-directed unit normal at edge of $\alpha\beta$ interface surface, —
P	specified pressure distribution within a domain, $M/(LT^2)$
P^b	specified pressure distribution on $\mathbf{x}_b^w$, the domain boundary, $M/(LT^2)$
P_{ext}	external pressure, $M/(LT^2)$
P	microscale pressure, $M/(LT^2)$
p_{atm}	atmospheric pressure, $M/(LT^2)$
p_c	microscale capillary pressure, $M/(LT^2)$
p^c	macroscopic capillary pressure($= p^n - p^w$), $M/(LT^2)$
p^s	macroscale force per unit area exerted on the solid by the adjacent fluids, $M/(LT^2)$
p_α	microscale pressure in phase α, $M/(LT^2)$
p^a	macroscale pressure in phase α, $M/(LT^2)$
Q_W^α	volumetric pumping rate of phase α at well W, L^3/T
$Q_W^{\alpha\prime}$	volumetric pumping rate of phase α at well W per cross-sectional area, L/T
q_B	combined velocity component used in fractional flow formulation ($= q^w + q^n$), L/T
q^b	specified normal flux of phase w at the boundary location $\mathbf{x}_b^w$, L/T
$\mathbf{q}^{w\prime}$	Darcy velocity of phase w in the horizontal plane, L/T
$\mathbf{q}^{w*}$	Darcy velocity of phase w at the top surface of an unconfined aquifer, L/T
$\mathbf{q}^\alpha$	specific discharge (Darcy velocity) of phase α, L/T
q^α	magnitude of $\mathbf{q}^\alpha$, L/T
R	radius of curvature, L
R_c	geometric mean radius of curvature, L

$R^{i\alpha}$	molar rate of production of species i in phase α per unit volume ($= r^{i\alpha}/M^i$), $moles/(L^3 T)$
R_S^{kw}	standard retardation factor, —
R_T^{kw}	total retardation factor for species k in the w phase, —
r	radius of cylinder, L
$\mathbf{r}$	global coordinate system vector, L
r_i	microscale rate of mass production of species i per unit volume, $M/(L^3/T)$
$r^{i\alpha}$	macroscale rate of production of species i per unit volume of α phase, $M/(L^3 T)$
r_{max}	maximum pore radius contributing to flow, L
r_{min}	minimum pore radius contributing to flow, L
S	boundary surface of a volume, L^2
S	storage coefficient ($= S_s b$), —
S_s	specific storage, $1/L$
S_y	specific yield ($= S + \varepsilon_{eff}$), —
$S^{\alpha\beta}$	surface between phases α and β in the interior of an averaging volume δV, L^2
s	solid phase
s_r^n	residual saturation of the nonwetting phase, —
s_{rmax}^n	maximum value of residual saturation achievable in the porous medium, —
s_e^w	effective saturation of w phase ($= (s^w - s_i^w)/(1 - s_i^w)$), —
$s_{e_r}^w$	revised effective saturation of w phase ($= (s^w - s_t^w)/(1 - s_t^w - s_t^n)$), —
s_i^w	irreducible saturation of w phase, —
s^α	saturation of the α phase, —
s_t^α	trapped saturation of the α phase
T	temperature, $^\circ$
$\mathbf{T}^{w''}$	two-dimensional transmissivity tensor for the w phase in the lateral plane ($= b\mathbf{K}^{w''}$), L^2/T
T^v	transmissivity of an isotropic medium for the w phase, L^2/T
T^α	macroscopic temperature in phase α, $^\circ$
t	time, T
V	volume under consideration, L^3
V^α	subvolume of V occupied by α phase, L^3
V_C^α	volume of α phase in a cylinder, L^3
V_T^α	total volume of α phase, L^3
$V\infty$	total macrospace domain of the system under study, L^3
$\mathbf{v}$	velocity, L/T
v	magnitude of velocity, L/T
$\mathbf{v}_i$	microscale velocity of chemical species i, L/T
$\mathbf{v}_i|_\alpha$	microscale velocity of species i evaluated in the α phase, L/T
$\mathbf{v}^{i\alpha}$	macroscopic velocity of species i in the α phase, L/T
$\mathbf{v}^\alpha$	barycentric macroscale α phase velocity, L/T
$\mathbf{w}$	velocity of the boundary surface of a region, L/T
w	wetting phase

w_i	fractional weight of sand between diameters of sieve i and sieve $i + 1$, —
$\mathbf{x}$	macroscale coordinate system, also centroid of an REV, L
$\mathbf{x}_b^w$	spatial coordinates of the boundary of a study region, L
$\mathbf{x}^w$	well location, L
x_{mass}^α	fraction of total mass at a macroscale point that is α phase, —
z	vertical coordinate
z_i	elevation coordinate, L
z_0	reference elevation, L

Greek Letters

α	refers to a general phase
α	fitting parameter for Brooks-Corey or van Genuchten relationship, $1/L$
α^b	matrix compressibility, LT^2/M
a_{L}	longitudinal dispersivity, L
a_{T}	transverse dispersivity, L
β	refers to a general phase
$\beta^{i\alpha}$	concentration compressibility for species i in phase α, —
$\beta_\mu^{i\alpha}$	viscosity-composition coefficient with respect to species i in phase α, —
β^α	compressibility of phase α, LT^2/M
γ_{wns}	microscale line/curve tension of common line where w, n, and s phases meet, ML/T^2
γ_α	phase distribution function for phase α, —
$\gamma_{\alpha\beta}$	microscale interfacial/surface tension for interface between α and β phases, M/T^2
$\delta S^{\alpha\alpha}$	part of the boundary of δV intersected by the α phase, L^2
δV	volume of representative elementary volume (REV), L^3
δV^α	volume of the α phase in δV, L^3
$\delta(\mathbf{x} - \mathbf{x}^W)$	Dirac delta function acting at location $\mathbf{x}^W$, $1/L^3$
ε	porosity, —
ε_{eff}	effective porosity, —
ε^α	volume fraction of an REV occupied by the α phase, —
ζ	variable of integration
ς	pore connectivity parameter for the nonwetting phase relative permeability relationship, —
θ	water content $(= \varepsilon s^w)$, —
θ	contact angle, —
θ_r	residual water content, —
θ_s	saturated water content, —
κ	pore connectivity parameter in wetting phase relative permeability relationship, —
κ^b	measure of the permeability of the boundary to flow, L^2T/M
κ_g	geodesic curvature, $1/L$
$\kappa_{\alpha\beta}^{i\alpha}$	coefficient of mass transfer of species i between the α and β phases based on conditions in the α phase, $M/(L^3T)$

κ_n	normal curvature, $1/L$
κ^α	bulk modulus of phase α, $M/(LT^2)$
λ	fitting parameter for Brooks-Corey relationship, —
λ_d^k	radioactive decay coefficient for species k, $1/T$
μ	dynamic viscosity, $M/(LT)$
μ^α	macroscale average of the dynamic viscosity of the α phase, $M/(LT)$
ξ	local coordinate system relative to REV centroid, L
ρ	mass density, M/L^3
ρ_i	mass density of species i $(= \rho\omega_i)$, M/L^3
ρ_α	microscale mass density of phase α, M/L^3
ρ^α	macroscale mass density of α phase, M/L^3
ρ_W^α	macroscale mass density of phase α being pumped at well W, M/L^3
ρ_0^α	macroscale reference density for phase α, M/L^3
τ	tortuosity, —
Φ^w	Hubbert potential for w phase, L^2/T^2
χ	Bishop parameter, —
ψ	suction head, L
$\Omega^{i\alpha}$	specified macroscale spatial distribution of species i in α phase, —
ω_i	mass fraction of species i at a point in α phase, —
ω_{max}^{is}	constant in Langmuir isotherm, —
$\omega^{i\alpha}$	macroscale mass fraction of species i in phase α, —
$\omega_{eq\beta}^{i\alpha}$	mass fraction of species i in the α phase that would be in equilibrium with the actual mass fraction, $\omega^{i\beta}$, of i in the β phase (Note: if the β phase is pure species i, this quantity is the solubility limit of i in the α phase), —
$\omega_W^{i\alpha}$	macroscale mass fraction of species i in phase α pumped at well W, —
$\omega_0^{i\alpha}$	reference macroscale mass fraction of species i in phase α, —

Mathematical Operations

$D(\cdot)/Dt$	microscale material time derivative, $1/T$
$D^\alpha(\cdot)/Dt$	macroscale material time derivative taken with $\mathbf{v}^\alpha$ as the reference velocity, $1/T$
∇	three-dimensional spatial del, $1/L$
$\nabla_{\mathbf{x}}$	macroscale spatial derivative operator holding the ξ coordinates fixed, $1/L$
∇_ξ	microscale spatial derivative operator holding the centroid $\mathbf{x}$ of the REV fixed, $1/L$
∇'	two-dimensional del operator acting in a surface, $1/L$
$\nabla' \cdot \mathbf{n}_n$	mean curvature of the n phase surface, $1/L$
∇''	one-dimensional del operator acting along a curve, $1/L$

INDEX

1-D simulation of air-water flow 277
    drainage in a heterogeneous soil 286
    drainage in a homogeneous soil 282
    imbibition in homogeneous soil 288
1-D simulation of DNAPL-water flow 290
    density effect 292
    DNAPL drainage in homogeneous soil 292
    primary DNAPL imbibition in homogeneous soil 290
    primary imbibition in heterogeneous soil 296
    secondary drainage in homogeneous soil 295
    secondary imbibition of DNAPL in homogeneous soil 293
1-D two-phase flow and transport 305
2-D simulation of DNAPL-water flow 298
    DNAPL descent into a water-saturated reservoir 298
2-D single-phase flow and transport 315
    base case 323
    effect of a low transmissivity region 329
    effect of adsorption 328
    effect of inflow 325
    effect of rate of reaction 335
    impact of well discharge 327
2-D three-phase flow 341
2-D two-phase flow and transport 308
3-D single-phase flow and transport 337

adsorption 310
adsorption isotherm 257, 310
advective term 232
anisotropy 144

applications of transport 101
    integral analysis 103
    point analysis 106
averaging criteria 77
averaging theorems 86
    spatial averaging theorem 89, 91
    temporal averaging theorem 91, 92

balance on the common line 32
Bishop parameter 168
bond number 184
bounding loop 190
Brooks and Corey parametric model 193
Buckley-Leverett analysis 215
bulk compressibility 150
bulk density 191
bulk modulus 125
by-passing 181

capillarity. *See* surface tension 36
capillary depression 38
capillary fringe 187, 344
capillary head 171
capillary number 184
capillary pressure 27
    formulation 210
capillary rise 38
capillary tube. *See* interfacial tension 36
capillary wetting phase 181
chain rule 150
chemical reaction 232
    rates 250
closure relations for the dispersion vector 246
coefficient of curvature 15
coefficient of mass transfer 255
coefficient of uniformity 14
collinear vector 139

collocation finite element method 282
common line 23, 28, 29
compressibility 125
        of pure water 126
concentration 42
concentration compressibility 125
concept
        of concentration 42
        of pressure 17
        of saturation 15
confined aquifer 163
connectivity 4
conservation equations 50
conservation of mass, integral form of 59
        See also mass balance
conservative species 258
constitutive equations 117
constitutive relationships 17, 50
contact angle 32, 39, 199
continuum scale 52
convection. *See* mass balance 60
cumulative grain size distribution curve 192
curvature 31
cylindrical coordinate system 162

Darcy velocity 139, 167, 236
Darcy's experiments 118
Darcy's law 137
decay reaction 336
dense non-aqueous phase (DNAPL) flow 290
density effect on DNAPL flow 292
derivation
        of the Buckley-Leverett equation 218
        of flow equations 166
        of groundwater flow equation 146
derivatives of hydraulic head 134
diffusion 61, 246
diffusion coefficient 246
direct approach 233
Dirichlet conditions for flow 156
dispersion 61, 246
dispersion coefficient 306
dispersion vector 232
dispersivity 320
distribution approach 239
distribution coefficient 257, 310
distribution form of the species conservation
        equation 239
divergence theorem 26, 63
DNAPL descent into a water-saturated
        reservoir 298
DNAPL drainage in homogeneous soil 292

DNAPL experiment 343
DNAPL pooling 300
drainage
        in a heterogeneous soil 286
        in a homogeneous soil 282
        See also capillarity 40
drainage curve 178, 190
        main 190
        primary 190
        scanning 190
Dupuit assumption 160
dynamic viscosity 122, 127

effect of a high transmissivity region 332
effective porosity 163
effective saturation 185, 205
elevation head 132
entry pressure 176, 193
equation of state 123
equations of state for fluids 123
equilibrium formulation 254, 260
exchange term 255
extensive quantities 78

Fick's Law 246
first order reaction 335
first type conditions for flow 156
first-order chemical reaction 319
flow equations 115
fluid compressibility 126, 149
fluid properties 121
fluid viscosity 127
flux condition 157
fractional flow 215
Freundlich isotherm 257

gauge pressure 37
geodesic curvature 32
global and local coordinate systems 69
grain and pore size distributions 8
grain density 191
grain size 7, 9
        distribution 191
        distribution curve 13
        effective grain size 15
        grading 14
        sorting 14
        uniformity coefficient 14
granular soils 172

Haines jump 178
half life 335

head, hydraulic 119
Henry Darcy 118
homogeneous reaction rate 250
Hubbert potential 134
hydraulic conductivity 120, 152
hydraulic force and hydraulic head 130
hydraulic head
    derivatives 134
    and permeability 140
hydraulic potential 128
hydraulically disconnected 181
hydrometer method 12
hysteresis 179
hysteresis loop 179
hysteresis. *See* capillarity 41

ideal gas constant 223
imbibition, in homogeneous soil 288
    See also capillarity 40
imbibition curve 179
    main 190
    primary 190
    scanning 190
incomplete displacement 184
initial and boundary conditions for flow 154
initial conditions 348
integral analysis, transport 103
integral forms 99
    of mass conservation 55
integral theorems 62
    divergence theorem 63
    transport theorem 63
intensive variables 78
interfacial tension 22, 23, 199
interfacial transport 17
interphase mass transfer 318
interphase transfer terms 232, 253
    equilibrium formulation 260
    kinetic formulation 255
    kinetic vs. equilibrium formulations 269
interphase transport 17, 256
intrinsic permeability 141
intrinsic phase average 79
irreducible saturation 181, 183, 188, 193, 199
isotropy 139, 146

$k_{rel}^{\alpha} - s^w$ relationship 204
kinetic formulation 254
kinetic process 309
kinetic rate coefficient 310
kinetic vs. equilibrium formulations 269
Kozeny relationship 142

Land trapping model 190
Langmuir isotherm 258
Laplace equation for capillary pressure 28
    195
Leibnitz rule 135
Leverett J-function 195
linear adsorption isotherm 257, 318
local coordinate system 69, 71
longitudinal dispersivity 248

macroscale 53, 67
macroscale density 122
macroscale mass conservation 92
    integral forms 99
    macroscale point forms 93
macroscale perspective 67
    definitions of macroscale quantities 78
    global and local coordinate systems 69
    macroscopic variables 74
    representative elementary volume 68
macroscale point forms 93
macroscale quantities
    definitions 78
    summary 85
macroscopic level of observation 76
macroscopic variable 74
manometer 21, 119
mass balance equation 116
mass conservation equations 49
mass conservation of the solid phase 236
mass density and pressure 125
mass dispersion vector 246
mass fraction 124, 232, 252
mass fraction solubility limit 309
mass rate of production 251
mass transfer 309
mass transport equations 229
material derivative 110, 147, 149, 167, 236
matrix compressibility 169
mean curvature 27
mean radius of curvature. *See* capillarity
    27
mechanical dispersion 247
microcopic level of observation 76
microscale 52
microscale mass conservation 53
mixed form of Richards equation 212
mixed formulation 210
moisture retention curve 187, 192
molar concentrations 251
molar rate of production 251
multiphase simulation 277

NAPL model 349
negative pressure 37
Neumann condition for flow 157
no flow boundary 157
non-conservative species 258
non-wetting fluid 24
 See also interfacial tension 33
normal curvature 32
normal flux 157

packing, cubic, grain packing, rhombohedral
 8
packing factor 143
parametric models 192
partitioning 254
p^c-s^w relationship 173
p^c-s^w relationship formulas 185
perchloroethylene 342
phase 2
 density 279
 distribution function 75
 transformation 17
phases and porous media 2
phreatic surface 187
pipette method 12
plunging plume 337
point analysis, transport 106
point forms of mass conservation 64
pore 5
 connectivity parameter 203
 diameter 195
 doublet 182
 high aspect-ratio 182
 radius 191
 size distribution function 191
 size distribution index 193
 velocity 140
porosity 7, 136, 279
porous media 2
pressure 17
 negative 37
pressure gauge 37
pressure head 132, 151
primary DNAPL imbibition in homogeneous
 soil 290
primary drainage 194
 curve 187, 201
primary imbibition
 of DNAPL 291
 in heterogeneous soil 296
primary-secondary DNAPL imbibition
 comparison 294

product rule 209
propagation of concentration front
 307

rate limited model 254
reaction term 309
reference pressure 131
relative permeability 170, 197
relative permeability-saturation
 relationship 197
Representative Elementary Volume 68
residual saturation 17, 183, 199
residual water content 214
retardation coefficient 310
REV 68
Richards equation 212
rigorous approach 235
Robin condition 158

sand shape factor 143
saturated water content 214
saturation 16, 279
 formulation 210
 residual saturation 17
scaling relation 194
scanning curve 190, 180
second type boundary condtion for flow
 157
secondary DNAPL drainage in homogeneous
 soil 295
secondary drainage curve 201
secondary imbibition 293
secondary imbibition of DNAPL in
 homogeneous soil 293
sieve, mesh size 9
simulation 277
 of multiphase flow and transport 305
  1-D two-phase flow and transport 305
  2-D two-phase flow and transport 308
single-phase fluid flow 136
snap-off 181
soil retention curve 187
soil water diffusivity 213, 214
solid compressibility 169
solid phase deformation 237
solution of the Buckley-Leverett equation
 219
spatial averaging theorem 89
special cases of multiphase flow 209
specific discharge 139
specific storage 151
specific yield 164

standard retardation factor 267
steady state flow 156
storage coefficient 161
storativity 161
suction head 212
suction head form of Richards equation
    213
surface tension 21, 23, 36, 195
    considerations 21
    capillarity 36
    See also interfacial tension 22

temporal averaging theorem 91
third type boundary condition for flow
    158
three-phase flow 341
threshold pressure 176
tortuosity 142
total mass conservation 241
total retardation factor 266
transient flow 155
transmissivity 161, 316, 329
transport equation 309
transport theorem 63
transverse dispersivity 248
trapped-phase volume 181
two-dimensional flow 159
two-phase flow and transport 305
two-phase immiscible flow 165

unconfined aquifer 163
uniform soil 15

uniformity coefficient 14
unsaturated zone 185

vadose zone 185
van Genuchten parametric model 193
velocity of the species transport equations 232
    direct approach 233
    distribution approach 239
    rigorous approach 235
velocity, barycentric 60
vertical averaging 337
vertically averaged transport equation 318
viscosity ratio 199
viscosity-composition coefficient 128
viscous forces 184
void fraction 7
void ratio 44
volume fraction 76

water capacity function 213
water content 212
    form of Richards equation 213
    See also saturation 16
water table 36, 163, 187
wet methods 12
wettability 34, 39
wetting fluid 24
    See also interfacial tension 33
wetting phase entrapment 181
wetting phenomenon 39

Young's equation: interfacial tension 33